Environmental Compliance Handbook

Environmental laws and regulations are extremely complex and difficult to understand. In order for people to comply with them, they need to be explained in layperson's terms. This book identifies many changes in regulations and recommends ways to apply and implement them. Containing the latest environmental information, the book addresses environmental compliance with land and provides a historical perspective to help follow the logical growth and increased complexity of land regulations through time. Structured as a "step-by-step how-to" book, readers will find real-life examples for the most important aspects of language, permit terms, demonstrating compliance, and organization for land projects

Features include:

- Introduces all land pollution control regulations and the requirements of any land pollution control permits available to date.
- Answers in depth all practical questions that arise when working on compliance projects in a "how-to" method.
- Addresses a wider spectrum of issues that go beyond chemical-based contamination and environmental regulations and examines the impacts of climate change.
- Includes many real-life examples and case studies from industry and institutions that comply with land use regulations.
- Is global in coverage and very useful to companies that have expanded operations outside their country of origin.

This book will be of benefit to professionals, engineers, scientists, senior under-graduates, and graduate students in the fields of environmental sciences, laws, and management.

Environmental Compliance Handbook

Land, Volume III

Daniel T. Rogers

CRC Press
Taylor & Francis Group
Boca Raton London

CRC Press is an imprint of the
Taylor & Francis Group, an **informa** business

First edition published 2023
by CRC Press
6000 Broken Sound Parkway NW, Suite 300, Boca Raton, FL 33487–2742

and by CRC Press
4 Park Square, Milton Park, Abingdon, Oxon, OX14 4RN

CRC Press is an imprint of Taylor & Francis Group, LLC

© 2023 Taylor & Francis Group, LLC

Reasonable efforts have been made to publish reliable data and information, but the author and publisher cannot assume responsibility for the validity of all materials or the consequences of their use. The authors and publishers have attempted to trace the copyright holders of all material reproduced in this publication and apologize to copyright holders if permission to publish in this form has not been obtained. If any copyright material has not been acknowledged please write and let us know so we may rectify in any future reprint.

Except as permitted under U.S. Copyright Law, no part of this book may be reprinted, reproduced, transmitted, or utilized in any form by any electronic, mechanical, or other means, now known or hereafter invented, including photocopying, microfilming, and recording, or in any information storage or retrieval system, without written permission from the publishers.

For permission to photocopy or use material electronically from this work, access www.copyright.com or contact the Copyright Clearance Center, Inc. (CCC), 222 Rosewood Drive, Danvers, MA 01923, 978-750-8400. For works that are not available on CCC please contact mpkbookspermissions@tandf.co.uk

Trademark notice: Product or corporate names may be trademarks or registered trademarks and are used only for identification and explanation without intent to infringe.

ISBN: 978-0-367-70601-2 (hbk)
ISBN: 978-0-367-71274-7 (pbk)
ISBN: 978-1-003-15010-7 (ebk)

DOI: 10.1201/9781003150107

Typeset in Times
by Apex CoVantage, LLC

*To Hannah and Matt, the two greatest
children, who love science and inspire me*

Contents

Preface xvii
Acknowledgments xix
About the Author xxi
Acronyms, Elements, Symbols, Molecules, and Units of Measure xxiii

Chapter 1   Themes and Overview 1
    1.1   Introduction 1
    1.2   Major Themes of This Book 2
    1.3   Theme 1: No Matter Where You Are, It's All the Same 3
    1.4   Theme 2: Pollution and Its Behavior in Nature 3
    1.5   Theme 3: Land Use and Where Cities Are Built 5
    1.6   Theme 4: History of Environmental Regulations 7
    1.7   Theme 5: Complexity and Effectiveness of Environmental Regulations 8
    1.8   Theme 6: Environmental Regulations and Pollution of the World 10
    1.9   Theme 7: Maintaining Compliance with Environmental Regulations 11
    1.10 Summary 12
    1.11 References 13

Chapter 2   History of Land Pollution and the Beginning of Environmental Regulations in the United States 15
    2.1   Introduction 15
    2.2   Overview of Environmental Regulations for Land in the United States 16
    2.3   The Birth of the Environmental Movement and Regulations in the United States 18
    2.4   Formation of National Policy and the Environmental Protection Agency 19
    2.5   Environmental Laws of the United States 22
    2.6   National Environmental Protection Act 23
    2.7   Environmental Justice 24
    2.8   Environmental Enforcement 25
    2.9   Current Status of Land Pollution in the United States 26
    2.10 Summary of Land Pollution Regulations in the United States 27
    2.11 References 27

Chapter 3   Land Contamination 29
    3.1   Introduction 29
    3.2   The Need for Regulation 29

vii

viii   Contents

3.3   Toxicity of Contamination 32
3.4   Volatile Organic Compounds 34
    3.4.1   Light Nonaqueous Phase Liquids 36
    3.4.2   Dense Nonaqueous Phase Liquids 37
    3.4.3   Trihalomethanes 39
3.5   Polynuclear Aromatic Hydrocarbons 40
3.6   Polychlorinated Biphenyls 40
3.7   Semi-Volatile Organic Compounds 42
    3.7.1   Phthalates 43
    3.7.2   Phenol 43
    3.7.3   Amines 44
    3.7.4   Esters 45
3.8   Heavy Metals 45
3.9   Pesticides and Herbicides 47
3.10 Dioxins 49
3.11 Fertilizers 49
3.12 Cyanide 50
3.13 Asbestos 50
3.14 Acids and Bases 52
3.15 Radioactive Compounds 53
3.16 Greenhouse Gases 54
3.17 Carbon Dioxide 55
3.18 Carbon Monoxide 56
3.19 Ozone 56
3.20 Sulfur Dioxide 57
3.21 Particulate Matter 57
3.22 Bacteria, Parasites, and Viruses 59
3.23 Invasive Species 60
3.24 Polyfluoroalkyl Substances 61
3.25 Emerging Contaminants 62
3.26 Summary and Conclusion 63
3.27 References 64

Chapter 4   Nature's Response to Land Contamination 71
4.1   Introduction 71
4.2   Release of Contamination into the Environment 72
4.3   Principles of Contaminant Fate and Transport 73
    4.3.1   Basic Contaminant Transport Concepts 76
    4.3.2   Basic Contaminant Degradation Concepts 80
        4.3.2.1   Biotic Degradation 82
        4.3.2.2   Abiotic Degradation 82
    4.3.3   Transport and Fate of Contaminants in Soil 83
    4.3.4   Transport and Fate of Contaminants in Surface
        Water 85
    4.3.5   Transport and Fate of Contaminants in Groundwater 87

Contents

ix

	4.3.6	Transport and Fate of Contaminants in the Atmosphere	91
4.4		Fate and Transport of Contaminants	96
	4.4.1	VOCs	96
	4.4.2	PAHs	100
	4.4.3	PCBs	100
	4.4.4	SVOCs	100
	4.4.5	Heavy Metals	101
	4.4.6	Pesticides and Herbicides	104
	4.4.7	Dioxins	104
	4.4.8	Fertilizers	105
	4.4.9	Cyanide	105
	4.4.10	Asbestos	105
	4.4.11	Acids and Bases	106
	4.4.12	Radioactive Compounds	106
	4.4.13	Greenhouse Gases	106
	4.4.14	Carbon Dioxide	106
	4.4.15	Carbon Monoxide	107
	4.4.16	Ozone	107
	4.4.17	Sulfur Dioxide	107
	4.4.18	Particulate Matter	108
	4.4.19	Bacteria, Parasites, and Viruses	108
	4.4.20	Invasive Species	108
	4.4.21	Polyfluoroalkyl Substances	108
	4.4.22	Emerging Contaminants	108
4.5		Summary and Conclusion	109
4.6		References	109

Chapter 5 Resource, Conservation, and Recovery Act, Medical Waste, and Low-Level Radioactive Waste 115

5.1		Introduction	115
5.2		Subtitle D—Non-Hazardous Waste	117
5.3		Subtitle C—Hazardous Waste	117
	5.3.1	Characteristic Hazardous Waste	118
	5.3.2	Listed Hazardous Waste	122
	5.3.3	Universal Hazardous Waste	124
	5.3.4	Used Oil	124
	5.3.5	Contained-In Policy and Derived-From Rule	124
	5.3.6	Land Disposal Restrictions	125
	5.3.7	Analytical Methods	126
	5.3.8	Hazardous Waste Manifests	127
5.4		Subtitle I-Petroleum Underground Storage Tanks	127
5.5		Summary of the Resource Conservation and Recovery Act	129
5.6		Medical Waste	129
5.7		Low-Level Radioactive Waste	130
5.8		References	130

x

Contents

Chapter 6 Comprehensive Environmental Response, Compensation, and
Liability Act.............. 133
6.1 Introduction 133
6.2 Potential Responsible Parties............... 133
6.3 CERCLA History 134
    6.3.1 Times Beach........... 134
    6.3.2 Love Canal 135
    6.3.3 Valley of the Drums 136
    6.3.4 Wells G and H Superfund Site in Woburn,
        Massachusetts..................... 137
    6.3.5 Ciba-Geigy and Reich Farm Site in Toms River,
        New Jersey 137
    6.3.6 PG&E Hinkley Site in California......... 138
    6.3.7 Summary of CERCLA History 139
6.4 Site Scoring 139
6.5 Remediation Criteria 140
6.6 Summary of Comprehensive Environmental Response,
Compensation, and Liability Act................. 145
6.7 References 145

Chapter 7 Hazardous Materials Transportation Act, Superfund
Amendments and Reauthorization Act, Emergency Planning and
Community Right to Know Act, Toxic Substance Control Act,
Pollution Prevention Act, Brownfield Revitalization Act............... 147
7.1 Introduction 147
7.2 Hazardous Materials Transportation Act................ 147
    7.2.1 Material Designation and Labeling........ 150
7.3 Toxic Substance Control Act 153
    7.3.1 Restricting Chemicals 154
    7.3.2 Regulating PCBs under TSCA............. 156
7.4 Superfund Amendments and Reauthorization Act and
Emergency Planning and Community Right-to-Know Act ... 156
7.5 Pollution Prevention Act.................. 163
7.6 Brownfield Revitalization Act 163
7.7 Summary and Conclusions................. 167
7.8 References 168

Chapter 8 Land-Related Environmental Regulations of the United States....... 171
8.1 Introduction 171
8.2 Insecticide Act; Feed, Drug, and Cosmetic Act; and
Federal Insecticide, Fungicide, and Rodenticide Act 172
    8.2.1 The Insecticide Act of 1910 172
    8.2.2 The Food, Drug, and Cosmetic Act........ 172
    8.2.3 The Insecticide, Fungicide, and Rodenticide Act
        of 1947 173

Contents   xi

	8.3	Occupational Safety and Health Act	174
	8.4	Lacey Act of 1900	176
	8.5	Antiquities Act of 1906	176
	8.6	Migratory Bird Treaty Act 1916	177
	8.7	National Parks Act of 1916	177
	8.8	Mineral Leasing Act of 1920	177
	8.9	Federal Power Act of 1935	178
	8.10	Atomic Energy Act of 1954	178
	8.11	National Historic Preservation Act of 1966	178
	8.12	Endangered Species Act of 1973	178
	8.13	National Forest Management Act of 1976	180
	8.14	Surface Mining Control and Reclamation Act of 1977	182
	8.15	Nuclear Waste Policy Act of 1982	183
	8.16	Energy Policy Act of 1992 and 2005	184
	8.17	Food Quality Protection Act of 1996	184
	8.18	Summary of Environmental Regulations of the United States	186
	8.19	References	187

Chapter 9 Review of Global Assessments and Standards for Land Contamination 189

	9.1	Introduction	189
	9.2	Climate and Geographic Influences	189
	9.3	Global Assessments and Environmental Standards for Land Pollution	192
	9.4	Summary of Global Assessments and Standards	192
	9.5	References	194

Chapter 10 Land Pollution Regulations of North America 197

	10.1	Introduction to North America		197
	10.2	Canada		197
		10.2.1	Environmental Regulatory Overview of Canada	198
		10.2.2	Solid and Hazardous Waste	200
		10.2.3	Remediation Standards	200
		10.2.4	Summary of Environmental Regulations of Canada	201
	10.3	Mexico		201
		10.3.1	Environmental Regulatory Overview of Mexico	202
		10.3.2	Solid and Hazardous Waste	202
		10.3.3	Remediation Standards	203
		10.3.4	Summary of Environmental Regulations in Mexico	206
	10.4	Summary and Conclusion		206
	10.5	References		207

xii Contents

Chapter 11 Land Pollution Regulations of Europe 209
    11.1   Introduction 209
    11.2   European Union... 209
          11.2.1   European Union Environmental Regulatory
                   Overview 209
          11.2.2   Solid and Hazardous Waste................................. 212
          11.2.3   Remediation 213
          11.2.4   Summary of Environmental Regulations of the
                   European Union 214
    11.3   Russia 214
          11.3.1   Environmental Regulatory Overview and
                   History of Russia... 215
          11.3.2   Solid and Hazardous Waste................................. 217
          11.3.3   Environmental Impact Assessments...................... 218
          11.3.4   Summary of Environmental Regulations and
                   Protection in Russia... 219
    11.4   Norway 219
          11.4.1   Environmental Regulatory Overview of Norway.... 220
          11.4.2   Solid and Hazardous Waste................................. 220
          11.4.3   Remediation 221
          11.4.4   Summary of Environmental Regulations of
                   Norway 221
    11.5   Switzerland 221
          11.5.1   Environmental Regulatory Overview of
                   Switzerland.. 222
          11.5.2   Solid and Hazardous Waste 223
          11.5.3   Remediation 223
          11.5.4   Summary of Environmental Regulations of
                   Switzerland.. 224
    11.6   Turkey 224
          11.6.1   Environmental Regulatory Overview of Turkey 224
          11.6.2   Solid and Hazardous Waste................................. 225
          11.6.3   Remediation 225
          11.6.4   Summary of Environmental Regulations
                   in Turkey 225
    11.7   Summary and Conclusions.. 226
    11.8   References 226

Chapter 12 Land Pollution Regulations of Africa.. 229
    12.1   Introduction 229
    12.2   South Africa... 231
          12.2.1   Environmental Regulatory Overview of South
                   Africa... 231
          12.2.2   Waste Management 234
          12.2.3   Remediation Standards.. 235
          12.2.4   Summary of South Africa Land Environmental
                   Regulations.. 236

Contents   xiii

12.3 Kenya.. 236
      12.3.1 Overview of Kenya Environmental Regulations..... 237
      12.3.2 Solid and Hazardous Waste................................... 237
      12.3.3 Remediation Standards.. 238
      12.3.4 Summary of Land Environmental Regulations
              of Kenya 239
12.4 Tanzania... 239
      12.4.1 Environmental Regulatory Overview
              of Tanzania 240
      12.4.2 Solid and Hazardous Waste................................... 240
      12.4.3 Remediation Standards.. 241
      12.4.4 Summary of Land Environmental Regulations
              of Tanzania 241
12.5 Egypt.. 241
      12.5.1 Environmental Regulatory Overview of Egypt....... 242
      12.5.2 Solid and Hazardous Waste................................... 242
      12.5.3 Remediation Standards.. 243
      12.5.4 Summary of Egyptian Land Environmental
              Regulations... 243
12.6 Summary and Conclusion.. 243
12.7 References 243

Chapter 13 Land Pollution Regulations of Asia................................. 247
13.1 Introduction 247
13.2 China.. 247
      13.2.1 Environmental Regulatory Overview of China....... 249
      13.2.2 Solid and Hazardous Waste................................... 250
      13.2.3 Remediation Standards.. 251
      13.2.4 Summary of Environmental Regulations
              in China 252
13.3 Japan 253
      13.3.1 Environmental Regulatory Overview of Japan 254
      13.3.2 Solid and Hazardous Waste................................... 254
      13.3.3 Remediation Standards.. 255
      13.3.4 Summary of Environmental Regulations
              of Japan 256
13.4 India.. 256
      13.4.1 Environmental Regulatory Overview of India 257
      13.4.2 Solid and Hazardous Waste................................... 257
      13.4.3 Remediation Standards.. 258
      13.4.4 Summary of Environmental Regulations
              of India 259
13.5 South Korea 259
      13.5.1 Environmental Regulatory Overview of
              South Korea.. 260
      13.5.2 Solid and Hazardous Waste................................... 261

xiv Contents

13.5.3 Remediation Standards.................................... 262
13.5.4 Summary of Environmental Regulations
of Korea... 263
13.6 Saudi Arabia 263
13.6.1 Environmental Regulatory Overview of
Saudi Arabia... 264
13.6.2 Solid and Hazardous Waste........................ 264
13.6.3 Remediation Standards.............................. 265
13.6.4 Summary of Environmental Regulations of
Saudi Arabia... 266
13.7 Indonesia... 266
13.7.1 Environmental Regulatory Overview
of Indonesia... 267
13.7.2 Solid and Hazardous Waste........................ 267
13.7.3 Remediation Standards.............................. 267
13.7.4 Summary of Environmental Regulations
in Indonesia... 268
13.8 Malaysia 268
13.8.1 Environmental Regulatory Overview
of Malaysia.. 268
13.8.2 Solid and Hazardous Waste........................ 268
13.8.3 Remediation Standards.............................. 268
13.8.4 Summary of Environmental Regulations
of Malaysia.. 269
13.9 Summary and Conclusion...................................... 269
13.10 References 269

Chapter 14 Land Pollution Regulations of Oceania...................... 273
14.1 Oceania... 273
14.2 Australia 273
14.2.1 Environmental Regulatory Overview
of Australia.. 274
14.2.2 Solid and Hazardous Waste........................ 274
14.2.3 Remediation Standards.............................. 275
14.2.4 Summary of Environmental Regulations of
Australia... 275
14.3 New Zealand.. 275
14.3.1 Environmental Regulatory Overview of New
Zealand.. 276
14.3.2 Solid and Hazardous Waste........................ 276
14.3.3 Remediation Standards.............................. 278
14.3.4 Summary of Environmental Regulations of
New Zealand 278
14.4 Summary and Conclusion...................................... 279
14.5 References 279

Contents   xv

Chapter 15 Land Pollution Regulations of South America.............................. 281
      15.1   Introduction 281
      15.2   Argentina... 281
            15.2.1   Environmental Regulatory Overview
                  of Argentina.. 281
            15.2.2   Solid and Hazardous Waste...................... 282
            15.2.3   Remediation Standards............................. 283
            15.2.4   Summary of Environmental Regulations
                  of Argentina.. 283
      15.3   Brazil 283
            15.3.1   Environmental Regulatory Overview of Brazil....... 284
            15.3.2   Solid and Hazardous Waste...................... 285
            15.3.3   Remediation Standards............................. 285
            15.3.4   Summary of Environmental Regulations
                  of Brazil.. 285
      15.4   Chile 286
            15.4.1   Environmental Regulatory Overview of Chile........ 286
            15.4.2   Solid and Hazardous Waste...................... 287
            15.4.3   Remediation Standards............................. 288
            15.4.4   Summary of Environmental Regulations
                  of Chile... 288
      15.5   Peru... 289
            15.5.1   Environmental Regulatory Overview of Peru 289
            15.5.2   Solid and Hazardous Waste...................... 290
            15.5.3   Remediation Standards............................. 290
            15.5.4   Summary of Environmental Regulations of Peru.... 291
      15.6   Summary and Conclusion.. 291
      15.7   References 291

Chapter 16 Current Status of Land Pollution and Regulations of the World 293
      16.1   Introduction 293
      16.2   Antarctica 293
            16.2.1   Pollution in Antarctica............................. 293
            16.2.2   Environmental Regulations of Antarctica 294
            16.2.3   Summary of Environmental Regulations in
                  Antarctica 295
      16.3   Oceans 295
            16.3.1   Ocean Pollution.. 296
            16.3.2   Environmental Regulations of the Oceans.............. 296
            16.3.3   Summary of Environmental Regulations of the
                  Oceans 297
      16.4   Summary of Environmental Regulations and Water
           Pollution of the World 297
            16.4.1   Evaluating and Ranking Each Country's
                  Regulatory Effectiveness for Land.......................... 298
            16.4.2   Environmental Challenges of Each Country........... 301

xvi   Contents

16.5 Summary and Conclusion.. 302
16.6 References 303

Chapter 17 Achieving and Maintaining Compliance with Land Regulations 305
17.1 Introduction 305
17.2 USEPA Audit Policy.. 306
17.3 Starting the Process 309
17.4 Conducting an Environmental Audit 310
    17.4.1 Categories of Findings 310
    17.4.2 Document List.. 311
    17.4.3 Opening Meeting... 312
    17.4.4 Environmental Audit Checklist 313
        17.4.4.1 Hazardous Waste 313
        17.4.4.2 Solid Waste... 317
        17.4.4.3 Used Oil... 319
        17.4.4.4 Universal Waste 319
        17.4.4.5 Company Owned or Operated
                 Landfills.. 320
        17.4.4.6 Spills... 321
        17.4.4.7 PCBs.. 322
        17.4.4.8 Storage Tanks 322
        17.4.4.9 Hazardous Materials................................. 323
        17.4.4.10 Toxic Substance Control Act.................. 323
        17.4.4.11 Emergency Planning and Community
                 Right-To-Know-Act 324
        17.4.4.12 Site Inspection 324
        17.4.4.13 Closing Meeting 326
        17.4.4.14 Report Preparation................................. 327
17.5 Agency Inspections.. 328
17.6 Environmental Audits and Sustainability 329
17.7 Summary of Fundamental Concepts of Environmental
     Compliance.. 329
17.8 References 333

Index.. 335

Preface

For the last 10,000 years, humans have modified the Earth through agriculture, building cities, resource exploitation, and other related activities. Now with more than 7 billion people on the planet, we are feeling the effects of our own actions by running out of habitable land, polluting the entire planet, and ultimately being responsible for the sixth major extinction in Earth's 4.54 billion year history. Modifications to the environment have been conducted to suit our needs, make our lives safer and more comfortable, and support our ever-growing population. To avoid feeling guilty and taking responsibility for our own actions, we have fooled ourselves and tried to justify the destruction of the natural environment by creating clever labels such as taming nature, western expansion, progress, sustainable development, and falsely claiming that Earth was created just for us.

Development of the land has many negative impacts on the environment that go well beyond the introduction of what we have come to associate with pollution. Many of these other forms do not introduce chemicals into the natural world. For instance, there are any number of different land uses where humans inflict enormous harm on our environment, including mining, petroleum development, urban land development, wetlands destruction, deforestation, forest management techniques, and agriculture. As we are discovering, these activities do not just affect a centralized area where the activity took place. These types of activities often have a rippling effect that negatively impacts surrounding areas and in some cases spreads like a disease that affects the entire planet.

Although there is much we can be proud of in limiting pollutant effects on the land through environmental regulation in the last 50 years, it clearly has not been enough, and our species is now at a critical crossroads. What we choose to do in the next few years to address not only climate change but other adverse impacts to water, land, and other living species will have a profound impact on how we and other species survive in the future.

One thing is for sure: things are going to change, and it's time to re-think how we have treated nature and adapt to a different world in the future. We have learned a hard lesson: instead of trying to change and tame nature, we must learn to understand and live in productive harmony with nature. This will mean that there will likely be a future increased need for additional environmental scientists and engineers to address and help solve anthropogenic impacts of pollution to the air, water, land, and living organisms. One very significant aspect of our technological advances in the last few centuries, which have introduced a multitude of pollutants, is that they have greatly outpaced nature's ability to evolve and adapt to humans' destructive behavior.

This book addresses environmental compliance with environmental regulations that apply to land. In addition to the characteristics of pollutants and associated regulations, the study of pollution is an interdisciplinary journey through the chemistry, physics, geology, and hydrology of our planet and its interconnectedness with geography, the oceans, land, biosphere, and even our nearest star, the sun. All of these factors and still others affect pollution. Therefore, we must thoroughly understand these

xvii

effects and influences if we desire to understand why and how pollution causes harm to us and the environment. This is where science meets environmental regulation. The purpose of environmental regulation is to protect human health and the environment. A major theme of this book and really any book that addresses pollution is that "no matter where you are, it's all the same." The reason is simple: pollution does not respect political boundaries between countries. There is a common saying in the geological world that states, "everything nature creates eventually ends up in the oceans."

The goal of this book is to provide the user with enough information so that compliance with environmental regulations that address land is achieved and maintained. To achieve this goal, several subject areas are addressed involving the science of pollution, the history of land pollution, land pollution regulations of the United States and the world, other land-related environmental regulations, worldwide assessments of land pollution and current status, and conducting an environmental audit of land pollution regulation compliance and applicability.

If we are successful at effectively addressing the harmful effects of climate change, we can then move on to correcting and restoring the impacts from many other pollutants from anthropogenic sources that are discharged into the air, water, and land and affect all life forms on Earth. Some of these other pollutants include ozone, smog, acid rain, fugitive emissions, volatile organic compounds, polychlorinated biphenyls (PCBs), polyaromatic hydrocarbons, heavy metals, criteria pollutants, fertilizers, pesticides, herbicides, semi-volatile organic compounds, and a plethora of hazardous pollutants, many of which are discussed through the regulatory framework in this book.

Sadly, in conclusion, climate change is just one of many examples of how humans have negatively impacted Earth through pollution in its many forms at a global scale. Other impacts caused by pollution that may not be in the public eye but are just as significant include impacts to the land, oceans, surface water, groundwater, and the biosphere. Each is related, and they affect each other. These impacts involve the physical and chemical composition and dynamics of pollutant behavior within the natural world. We must now accept that humans have adversely impacted the entire Earth and that our efforts to improve our environment since the enactment of environmental regulations in the United States and worldwide have been largely inadequate. To underscore this point, Earth scientists are convinced that we have now moved the needle of geologic time into a new period called the Anthropocene, which is defined as *the age of humans*. This is significant and means that scientists believe we are now in a time period that affects the entire Earth in a way that cannot be erased.

Acknowledgments

First and foremost, I wish to thank Dr. Jack Bregman and Robert Edell who together wrote the second edition of the Environmental Compliance Handbook. Their contributions to environmental science and education and their military service to the United States will be remembered and respected by all of us.

I wish to thank many regulatory agencies and scientific research organizations that assisted over the years providing research, opinions, and advice to many tough questions concerning protecting human health and the environment that have contributed to the development of this book. International regulatory agencies and countries I wish to thank include the World Health Organization, the United Nations, European Environmental Agency, Environment Canada, Mexico Environmental and Natural Resource Ministry, Norwegian Environment Agency, Turkey Ministry of Environment and Urbanization, India Ministry of Environment, Forestry, and Climate Change, Indonesia Ministry of Environment, Japan Ministry of the Environment, Korea Ministry of the Environment, Malaysia Department of Environment, Egyptian Environmental Affairs Agency, Kenya National Environment Management Authority, South Africa Department of Environmental Affairs, Tanzania Department of Environment, Brazil Ministry of the Environment, Chile Ministry of the Environment, Peru Ministry of the Environment, Argentine Secretariat for the Environment, Australia Department of Environment and Energy, and New Zealand Ministry for the Environment.

Other organizations and state agencies include International Union of Geological Sciences, Geological Society of America, International Association of Hydrogeologists (IAH), United States Environmental Protection Agency, Intergovernmental Panel on Climate Change (IPCC), United States Geological Survey (USGS), British Geological Survey, National Oceanic and Atmospheric Administration, Illinois State Geological Survey, California Environmental Protection Agency, California Department of Toxic Substances and Control, Delaware Department of Natural Resources and Environmental Control, Illinois Environmental Protection Agency, Indiana Department of Environmental Management, Iowa Department of Natural Resources, Kansas Department of Health and Environment, Maryland Department of Environment, Michigan Department of Environment, Great Lakes, and Energy, New Jersey Department of Environmental Protection, Ohio Environmental Protection Agency, Oregon Department of Environmental Quality, and the Wisconsin Department of Natural Resources.

Individuals who have provided guidance and advice from the research and academic field over the years include my colleagues at the University of Michigan, Dr. Martin Kaufman and Dr. Kent Murray and Dr. Ken Howard at the University of Toronto, Dr. Jack Sharp at the University of Texas, Dr. Richard Berg at the Illinois State Geological Survey, Dr. Robert Bobrowsky at the University of British Columbia, Dr. Peter Kolesar at Utah State University, Dr. Bill Farrand at the University of Michigan, Dr. Robert Oaks at Utah State University, Dr. Krause formerly at Henry Ford Community College, Dr. Derek Wong at APEX Environmental,

xix

Dr. Fred Payne with Arcadis, Dr. Colin Booth at Northern Illinois University, Dr. Mike Barcelona at Western Michigan University, Dr. Rebecca Spearot formerly with Clayton Environmental, Dr. Mary Ann Thomas at USGS, Dr. Garth van der Kamp at Environment Canada, Dr. Vladimir Kovalevsky at the Russian Academy of Sciences, Dr. John Moore with IAH, Dr. Hugo Loaiciga with the University of California, and Dr. John Chilton and Dr. Craig Foster with the British Geological Survey.

When dealing with environmental regulations of the United States and the world, one must also have guidance from the legal field. I wish to thank those special individuals who have been a source of inspiration, counsel, and debate concerning environmental regulations, who include Chris Athas, Esq.; Ed Brosius, Esq.; James Enright, Esq.; Rick Glick, Esq.; Geneva Halliday, Esq.; Brian Houghton, Esq.; Michael Krautner, Esq.; Stephen Lewis, Esq.; Michael Maher, Esq.; Chris McNevin, Esq.; Granta Y. Nakayama, P. C., Esq.; Jeryl Olsen, Esq.; Ron Paterson, Esq.; Tom Petermann, Esq.; Doug Schleicher, Esq.; Stephen Smith, Esq.; and Tom Wilczak, Esq.

I wish to thank the following professional colleagues for their assistance, cooperation, and ideas over the many years that it took to prepare this book: Lauren Alkadis, Cathee Andrews, Tony Anthony, Ann Barry, Gary Blinkiewicz, Tom Buggey, Kristine Casper, Chris Christensen Russ Chadwick, Bob Cigale, Tom Cok, Marc DeLoecker, Sheryl Doxtader, Ian Drost, Rob Ellis, Rose Ellison, Rob Ferree, Jennifer Formoso, Orin Gelderloos, Steve Hoin, Heather Hopkins, George Karalus, Steve Kitler, Shannon Lian, Rick Linnville, Daniel J. Lombardi, Bill Looney, Kim Myers, Ernie Nimister, David O'Donnell, Tom O'Hara, Ray Ostrowski, Paul Owens, Bruce Patterson, Sara Pearson, Scott Pearson, Mark Penzkover, Robert Ribbing, Eric Ross, Jeanne Schlaufman, Dave Slayton, Dave Smith, Ed Stewart, Patricia Thornton, Mary Vanderlaan, Chris Venezia, Nick Welty, Cheryl Wilson, and Tom Wenzel.

About the Author

Daniel T. Rogers is the Director–Environmental Affairs at Amsted Industries Incorporated in Chicago, Illinois. Amsted Industries is a diversified manufacturing company of industrial components serving railroad, vehicular, construction and building markets. Amsted Industries has more than 70 manufacturing locations in 15 countries. Mr. Rogers participates in environmental due diligence for acquisitions and divestitures; investigation and remediation of contamination; creating and evaluating environmental compliance and sustainability programs; and providing environmental advice, oversight, training, and negotiation strategies at various levels within the organization.

Over the last few decades, Mr. Rogers has published nearly 100 research papers in professional and academic publications and peer-reviewed journals on subjects including environmental geology, hydrogeology, geologic vulnerability and mapping, contaminant fate and transport, urban geology, environmental site investigations, contaminant risk, brownfield re-development, remediation, pollution prevention, environmental compliance, and sustainability. He has authored *Urban Watersheds: Geology, Contamination, Environmental Regulations, and Sustainability* Second Edition (2020); *Environmental Compliance and Sustainability: Global Challenges and Perspectives* (2019); and *Environmental Geology of Metropolitan Detroit* (1996). Mr. Rogers has published surficial geologic and contaminant vulnerability maps of the Rouge River watershed in southeastern Michigan. In addition, he is a contributing author of *The Encyclopedia of Global Social Issues* (2013), *Urban Groundwater* (2007), *Geoenvironmental Mapping* (2002), and *Groundwater in the Urban Environment* (1997).

He has taught geology and environmental chemistry at Eastern Michigan University and the University of Michigan and has presented guest lectures at numerous colleges and universities both in the United States and internationally.

Acronyms, Elements, Symbols, Molecules, and Units of Measure

ACRONYMS

ACGIH	American Conference of Governmental Industrial Hygienists
ADP	Atmospheric Decontamination Plan
APCP	Air Prevention and Control of Pollution Act
APP	Atmospheric Prevention Plan
AST	Aboveground storage tank
ASTM	American Society for Testing Materials
ATSDR	Agency for Toxic Substances and Disease Registry
BACT	Best available control technology
BLM	Bureau of Land Management
BMP	Best management practice
BPA	Bisphenol A
BTEX	Benzene, toluene, ethyl benzene, and xylenes
CAA	Clean Air Act
CAAC	Clean Air Alliance of China
CAS	Chemical Abstract Service
CalEPA	California Environmental Protection Agency
CEPA	Canadian Environmental Protection Act
CESQG	Conditionally exempt small quantity generator
CFC	Chlorofluorocarbons
CMEE	China Ministry of Ecology and Environment
CONAGUA	Mexico National Water Commission
CERCLA	Comprehensive Environmental Response, Compensation and Liability Act
CFR	Code of Federal Regulation
CRF	Contaminant or pollution risk factor
CRF_{air}	Contaminant or pollution risk factor for air
CRF_{gw}	Contaminant or pollution risk factor for groundwater
CRF_{soil}	Contaminant or pollution risk factor for soil
CSO	Combined sewer overflow
CVOC	Chlorinated volatile organic hydrocarbon
CWA	Clean Water Act
DBP	Di-n-butyl phthalate
DCE	dichloroethylene, dichloroethane
DEHP	Bi or Bis(2-ethylhexyl) phthalate
DEP	Diethyl phthalate

xxiii

DDT	Dichlorodiphenyltrichloroethane
DIVs	Dutch Intervention Values
DNAPL	Dense nonaqueous phase liquid
DOT	Department of Transportation
EEA	European Environment Agency
EEC	European Economic Community
EIA	Environmental impact assessment
EIS	Environmental impact statement
EHS	Extremely hazardous substance
EH&S	Environmental health and safety
EINETICS	European Information and Observation Network
EPCRA	Emergency Planning and Community Right-to-Know Act
EPA	Environmental Protection Agency
ESA	Environmental site assessment
ESPM	Elimination, substitution, prevention, and minimization
ETS	Emission Trading System
EU	European Union
FDA	Food and Drug Administration
FEPCA	Federal Environmental Pesticide Control Act
FFDCA	Federal Feed, Drug, and Cosmetic Act
FIFRA	Federal Insecticide, Fungicide, and Rodenticide Act
GACT	Generally available control technology
GHG	Greenhouse gas
GWP	Global warming potential
HAP	Hazardous air pollutant
HAZMAT	Hazardous materials
HCFC	Hydrochlorofluorocarbons
HFC	Hydrofluorocarbon
HMTA	Hazardous Materials and Transportation Act
HRS	Hazard ranking system
IED	Industrial emissions directive
IPCC	Intergovernmental Panel on Climate Change
KDNREP	Kentucky Department of Natural Resources and Environmental Protection
LEPC	Local emergency planning committee
LNAPL	Light nonaqueous phase liquid
LOAEL	Lowest-observed-adverse-effect level
LQG	Large quantity generator
MACT	Maximum achievable control technology
MCL	Maximum contaminant level
MCLG	Maximum contaminant level goal
MDEP	Massachusetts Department of Environmental Protection
MDEQ	Michigan Department of Environmental Quality
MEK	Methyl ethyl ketone
MEP	China Ministry of Environmental Protection
MNR	Russian Ministry of Natural Resources

MS4S	Large municipal separate storm sewer systems
MSDS	Material safety data sheet
MTBE	Methyl tertiary butyl ether
NAAQS	National Ambient Air Quality Standards
NAFTA	North American Free Trade Agreement
NAICS	North American Industry Classification System
NASA	National Aeronautics and Space Administration
NATO	North Atlantic Treaty Organization
NCZMP	National Coastal Zone Management Program
NERRS	National Estuarine Research Reserve System
NEA	Norwegian Environment Agency
NEPA	National Environmental Protection Act
NEPC	Australian National Environmental Protection Council
NESHAPS	National Emission Standards for Hazardous Air Pollutants
NGS	National Geographic Society
NHTSA	National Highway Traffic Safety Administration
NJDEP	New Jersey Department of Environmental Protection
NOAA	National Oceanic and Atmospheric Administration
NOM	Mexico Standards or Normas
NOV	Notice of violation
NPDES	National Pollutant Discharge Elimination System
NPDWS	National Primary Drinking Water Standards
NPL	National Priority List
NPS	Non-point source
NRC	National Response Center
NRC	Nuclear Regulatory Commission
ODS	Ozone-depleting substance
OECD	Organization for Economic Co-operation and Development
OPP	Office of Pesticide Programs (USEPA)
OSHA	Occupational Safety and Health Administration
OSPA	Oil Spill Prevention Act
PAH	Polycyclic aromatic hydrocarbon
PFC	Perfluorocarbon
PBT	Persistent bio-accumulative toxic
PCB	Polychlorinated biphenyl
PCE	Tetrachloroethylene, tetrachloroethene, perchloroethylene, perchloroethene
PCP	Pentachlorophenol
PERC	Tetrachloroethylene, tetrachloroethene, perchloroethylene, perchloroethene
PEL	Permissible exposure limit
PFOA or PFAS	Per- and polyfluoroalkyl substances
PM	Particulate matter
PNA	Polynuclear aromatic hydrocarbon
POTW	Publicly owned treatment works
PPA	Pollution Prevention Act

PPE	Personal protective equipment
PREFEPA	Mexico Federal Attorney Generalship of Environmental Protection
PRP	Potential responsible party
PS	Point source
PSD	Prevention of significant deterioration
PTE	Potential to emit
PVC	Polyvinyl chloride
RCI	Reactivity, corrosivity, and ignitability
RCRA	Resource, Conservation, and Recovery Act
REC	Recognized environmental condition
RQ	Reportable quantity
SARA	Superfund Amendments and Reauthorization Act
SDWA	Safe Drinking Water Act
SDS	Safety data sheet
SEMARNAT	Mexico Secretariat of the Environment and Natural Resources
SEP	Supplemental environmental project
SERC	State Emergency Response Commission
SIC	Standard Industrial Code
SMG	Small quantity generator
SPCC	Spill Prevention Control and Countermeasures Plan
SPLP	Synthetic precipitant leaching procedure
SRP	Spill response plan
SVOC	Semi-volatile organic compound
SWPPP	Stormwater pollution prevention plan
TCA	Trichloroethane
TCE	Trichloroethylene, trichloroethene
TCLP	Toxic characteristic leaching procedure
THM	Trihalomethane
UK	United Kingdom
TPQ	Threshold planning quantity
TSCA	Toxic Substance and Control Act
TSP	Total suspended particulates
UN	United Nations
UNFCCC	United Nations Framework Convention on Climate Change
UNESCO	United Nations World Heritage Site
USCB	United States Census Bureau
USCIA	United States Central Intelligence Agency
USDA	United States Department of Agriculture
USDC	United States Department of Commerce
USDOE	United States Department of Energy
USDHH	United States Department of Health and Human Services
USDOT	United States Department of Transportation
USEPA	United States Environmental Protection Agency
USFS	United States Forest Service
USGS	United States Geological Survey
USNRC	United States Nuclear Regulatory Commission

Acronyms, Elements, Symbols

UST	Underground storage tank
VC	Vinyl chloride
VOC	Volatile organic compound
VSQG	Very small quantity generator
WDNR	Wisconsin Department of Natural Resources
WHO	World Health Organization

ELEMENTS

Ag	Silver
As	Arsenic
Ba	Barium
Br	Bromine
C	Carbon
Cd	Cadmium
Cl	Chlorine
Cr	Chromium
Cr III	Trivalent chromium
Cr VI	Hexavalent chromium
$Cr+^3$	Trivalent chromium
$Cr+^6$	Hexavalent chromium
Cu	Copper
F	Fluorine
H	Hydrogen
Hg	Mercury
I	Iodine
K	Potassium
N	Nitrogen
O	Oxygen
P	Phosphorus
Pb	Lead
S	Sulfur
Se	Selenium
Zn	Zinc

SYMBOLS

η	Effective porosity
ρb	Bulk density
W_s	Water solubility
ATM	Atmosphere
$°C$	Degrees Celsius
$°F$	Degrees Fahrenheit
Foc	Organic carbon partition coefficient
H	Henry's law constant
Kd	Distribution coefficient

Koc	Fraction of total organic carbon
M	Mobility
MW	Molecular weight
P	Persistence
R	Retardation factor
T	Toxicity
VP	Vapor pressure

MOLECULES

CO	Carbon monoxide
CO_2	Carbon dioxide
CH_4	Methane
H_2O	Water
NO^x	Nitrogen oxides
N_2O	Nitrous oxide
O_3	Ozone
SO_2	Sulfur dioxide
SF_6	Sulfur hexafluoride

UNITS OF MEASURE

mg/kg	milligram per kilogram
ug/kg	microgram per kilogram
mg/l	milligram per liter
ug/l	microgram per liter
g/l	gram per liter
mg/m^3	milligram per cubic meter
ug/m^3	microgram per cubic meter
PM2.5	particulate matter less than 2.5 microns
PM10	particulate matter less than 10 microns
ppm	part per million
ppb	part per billion
TPY	tons per year
BTU	British thermal unit

1 Themes and Overview

1.1 INTRODUCTION

Land makes up 30% of the Earth's surface. Of that proportion, approximately 71% is considered habitable, with 10% covered by glaciers and the remaining 19% either being mountainous or extreme desert. Of the 71% of land that is habitable, 50% is used for agriculture, 36% is forest, 11% is shrubland, 2% is urban, and 1% is surface water, composed of lakes and streams (United Nations 2016a). Figure 1.1 shows the breakdown for land and land use on Earth. What should be apparent is that the amount of land on Earth that is already developed is significant, and this will have significant implications on environmental regulations and sustainability now and in the future.

For thousands of years, human civilization has been busy changing the environment for its own benefit, with little or no regard for the natural environment. Historically the focus was on exploitation of the environment rather than protecting the natural environment. This disregard for the natural environment even extended to the United States, where numerous examples of polluting land from anthropogenic sources were common and even accepted practices. Now that climate change has caught our attention, we are noticing and experiencing several of the detrimental effects, including higher temperature; severe drought; flooding; and more frequent, stronger, and slower-moving storm systems. This has raised our collective awareness to the fact that we can impact our environment on a global scale, and it does not include just the atmosphere.

The purpose of environmental regulation is to protect human health and the environment. To achieve this goal, nations of the world have learned to concentrate their efforts on environmental regulations that protect the air, water, land, and biosphere. The goal of this book is to provide the user with enough technical information and practical experience so that compliance with environmental regulations that address land is achieved and maintained.

A rhetorical theme of this book is, "no matter where you are, it's all the same." Land is the same throughout the world and must be protected with a sense of consistency and care, because without habitable land, our species and many others are at risk. This approach is critical if humans are to restore a sustainable balance with Earth, because, as we shall discover, land pollution does not respect political boundaries between countries. Land pollution caused by one country often pollutes neighboring countries and sometimes the entire planet when the pollutants released are at the right place, at the right time, and in the right amount. There is a saying that Earth scientists use to describe the erosional effects of water: "eventually everything that nature makes, eventually ends up in the oceans." This is an interesting concept and is also factual given enough time. What is interesting is that it also applies to pollution.

Environmental regulations that address land are most often reactive and are born out of a historical incident that results in an environmental disaster that all too often caused

DOI: 10.1201/9781003150107-1

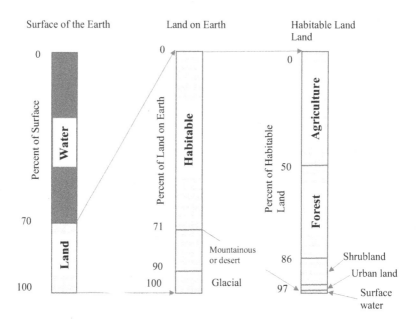

FIGURE 1.1 Breakdown of land and land use on Earth (from United Nations. *Human Development Report*. United Nations Development Programme. www.hdr.undr.org/sites/2017 [accessed July 25, 2021] 2016a).

enormous harm to humans and the environment. We shall explore many environmental incidents that have been significant in shaping environmental regulations that address land. We shall also go into detail about environmental regulations that deal with land to learn how to navigate the regulatory landscape and also learn the spirit and background of certain requirements under land regulations. This book will also provide guidance on land-related environmental regulations so that the user has practical information on technical information for almost all situations that deal with land regulations.

In addition to the environmental regulations that deal with land in the United States, this book also explores the many aspects of how humans have caused harm to the environment and examines how more than 50 countries choose to protect land through environmental regulations. This book concludes with a chapter dedicated to conducting an environmental audit at any facility with a focus on environmental regulations that address land.

1.2 MAJOR THEMES OF THIS BOOK

The following sections provide a brief introduction to the central themes that are covered in this book:

1. No matter where you are, it's all the same.
2. Pollution and its behavior in nature.

Themes and Overview

3. Land use and where cities are built.
4. History of environmental regulations.
5. Complexity and effectiveness of environmental regulations.
6. Environmental regulations and land pollution of the world.
7. Maintaining compliance with environmental regulations.

1.3 THEME 1: NO MATTER WHERE YOU ARE, IT'S ALL THE SAME

Of any theme presented in this book, "No matter where you are, it's all the same," is probably the most significant. As stated in the introduction, pollution does not care about country borders. As we shall discuss, land pollution can and does pollute the entire planet. Therefore, it makes sense that when dealing with many types of land pollutants, there should be consistency in environmental regulations that deal with land pollution throughout the world. This becomes most apparent when we discuss environmental regulations of the world.

1.4 THEME 2: POLLUTION AND ITS BEHAVIOR IN NATURE

Air, water, and land are essential for life on Earth. Life becomes threatened when any one of these becomes contaminated. As the human population continues to increase, we continue to modify and impact our environment. In many instances these actions result in the introduction of pollutants into the environment that degrade not just the local environment but can now be observed on a global scale and affect all life on Earth. Therefore, we must evaluate how these actions are impacting our planet and either eliminate, modify, or reduce the harmful impacts if we and other organisms are to survive. Protection of these three essentials for life on Earth are the environmental management fundamentals, along with directly protecting living organisms.

There would be no need for environmental laws and regulations or the United States Environmental Protection Agency if there were no pollution. Therefore, it seems logical that we begin our journey with examining pollution in its many forms. However, it does not just begin and end with pollution. There are other activities that impact our environment that go beyond what is commonly termed pollution. Many of these other forms do not introduce chemicals into the natural world. For instance, there are any number of different land uses where humans inflict enormous harm on our environment, including mining, petroleum development, urban land development, wetland destruction, deforestation, forest management techniques, and agriculture. As we are discovering, these activities do not just affect a centralized area where the activity took place. These types of activities often have a rippling effect that negatively impacts surrounding areas and in some cases spreads like a disease that affects the entire planet. As we will discover in the course of this book, this is where sustainability then plays an important and central role. Sustainability goes beyond addressing pollution and examines how humans, albeit unintentionally, adversely impact our planet as though we ourselves are a form of pollution.

4   Environmental Compliance Handbook

In order to fully understand pollution and its effects, we need to address four aspects:

- Physical chemistry of pollution
- Behavior of pollution in nature
- Nature's response to pollution
- The effects of pollution on humans and the environment

There are thousands of different types of pollution. To understand the behavior of pollution in nature, one must evaluate the physical chemistry of pollution and how pollution behaves in the geologic realms beneath the surface, in water, and in the atmosphere.

Sources of pollution include many human activities performed primarily at the surface resulting in the release of toxic substances into the environment. These toxic substances may be transported over time or remain relatively close to their source before they are degraded, transformed, or destroyed. Pollutants released into the environment are often mixtures, each contaminant is unique chemically, and the geologic environment in which the pollutants arc released is also unique. Understanding the physical chemistry of these chemicals and the geologic environment into which they are released becomes necessary for characterizing their fate and transport in nature (Rogers et al. 2007).

During their transport and before reaching their final sink, certain pollutants may reside at multiple intermediate sinks for different periods of time. Intermediate sinks include surface water, groundwater, and the atmosphere, because the contaminants held within these water- and land-containing sinks will flow and ultimately reach the oceans. Aquifers with a very low hydraulic conductivity are for practical purposes considered final sinks, as are inland bogs and some wetlands. Sediment and soil can function as intermediate or as final sinks, since erosion may move both of these unconsolidated materials. The oceans are almost always a final sink of pollution, but wave action occasionally brings pollutants onshore. To a large extent, the level of human health and/or environmental risk is a function of two fundamental concepts: *mobility* and *persistence*. Mobility is a measure of a substance's potential to migrate. Persistence is a measure of a substance's ability to remain in the environment before being degraded, transformed, or destroyed (Rogers et al. 2007). Substances with higher toxicity pose greater health risks, but the risk to humans and the environment grows exponentially if the chemical is both mobile and persistent. For example, a highly toxic but immobile chemical may affect a few people in a warehouse through inhalation, whereas a mobile and persistent chemical of moderate toxicity can contaminate a public water supply or migrate to a different sink where the potential for widespread human exposures and ecosystem damages is much higher.

We have control over the chemicals we use and where and how we use them. Control over the geologic environment, however, is beyond our means. Therefore, we must understand the geologic environment where our urban areas are located and develop methods to minimize or eliminate the potential harmful effects of contaminants upon human health and the environment.

Themes and Overview

Throughout the world, the largest cities share a geologic environment dominated by unconsolidated sedimentary deposits and are located near water (see Table 1.1). Most of those sedimentary deposits are saturated with water very near the surface and function as sources of drinking water and/or as hydraulic connections to surface water and ecosystems. Moreover, water is considered the universal solvent, so any pollution released into the environment from anthropogenic or natural sources has the potential to migrate and cause harm. Scientific factors that control the severity of harm to the environment from pollutant releases are: (1) the geologic and hydrogeologic environment, (2) the physical chemistry of the contaminants and amounts released, and (3) the mechanism in which the release occurs (Rogers 1996; Murray and Rogers 1999; Kaufman et al. 2005; Rogers 2018). There are numerous but costly techniques available for investigating, managing, and reducing the harmful effects of land pollution (Rogers 2016a, 2016b).

The largest polluting activity of land worldwide is agriculture. The largest polluter of air worldwide is motor vehicles, including automobiles, diesel trucks, trains, buses, marine vessels, and aircraft. The largest polluter of land worldwide is biological pollutants, largely from human waste and from agricultural activities that include pesticides and herbicides, fertilizers, and erosion (United Nations 2016a).

The final resting place for much of the pollution of the world are the oceans. With respect to pollution, we will learn that:

- No country, continent, or ocean is pollution free.
- Pollution caused by humans is everywhere.
- Significant environmental degradation caused by humans has touched every country.
- Pollution does not respect country borders.
- Pollution will continue to migrate in the air and water and on land.
- The oceans have been treated as the human privy and garbage disposal of the world.

1.5 THEME 3: LAND USE AND WHERE CITIES ARE BUILT

For nearly 10,000 years, humans have applied themselves toward changing their environment for their own benefit and enjoyment. Over the past few decades, there has been a realization that perhaps changing our environment may not be a wise choice. John Muir once said that "changing nature cannot be attained by force but by understanding." This phrase forms one of the focal points of this book, as it relates to environmental regulations and also has significant implications for sustainability. Smart land use in concert with nature can play a significant role in shaping a sustainable future in preventing pollution and minimizing its harm on the environment.

The surface of the Earth is approximately 510,100,000 square kilometers (Coble et al. 1987). Approximately 70%, or 361,800,000 square kilometers, is covered by water, and 30%, or 148,300,000 square kilometers, is land. Of that proportion, approximately 71% is considered habitable, with 10% covered by glaciers and the remaining

6   Environmental Compliance Handbook

19% either being mountainous or extreme desert. Of the proportion of the 71% that is habitable, 50% is used for agriculture, 36% is forest, 11% is shrubland, 2% is urban, and 1% is surface water, composed of lakes and streams (United Nations 2016a).

Despite the availability of specific methods and procedures, the natural environment of most urban areas is not well understood. A common thread of all major cities of the world is that they are all located near water and on unconsolidated sediments (see Tables 1.1 and 1.2). Modifications to land use can have a significant impact on improving the environment and greatly increasing sustainability measures. Repurposing the land we have already developed can potentially significantly reduce the amount of developed land, even with an expanding human population, as we shall see when employing several techniques and options for smart land use that include numerous available options we shall explore in this book.

TABLE 1.1
Major Urban Areas of the United States and Associated Surface Water Features

Urban Area	Surface Water Features
Atlanta	Chattahoochee River
Baltimore	Atlantic Ocean
Boston	Atlantic Ocean
Chicago	Lake Michigan and Chicago River
Cincinnati	Ohio River
Cleveland	Lake Erie
Dallas	Trinity River and White Rock River
Detroit	Detroit River
Denver	South Platte River
Houston	Gulf of Mexico
Indianapolis	White River
Kansas City	Missouri River
Las Vegas	Colorado River
Los Angeles	Pacific Ocean
Miami	Atlantic Ocean
Minneapolis	Mississippi River
New Orleans	Mississippi River and Gulf of Mexico
New York	Atlantic Ocean and Hudson River
Philadelphia	Atlantic Ocean and Delaware River
Phoenix	Gila River and Salt River
Pittsburgh	Ohio River
Portland	Pacific Ocean, Columbia River and Willamette River
Saint Louis	Mississippi River and Missouri River
Salt Lake City	Great Salt Lake and Jordan River
San Antonio	Salado River, San Antonio River, and Olmos River
San Francisco	Pacific Ocean
Seattle	Pacific Ocean
Tampa	Gulf of Mexico
Washington, D. C.	Potomac River

Themes and Overview

TABLE 1.2

Water Bodies and Near-Surface Geology near the Major Cities of the World

City	Location	Estimated Metropolitan Population (millions)	Dominant Surficial Geology	Water Body
Tokyo	Japan	37.8	Unconsolidated—fluvial	Pacific Ocean
Shanghai	China	34.8	Unconsolidated—fluvial	Pacific Ocean
Jakarta	Indonesia	31.7	Unconsolidated—fluvial	Pacific Ocean
Delhi	India	26.4	Unconsolidated—fluvial	Yamuna River
Seoul	Korea	25.5	Unconsolidated—fluvial, lacustrine	Han River
Beijing	China	24.9	Unconsolidated—fluvial, lacustrine	Yongding River
New York City	USA	23.8	Unconsolidated—glacial, alluvial	Atlantic Ocean
Mexico City	Mexico	21.6	Unconsolidated—lacustrine	Lerma River
Sao Paulo	Brazil	21.2	Unconsolidated—fluvial, alluvial	Atlantic Ocean
Cairo	Egypt	20.5	Unconsolidated—fluvial, eolian	Nile River
Los Angeles	USA	18.7	Unconsolidated—fluvial, alluvial	Pacific Ocean
Moscow	Russia	16.9	Unconsolidated—fluvial, glacial	Moskve River
Istanbul	Turkey	15.2	Unconsolidated—fluvial, eolian, alluvial	Turkish Straits
London	England	14.2	Unconsolidated—fluvial	Thames River
Buenos Aires	Argentina	13.1	Unconsolidated—fluvial, alluvial	Atlantic Ocean
Paris	France	12.6	Unconsolidated—fluvial	Seine River
Rio de Janeiro	Brazil	12.3	Unconsolidated—fluvial, alluvial	Atlantic Ocean
Chicago	USA	9.8	Unconsolidated—fluvial, glacial, lacustrine	Lake Michigan
Johannesburg	South Africa	9.6	Unconsolidated—fluvial, lacustrine	Jukskei River
Riyadh	Saudi Arabia	7.7	Unconsolidated—eolian, alluvial	Simbacom River
Santiago	Chile	6.7	Unconsolidated—fluvial, alluvial, lacustrine	Pacific Ocean
Berlin	Germany	6.1	Unconsolidated—fluvial	Spree River
Toronto	Canada	5.9	Unconsolidated—fluvial, glacial, lacustrine	Lake Ontario
Sydney	Australia	5.0	Unconsolidated—fluvial, alluvial	Pacific Ocean

Source: United Nations, 2021. World Population Prospects. United Nations Department of Economic and Social Affairs. Population Division. https://esa.un.org/unpd/wpp/data. (Accessed August 20, 2021.)

1.6 THEME 4: HISTORY OF ENVIRONMENTAL REGULATIONS

In order to know where we are heading into the future as a species, we often must reflect back at our history for perspective. From a historical point of view, our treatment of the environment has not been kind. As recently as the early portion of the 20th century, environmental awareness and legislation were generally lacking not only in the United States but worldwide. There were a few international agreements that primarily focused on boundary waters, navigation, and fishing rights along shared waterways between countries. However, they ignored pollution and ecological issues. In fact, the most convenient and least costly method of waste disposal up until the middle of the 20th century was "up the stack or down the river" (Haynes 1954).

The 1962 Rachel Carson book *Silent Spring* described the environmental effects of DDT on birds and catapulted the environmental movement in the United States into the public eye (Carson 1962). The 1960s and 1970s are generally described as the age of environmentalism in the United States. This is when the United States realized that environmental degradation had significantly affected the air and water quality of many urban areas, especially the air in Los Angeles and water in Lake Erie, which is one of the Great Lakes. Starting on Earth Day in 1971, the Ad Council and Keep America Beautiful campaign aired public awareness commercials on television depicting a Native American shedding a tear when looking out over the polluted landscape of the United States. The commercial stated, "people start pollution, people can stop pollution" (United States Advisory Council on Historic Preservation 2021).

Although laws focused on protecting human health or the environment can be traced back more than 2,000 years, it took the United States in the 1970s with the passage of the most significant environmental regulations and laws to set the example for the rest of the world (Hahn 1994).

Environmental regulations of the United States and the world are largely based on reactions to an incident that caused enormous harm to human health and the environment. A few of the many examples of incidents that resulted in enactment of environmental regulations include Love Canal in New York; Times Beach in Missouri; the Bhopal, India, gas tragedy; the fire on the Cuyahoga River near Cleveland; the *Exxon Valdez* oil spill; and the recent incident on the *Deepwater Horizon* oil drilling platform in the Gulf of Mexico.

1.7 THEME 5: COMPLEXITY AND EFFECTIVENESS OF ENVIRONMENTAL REGULATIONS

Factors that influence the effectiveness of environmental regulations vary from country to country and within each country as well. The United States is no different. Factors that influence the degree to which a country has effective environmental regulations include (Bates and Ciment 2013):

- Political will
- Environmental tragedies, harm to the environment, and lessons learned
- Cost
- Geography and climate
- Enforcement
- Incentives
- Lack of overriding social factors such as war, political structure, greed, poverty, hunger, lack of infrastructure, educational awareness, and corruption

Environmental laws and regulations in the United States and the world have continually become more numerous and complex to the point that even many scientists and other professionals need assistance at interpreting, understanding, and applying many environmental laws and regulations.

Themes and Overview

The growth of the number and complexity of environmental laws and regulations can be more easily understood by simply applying the principle of the scientific term called entropy. Entropy is a principle that states that as time progresses, there is more disorder, which increases complexity. This can be applied to our human civilization as well. Our civilization has become more complex with time as a result of our technological advances, population increase, and other factors. This in turn has put pressure on the natural world, and our response has been the passage of environmental laws and regulations as an attempt to establish order in the disordered and out-of-equilibrium environment that humans caused. To make this point more clear, a list of laws that are commonly credited with protecting human health and the environment just in the United States include (USEPA 2021):

- Rivers and Harbors Act of 1899
- Antiquities Act of 1906
- Atomic Energy Act of 1946
- Atomic Energy Act of 1954
- Brownfield Revitalization Act of 2002
- Clean Air Act (CAA) of 1970
- Clean Water Act (CWA) of 1972
- Coastal Zone Management Act of 1972
- Comprehensive Environmental Response, Compensation, and Liability Act (CERCLA) of 1980
- Emergency Planning and Community Right-to-Know Act of 1986
- Endangered Species Act of 1973
- Energy Policy Act of 1992
- Energy Policy Act of 2005
- Federal Power Act of 1935
- Federal Feed, Drug, and Cosmetic Act (FFDCA) of 1938
- Federal Insecticide, Fungicide, and Rodenticide Act (FIFRA) of 1947
- Fish and Wildlife Coordination Act of 1968
- Food Quality Protection Act of 1996
- Fisheries Conservation and Management Act of 1976
- Global Climate Protection Act of 1987
- Hazardous Materials Transportation Act (HMTA) of 1975
- Insecticide Act of 1910
- Lacey Act of 1900
- Low-Level Radioactive Waste Policy Act of 1980
- Marine Protection Act of 1972
- Marine Mammal Protection Act of 2015
- Medical Waste Tracking Act of 1980
- Migratory Bird Treaty Act of 1916
- Mineral Leasing Act of 1920
- National Environmental Policy Act of 1969
- National Forest Management Act of 1976
- National Historic Preservation Act of 1966
- National Parks Act of 1980

10 Environmental Compliance Handbook

- Noise Control Act of 1974
- Nuclear Waste Policy Act of 1982
- Ocean Dumping Act of 1988
- Occupational Safety and Health Act (OSHA) of 1970
- Oil Spill Prevention Act of 1990
- Pollution Prevention Act (PPA) of 1990
- Refuse Act of 1899
- Resource, Conservation, and Recovery Act (RCRA) of 1976
- Rivers and Harbors Act of 1899
- Safe Drinking Water Act (SDWA) of 1974
- Superfund Amendments and Reauthorization Act (SARA) of 1986
- Surface Mining Control and Reclamation Act of 1977
- Toxic Substance Control Act of 1976
- Wild and Scenic Rivers Act 1968

1.8 THEME 6: ENVIRONMENTAL REGULATIONS AND POLLUTION OF THE WORLD

Some of the most important conclusions reached in this book concerning environmental regulations of the world include:

- Each country has environmental regulations
- No two countries are exactly the same
- Each country has a unique set of attributes that influence the effectiveness of environmental protection that include climate, geography, geology, social and economic concerns, and politics
- The platform of environmental regulations of each country is similar to that of the United States or the European Union
- Although some countries are doing a better job than others, each country struggles with protecting human health and the environment for various reasons, two of which are universal: a growing population and urban expansion
- Every country has been significantly affected by pollution of the air, water, and land

On Earth, the largest polluting activity is agriculture. The largest polluter of air is motor vehicles, including automobiles, diesel trucks, trains, buses, marine vessels, and aircraft. The largest polluter of land worldwide is biological pollutants, largely from human waste and from agricultural activities that include pesticides and herbicides, fertilizers, and erosion (United Nations 2016a).

The final resting place for much of the pollution of the world is the oceans.

International organizations such as the WHO and the UN have begun to assess, on a global scale, the quality of fresh water and air on Earth. An evaluation of land has yet to be completed. According to WHO (2013, 2015) and the United Nations (2016b, 2016c), approximately 2.8 billion humans lack access to basic sanitation and improved drinking water. While this number is improving, it still represents nearly

Themes and Overview 11

30% of the planet's human population. In addition, according to WHO (2021), 9 out of 10 people on Earth breathe air containing high levels of pollution. The United Nations estimates that most land pollution is caused by agricultural activities, such as grazing, use of pesticides, fertilizers, irrigation, plowing, and confined animal facilities (United Nations 2016a).

Some of these actions have affected an individual site or location, and others have affected the entire planet. This book explores how these actions have molded and influenced the development of environmental regulations worldwide and identifies and explores many new approaches that can be undertaken to improve our environment. These improvements and modifications range from the individual to global level through several avenues that include education, science-based assessments, improving existing environmental regulations, and enacting new environmental regulations with the goal of achieving a human population that is sustainable with Earth.

1.9 THEME 7: MAINTAINING COMPLIANCE WITH ENVIRONMENTAL REGULATIONS

Environmental compliance is sometimes perceived as just another set of rules and is easy to accomplish. Just follow the rules. This premise fails on several levels. First, the human population is ever increasing, and our thirst for land and modern conveniences increases as our standard of living is raised. Second, our technology is ever changing with new products and chemicals. Third, the Earth is dynamic. There is change, growing ever more complex as more information and science are discovered, learned, and shared. The weather changes constantly and is perhaps the best example of how to imagine environmental management and risk. It also changes, as does everything. Everything obeys the second law of thermodynamics, namely entropy, in that nature tends to become more complex as time marches on.

Understanding the natural setting of our urban and industrial areas, knowing the chemicals that are used and how they might cause harm to the environment and humans if they are released, what alternative chemicals are available, how to reduce chemical and energy use, preventing chemical releases to the environment, quantifying the costs involved with cleaning up the environment once a release occurs, and developing environmental stewardship are all subjects that should be taken into account when evaluating the environmental health of a country, province, state, city, manufacturing facility, or an individual household, and it all starts with conducting an environmental audit.

The significant lessons or principles of environmental compliance in this chapter include:

1. It all begins with conducting an environmental audit.
2. Being in compliance with environmental regulations does have sustainability implications.
3. Work yourself out of environmental regulations.
4. There is little we can control.
5. Limit chemical use when possible.

12 Environmental Compliance Handbook

6. When possible, eliminate chemicals that are very toxic or have a high chemical risk factor.
7. Fully understand environmental permits.
8. Never accept a permit term that the facility can't achieve.
9. Always be accurate and truthful.
10. When in doubt, it's usually always better to report a spill or other incident where reporting may be required.
11. The importance of waste characterization and points of generation.
12. Any operational changes may require notice and permit changes or a new permit.
13. Give yourself enough time for collecting additional compliance samples in case a data quality issue arises.
14. Keep well organized and communicate with management and employees regularly.
15. Housekeeping.
16. Signage.
17. Conduct spill drills.
18. Prepare a list of onsite chemicals with corresponding reportable quantities (RQs) and recommended cleanup procedures.
19. Conduct regular inspections.
20. Proper preparation prevents poor performance.
21. In many instances, proving a negative is required when evaluating a release.
22. When a question arises, consult management, in-house environmental counsel, outside environmental counsel, or the regulatory authority, as appropriate.
23. Last, remember that "no matter where you are, it's all the same."

We are learning that we live on one planet with no environmental boundaries and that contamination does not respect boundaries. Another point is that environmental regulations are built on the USEPA or EU platforms, and those platforms rely on basic principles that include:

1. Protect the air.
2. Protect the water.
3. Protect the land.
4. Protect living organisms.
5. Protect cultural and historic sites.

1.10 SUMMARY

The goal of this book is to provide the user with enough information so that compliance with environmental regulations that address land is achieved and maintained. To achieve this goal, we will navigate several subject areas involving the science of land pollution, the history of land pollution, land pollution regulations of the United States and the world, land-related environmental regulations, worldwide assessments of land pollution and current status, and conducting an environmental audit of land pollution regulation compliance and applicability.

Themes and Overview

1.11 REFERENCES

Bates, C. G. and Ciment, J. 2013. *Encyclopedia of Global Social Issues*. M.E. Sharpe Publishers. New York, NY. 1450p.

Carson, R. 1962. *Silent Spring*. Houghton Mifflin. Boston, MA.

Coble, C. R., Murray, E. G. and Rice, D. R. 1987. *Earth Science*. Prentice-Hall Publishers. Englewood Cliffs, NJ. 502p.

Hahn, R. W. 1994. United States Environmental Policy: Past, Present and Future. *Natural Resource Journal*. John F. Kennedy School of Government, Harvard University. Cambridge, MA. Vol. 34. No. 1. pp. 306–348.

Haynes, W. 1954. *American Chemical Industry—A History*. Vols. I–IV. Van Nostrand Publishers. New York, NY.

Kaufman, M. M., Rogers, D. T. and Murray, K. S. 2005. An Empirical Model for Estimating Remediation Costs at Contaminated Sites. *Journal of Water, Air and Soil Pollution*. Vol. 167. pp. 365–386.

Murray, K. S. and Rogers, D. T. 1999. Groundwater Vulnerability, Brownfield Redevelopment and Land Use Planning. *Journal of Environmental Planning and Management*. Vol. 42. No. 6. pp. 801–810.

Rogers, D. T. 1996. *Environmental Geology of Metropolitan Detroit*. Clayton Environmental Consultants, Novi, MI.

Rogers, D. T. 2016a. Next Generation of Urban Hydrogeologic Investigations. *Journal of the Italian Geological Society*. Rome, Italy. Vol. 39. No. 1. pp. 349–353.

Rogers, D. T. 2016b. Scientific Advancements That Improve the Conceptual Site Model in Urban Hydrogeological Site Investigations. *35th International Geological Congress*. Paper 3019. Cape Town, South Africa.

Rogers, D. T. 2018. Derivation of a Comprehensive Environmental Risk Model for Urban Groundwater Protection. *International Association of Hydrogeologists Congress*. Vol. 1. Daejeon, Korea.

Rogers, D. T., Murray, K. S. and Kaufman, M. M. 2007. Assessment of Groundwater Contaminant Vulnerability in an Urban Watershed in Southeast Michigan, USA. In: Howard, K. W. F. editor. *Urban Groundwater—Meeting the Challenge*. Taylor & Francis, London, England.

United Nations. 2016a. *Human Development Report*. United Nations Development Programme. www.hdr.undr.org/sites/2017. (Accessed September 16, 2017).

United Nations. 2016b. *Global Drinking Water Quality Index and Development and Sensitivity Analysis Report*. UNEP Water Programme Office. Burlington, Ontario, Canada. 58 pages.

United Nations. 2016c. *World Air Pollution Status*. United Nations News Center. New York, NY. www.un.org/sustainabledevelopment/2016/09. (Accessed July 25, 2021).

United Nations. 2021. World Population Prospects. United Nations Department of Economic and Social Affairs. Population Division. https://esa.un.org/unpd/wpp/data. (Accessed August 20, 2021).

United States Advisory Council on Historic Preservation. 2021. National History Preservation Act. www.achp.gov. (Accessed July 25, 2021).

United States Environmental Protection Agency. 2021. Summary of Environmental Regulations of the United States. https://epa.gov/summary-environmental-laws. (Accessed July 25, 2021).

World Health Organization. 2013. Water Quality and Health Strategy 2013–2020. www.who.int/water_sanitation_health/dwq/en/. (Accessed October 18, 2018).

World Health Organization. 2015. *Progress on Drinking Water and Sanitation*. World Health Organization. New York, NY. 90p.

World Health Organization. 2021. *Nine Out of Ten People Breathe Unhealthy Air*. Geneva, Switzerland. www.who.int/news-room/detail/02-02-2018. (Accessed July 25, 2021).

2 History of Land Pollution and the Beginning of Environmental Regulations in the United States

2.1 INTRODUCTION

Land pollution refers to the deterioration of the Earth's land surfaces, at or below ground level, from anthropogenic sources (NOAA 2021). Causes of anthropogenic land pollution are numerous and don't just include releases of chemical contaminants but also human development and activities that disturb the land and often destroy natural habitats. In this book, we will concentrate on chemical contaminants and leave the destruction of habitat and other negative environmental anthropogenic activities for the book that addresses sustainability. Land pollution is not a new phenomenon. Land pollution has been a problem for thousands of years and has become more and more serious as the human population increases. An increasing human population opens the door for exposure to chemical contaminants and to more bacteria and disease through accumulation of human waste, animal waste, and our garbage (NOAA 2021; USEPA 2021a).

Many believe that environmental regulations in the United States are the most comprehensive and protective on Earth. In many respects, this is true, but there are significant exceptions and gaps. The purpose of environmental regulations is to protect human health and the environment. However, protecting human health and the environment is difficult if the regulations are not comprehensive and all encompassing. To assist in explaining this point, it is best to demonstrate the need for comprehensive environmental regulations through example scenarios. For example, if a person were placed in a situation where there were ten detrimental environmental circumstances that may cause harm to that person and the person were only protected through regulations against nine of the items, that person would still be harmed. Some may conclude that the regulations were ineffective in this example because the person was still harmed. This is a common frustration of applying environmental regulations; there always seems to be something no one thought of that leads to an unacceptable exposure. This is one of the reasons environmental regulations are complex: they have to be in order to be truly protective.

DOI: 10.1201/9781003150107-2

Another challenge is that the environmental regulations of the United States only apply to the United States. What if the person in the previous example were protected from nine of the ten potential hazards, but the one hazard that the person was not protected from originated in a different country? This is more difficult to address because our laws only apply to our country. This scenario is central to one our themes in this book, which is *pollution does not respect boundaries*. This is another reason environmental regulations are so complex and, sometimes, ineffective.

A third challenge from our example would be that the person was harmed not because the regulations were inadequate but because something new was introduced into the situation that was not present when the regulations took effect and modified the situation enough to cause harm. This is why environmental regulations always seem to need to be updated and improved. New information and facts or new circumstances arise that must be accounted for and addressed. This is yet another example of why environmental regulations are so complex and become more complex with time; they need to be comprehensive.

Last, environmental regulations are in large part based on single incidents that cause large-scale harm to human health and the environment. Therefore, they are reactive rather than proactive, which tends to narrow the focus of regulations and does not fully address other circumstances that cause similar harm.

Factors that influence the effectiveness of environmental regulations vary from country to country and within each country as well. The United States is no different. Factors that influence the degree to which a country has effective environmental regulations include (Bates and Ciment 2013; United Nations 2021a):

- Political will
- Environmental tragedies, harm to the environment, and lessons learned
- Cost
- Geography and climate
- Enforcement
- Incentives
- Lack of overriding social factors such as war, political structure, greed, poverty, hunger, lack of infrastructure, educational awareness, and corruption

Along with our example situations, each of the previous listed factors either influences the effectiveness of environmental regulations in a positive way or undermines environmental regulations.

2.2 OVERVIEW OF ENVIRONMENTAL REGULATIONS FOR LAND IN THE UNITED STATES

To begin our journey through environmental regulations for land, we must first consider the meaning of the term **environment**. According to the United Nations (2021a), the environment is a broad term that encompasses all living and nonliving things on Earth or some regions thereof. This definition identifies the natural world existing within the atmosphere, hydrosphere, lithosphere, and biosphere and also includes the built or developed world (defined as the **anthrosphere**).

History of Land Pollution

Second, we must define the difference between a law and a regulation. A **law** is a written statute passed by an appropriate governing body. A **regulation**, which is often used interchangeably with a law, is a set of standards and rules adopted by administrative agencies that govern how laws will be enforced. In our case, that primarily means the United States Environmental Protection Agency. Regulations often have the same force as such laws, since, without them, regulatory agencies wouldn't be able to enforce laws. Therefore, the USEPA is called a regulatory agency because the United States Congress authorized the USEPA to write regulations that explain the technical, operational, and legal details necessary to implement laws passed by the United States Congress (USEPA 2021a, 2021b). Although laws focused on protecting human health or the environment can be traced back more than 2,000 years, it was the United States in the early 1970s that set the example when it enacted the most significant and comprehensive environmental regulations and laws in the world (Hahn 1994). Based on the example set by United States, over 50 countries have established laws that protect human health and the environment on the same platform as USEPA. Table 2.1 is a short list of the countries that have well-established laws that protect human health and the environment.

Environmental laws and regulations in the United States have methodically become more numerous and complex, to the point that even many scientists and other professionals need assistance interpreting, understanding, and applying many

---

TABLE 2.1

Select List of Countries with Established Environmental Regulations Based on the United States.

Africa	EUROPE	malta
Egypt	Austria	Netherlands
Kenya	Belgium	Norway
South Africa	Bulgaria	Poland
ASIA	Cyprus	Portugal
China	Czech Republic	Romania
India	Denmark	Russia
Japan	Estonia	Slovakia
Kyrgyzstan	Finland	Slovenia
Pakistan	France	Spain
NORTH AMERICA	Germany	Sweden
Canada	Greece	Switzerland
Mexico	Iceland	Turkey
United States	Ireland	United Kingdom
SOUTH AMERICA	Italy	**OCEANIA**
Bolivia	Latvia	Australia
Brazil	Lithuania	New Zealand
Chile	Luxembourg	Indonesia

Source: United Nations, 2021a. The United Nations Environment Programme. www.unenvironment.org/environment-you (accessed February 15, 2021).

18 Environmental Compliance Handbook

of these environmental laws and regulations. The current cost of protecting human health and the environment in the United States is generally considered 2% of the gross national product (USEPA 2021a).

As previously stated in the Chapter 1, the growth of the number and complexity of environmental laws and regulations can be more easily understood by simply applying the principle of the scientific term called entropy. Entropy is a principle that states that as time progresses, there is more disorder, which increases complexity. This can be applied to our human civilization as well. Our civilization has become more complex with time as a result of our technological advances, population increase, development, and other factors. This in turn has put pressure on the natural world, and our response has been the passage of environmental laws and regulations as an attempt to establish order in the disordered and out-of-equilibrium environment that humans caused.

The basic overriding principle of environmental law of the United States is to protect human health and the environment. This principle is broken down into categories that track some of the laws and how humans impact the environment and other living species with which we share our planet (Lazarus, R. 2004). These principles include:

- Environmental Impact Assessment
- Water Quality
- Contaminant Investigation
- Risk Assessment
- Chemical Transport
- Mineral Resources
- Wildlife
- Soil
- Hunting
- Equity
- Public Transparency
- Prevention

- Air Quality
- Waste Management
- Contaminant Cleanup
- Chemical Safety
- Water Resources
- Forest Resources
- Plants
- Fish
- Sustainable Development
- Transboundary Responsibility
- Public Participation
- Polluters Pay

2.3 THE BIRTH OF THE ENVIRONMENTAL MOVEMENT AND REGULATIONS IN THE UNITED STATES

The term **environmental regulation** refers to a set of laws, sometimes referred to as an amalgam of principles, treaties, laws, regulations, directives, ordinances, or policies, at the local, state, or federal level that are enacted to protect the environment in some shape or form (USEPA 2021a, 2021b). Environmental regulations or laws can be traced to the Roman Empire when the Roman Senate passed legislation to protect the city's supply of water that addressed concern over fresh water and sewage disposal (Journal of Roman Studies 2021).

In the 14th century, England prohibited the burning of coal in London and the disposal of waste into waterways, specifically the Thames River (Smithsonian 2021). In December 1682, in what would become the United States, William Penn ordered that one acre of forest be preserved for every five acres cleared for settlement (Pennsylvania Historical Association 2021). In addition, Benjamin Franklin widely wrote about the

History of Land Pollution

adverse effects of lead and dumping of waste (Environmental Education Associates 2021). However, it was not until 1899 that the United States government passed what is commonly referred to as the first environmental law in the United States, the Rivers and Harbors Act, which was intended to clear the harbors of garbage so that ships would not be impeded in transporting goods (USEPA 2021a, 2021b).

It was not until the industrial revolution, starting around 1760, that environmental degradation of the air, water, and land became a significant issue. The industrial revolution marked a time of rapid transition from making products by hand to making products with machines. The stationary steam engine invented by James Watt led to widespread use of steam power that used coal. New chemical manufacturing methods, iron production processes, the development of machine tools, and the rise of the factory system all came about in a relatively short time, which sparked the rise of the industrial age and resulted in the discharge of large amounts of waste into the air, into waterways, and discarded onto the land.

As recently as the first half of the 20th century, environmental awareness and legislation were generally lacking not only in the United States but worldwide. There were a few international agreements that primarily focused on boundary waters, navigation, and fishing rights along shared waterways between countries. However, they ignored pollution and ecological issues. In fact, the most convenient and least costly method of waste disposal at the time was "up the stack or down the river" (Haynes 1954). It was not until the 1962 publication of the Rachel Carson book *Silent Spring*, which described the environmental effects of DDT on birds, that the environmental movement in the United States catapulted into the public eye (Carson 1962).

The 1960s and 1970s are generally described as the age of environmentalism in the United States. **Environmentalism** is defined as a value system that seeks to redefine humankind's relationship with nature (Tarlock 2001). This is when the United States realized that environmental degradation had significantly affected the air and water quality of many urban areas, especially the air in Los Angeles and the water in Lake Erie, which is one of the Great Lakes. Starting on Earth Day in 1971, the Ad Council and Keep America Beautiful campaign aired public awareness commercials on television depicting a Native American shedding a tear when looking out over the polluted landscape of the United States. The commercial stated, "people start pollution, people can stop pollution" (see Figure 2.1).

Two significant sites of environmental contamination were discovered about this time: Times Beach in Missouri and Love Canal near Buffalo, New York. These two sites would become the first Superfund Sites in the United States. Currently, there are over 1,500 Superfund Sites (USEPA 2021b).

2.4 FORMATION OF NATIONAL POLICY AND THE ENVIRONMENTAL PROTECTION AGENCY

On December 2, 1970, then-President Richard Nixon signed an executive order that established the USEPA, and a cavalcade of environmental legislation ensued. The creation of the USEPA originated from the passage of the United States National Environmental Policy Act (NEPA) of 1969 (USEPA 2021a). NEPA required the

FIGURE 2.1 Pollution: it's a crying shame (from United States Advisory Council on Historic Preservation [2021]. (www.adcouncil.org/Our-Campaigns/The-Classics/pollution-keep-america-beautiful-iron-eyes-cody [accessed February 15, 2021], 2021).

federal government to give environmental issues priority when planning any type of large-scale projects and also provided for the establishment of councils and agencies to monitor and protect the environment. NEPA was developed to establish consistency in application of environmental laws and standards throughout the United States (Eccleston 2008; USEPA 2021a).

Prior to its passage and the creation of the USEPA, some states had pollution statutes, but many did not. For example, Ohio in particular had an active and highly regarded environmental program, but surrounding states did not. The result created significant confusion because what was considered permissible at a location in one state that resulted in the release of pollutants to the environment was not permissible in another state, and often there was a common border.

NEPA developed standards for conducting Environmental Impact Statements (EISs) (Eccleston 2008). An EIS describes investigation and remediation methods necessary in order to mitigate or minimize any and all potentially detrimental effects of any new development. Another method of evaluation is environmental auditing, modeled similarly to that of a financial audit, which examines processes and outcomes of environmental impacts.

Environmental laws, along with all other federal laws, are organized by subject into what is termed United States Code (USC). The United States Code contains

History of Land Pollution

all the current enacted statutory language. The official United States Code is maintained by the Office of the Law Revision Counsel in the United States House of Representatives (USEPA 2021a). USEPA is divided into ten regions (see Figure 2.2). Table 2.2 shows each region and the states within each region.

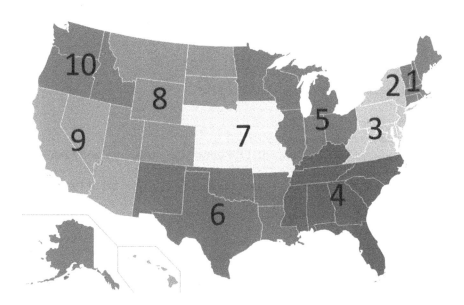

FIGURE 2.2 USEPA regions (from United States Environmental Protection Agency. Summary of Environmental Regulations of the United States. https://epa.gov/summary-environmental-laws. [accessed February 15, 2021], 2021b.)

TABLE 2.2
USEPA Regions

Region	States
Region 1	Maine, New Hampshire, Vermont, Rhode Island, Massachusetts, Connecticut
Region 2	New York, New Jersey
Region 3	Pennsylvania, West Virginia, Virginia, Maryland, Delaware, District of Columbia
Region 4	Tennessee, Kentucky, Georgia, North Carolina, South Carolina, Florida, Alabama, Mississippi
Region 5	Illinois, Wisconsin, Michigan, Ohio, Indiana, Minnesota
Region 6	Texas, Oklahoma, New Mexico, Arkansas, Louisiana
Region 7	Kansas, Nebraska, Iowa, Missouri
Region 8	North Dakota, South Dakota, Wyoming, Colorado, Utah, Montana
Region 9	California, Arizona, Nevada, Hawaii
Region 10	Washington, Oregon, Idaho, Alaska

Source: United States Environmental Protection Agency, 2021b. Summary of Environmental Regulations of the United States. https://epa.gov/summary-environmental-laws (accessed February 15, 2021).

2.5 ENVIRONMENTAL LAWS OF THE UNITED STATES

Laws that are commonly related to protecting human health and the environment include the following (USEPA 2021c) (those that appear in BOLD text are covered in this book):

- Rivers and Harbors Act of 1899
- Antiquities Act of 1906
- Atomic Energy Act of 1946
- Atomic Energy Act of 1954
- **Brownfield Revitalization Act of 2002**
- Clean Air Act of 1970
- Clean Water Act of 1972
- Coastal Zone Management Act of 1972
- **Comprehensive Environmental Response, Compensation, and Liability Act of 1980**
- **Emergency Planning and Community Right-to-Know Act of 1986**
- **Endangered Species Act of 1973**
- Energy Policy Act of 1992
- Energy Policy Act of 2005
- Federal Power Act of 1935
- **Federal Feed, Drug, and Cosmetic Act of 1938**
- **Federal Insecticide, Fungicide, and Rodenticide Act of 1947**
- Fish and Wildlife Coordination Act of 1968
- **Food Quality Protection Act of 1996**
- Fisheries Conservation and Management Act of 1976
- Global Climate Protection Act of 1987
- **Hazardous Materials Transportation Act of 1975**
- **Insecticide Act of 1910**
- **Lacey Act of 1900**
- **Low-Level Radioactive Waste Policy Act of 1980**
- Marine Protection Act of 1972
- Marine Mammal Protection Act of 2015
- **Medical Waste Tracking Act of 1980**
- Migratory Bird Treaty Act 1916
- **Mineral Leasing Act 1920**
- **National Environmental Policy Act of 1969**
- **National Forest Management Act of 1976**
- **National Historic Preservation Act of 1966**
- **National Parks Act of 1980**
- Noise Control Act 1974
- **Nuclear Waste Policy Act of 1982**
- Ocean Dumping Act of 1988
- **Occupational Safety and Health Act of 1970**
- Oil Spill Prevention Act of 1990
- **Pollution Prevention Act of 1990**

History of Land Pollution

- **Refuse Act of 1899**
- **Resource, Conservation, and Recovery Act of 1976**
- Rivers and Harbors Act of 1899
- Safe Drinking Water Act of 1974
- **Superfund Amendments and Reauthorization Act of 1986**
- Surface Mining Control and Reclamation Act of 1977
- **Toxic Substance Control Act (TSCA) of 1976**
- Wild and Scenic Rivers Act of 1968

The number of environmental laws and regulations is a very long list and encompasses hundreds of thousands of pages of rules. We shall discuss those that pertain to land, highlighted in bold type.

2.6 NATIONAL ENVIRONMENTAL PROTECTION ACT

The National Environmental Protection Act of 1969 is considered the "environmental Magna Carta" in that it established a pathway for the development of the national environmental policy and the emergence of the United States Environmental Protection Agency. NEPA grew out of the increased awareness and concern for the environment from human events that caused harm to the environment that included (USEPA 2021b):

- Rachel Carson's chronicled impact of DDT on the environment, especially birds, in her book published in 1962
- An oil spill off the coast of Santa Barbara in 1969
- Freeway revolts in the 1960s, which consisted of a series of protests in response to the destruction of many communities and ecosystems during the construction of the Interstate Highway System

The purpose of NEPA was to establish three goals (USEPA 2021b):

1. The nation's environmental policies and goals
2. Provisions for the federal government to enforce such policies
3. The Council of Environmental Quality in the executive branch of the president

The declaration of the national environmental policy contained in Section 2, Title I of the NEPA states (USEPA 2021b):

The Congress, recognizing the profound impact of man's activity on the interrelations of all components of the natural environment, particularly the profound influences on population growth, high-density urbanization, industrial expansion, resource exploitation, and new expanding technological advances and recognizing further the critical importance of restoring and maintaining environmental quality to the overall welfare and development of man, declares that is the continuing policy of the Federal Government in cooperation with State and local governments, and other concerned public and private organizations, to use all practicable means and measures, including financial and technical assistance, in a manner calculated to foster and promote the general welfare, to create and maintain conditions under which man and nature can exist in productive harmony, and fulfill the social, economic, and other requirements of present and future generations of Americans.

Of special note in the previous paragraph is the phrase "particularly the profound influences on population growth." This will be discussed in greater detail in subsequent chapters when we discuss the influence of continued population growth on sustainability. Specifically, the questions to be discussed are how humans can maintain sustainability with the environment with an ever increasing population and when the stresses caused by human population growth on the quality of air, water, land, and other species will become too great. Just since NEPA was written, our population in the United States has risen from 202 million (in 1969) to over 323 million (in 2019), an increase of more than 37% (United States Census Bureau 2021). During this same period of time, the world population has increased from 3.6 billion to more than 7.2 billion, or 100% (United Nations 2021a).

In essence, the purpose of NEPA was to ensure that environmental factors are weighted equally compared to other factors in the decision-making process undertaken by federal agencies and to establish a national environmental policy. NEPA also promotes the Council of Environmental Quality (CEQ) to advise the president in the preparation of an annual report on the progress of federal agencies in implementing NEPA. A significant environmental benefit as a result of NEPA is that it required federal agencies to prepare an **Environmental Impact Statement** to accompany reports and recommendations for Congressional funding. An EIS is a systematic scientific evaluation of the relevant environmental effects of a federal project or action, such as building a dam or an interstate highway (USEPA 2021b).

2.7   ENVIRONMENTAL JUSTICE

Environmental justice is the fair treatment and meaningful involvement of all people regardless or race, color, national origin, or income during environmental activities that include development, implementation, and enforcement of environmental laws, regulations, and policies (USEPA 2021d). Environmental justice had its beginning as far back as 1968 when sanitation workers in Memphis went on strike because of unfair treatment and environmental justice concerns. The strike advocated for fair pay and better working conditions for Memphis garbage workers. It was the first time that African Americans mobilized a national, broad-based group to oppose environmental injustices (USEPA 2021d). The position of the workers was summed up by Professor Robert Bullard, who stated, "whether by conscious design or institutional neglect, communities of color in urban ghettos, in rural poverty pockets, or on economically impoverished Native-American reservations face some of the worst environmental devastation in the nation."

A second incident worthy of note occurred in 1983 in Warren, North Carolina, where a polychlorinated biphenyl landfill was proposed to be constructed. Over 500 environmentalists and civil rights activists were arrested during a peaceful demonstration. The protests and demonstrations were ultimately unsuccessful, but this incident is widely considered the catalyst for the environmental justice movement (USEPA 2021d). Since 1983, there are perhaps hundreds of instances and examples

History of Land Pollution

of environmental justice throughout the United States. So many, in fact, that the USEPA created the Office of Environmental Justice (USEPA 2021d).

2.8 ENVIRONMENTAL ENFORCEMENT

Enforcing environmental laws is a central part of USEPA's strategic plan to protect human health and the environment. To ensure that environmental regulations are taken seriously by the regulated community, the USEPA has established an enforcement policy and has published several guidance documents that provide detail so that enforcement measures are applied equally and consistently (USEPA 2021e, 2021f). In addition, USEPA has a top priority to protect whole communities disproportionately affected by pollution through what is termed the environmental justice network, which was discussed in the previous section (USEPA 2021e).

In large part, the reason the United States has been so successful at improving the environment and protecting human health has been because USEPA has an effective enforcement program. Purposeful acts that harm the environment have been greatly reduced in large part because of education and the threat of enforcement. No longer is there a mentality that "It's not a violation until I get caught." In fact, it may be a good time to rethink enforcement policies that have not changed in 30 years.

There are generally two levels of enforcement penalties in the United States: civil and criminal (USEPA 2021g). The criminal enforcement program at USEPA was established in 1982; was granted full law enforcement authority by Congress in 1988; and employs special agents and investigators, forensic scientists and technicians, lawyers, and support staff. Civil enforcement arises most often through an environmental violation. Civil enforcement does not take into consideration what the responsible party knew about the law or regulation that was violated. Environmental criminal liability is triggered through some level of intent (USEPA 2021e). A simple explanation of the difference between a civil or criminal violation is that a criminal act is characterized as a "knowing violation" where a person or company is aware of the facts that create the violation. A conscious and informed action brought about the violation. In contrast, a civil violation may be caused by an accident or mistake (USEPA 2021e, 2021f, 2021g).

Types of enforcement results from a civil violation include settlements, civil penalties, injunctive relief, and supplemental environmental projects (SEPs) (USEPA 2021e). To evaluate the level of seriousness of an environmental violation, one must answer three basic questions:

1. Who first discovered the environmental violation? For the most part, environmental regulations are based on self reporting. It is generally a worse situation if the regulatory authority discovers a violation, especially one that should have been self reported.
2. Did the violation cause harm? It is generally a worse situation if the violation caused harm to human health or the environment. A permit violation or exceedance is generally considered a situation that has the potential to cause harm.
3. Was the violation a purposeful act? It will always be a worse situation if the violation was a purposeful act.

26 Environmental Compliance Handbook

To evaluate the level of a potential fine, the following are generally considered (USEPA 2021g):

- How long did the violation last? Under many circumstances, fines are determined per day the violation occurred and can range from $7,500 per day to as much as $37,500 per day.
- What was the economic benefit of the violation considered from a point of view of delayed and avoided costs? In addition compelling public concerns are considered.
- Did the violation cause harm to human health and the environment? To evaluate the level of risk, the toxicity of the pollutant, sensitivity of the environment, pollutant mass released, and length of time of the violation are each considered.
- Did the violation continue to occur after discovery?
- Size of the violator.
- Litigation risk.
- Ability to pay.
- Offsetting penalties such as those paid to state and local governmental agencies or citizen groups.
- Did more than one violation occur?

Settlement of violations often includes the following (USEPA 2021g):

- Payment of a fine. As stated, fines are assessed on how many days the violation occurred and the level of seriousness of the violation itself, such as whether the violation caused harm to human health and the environment.
- Permit modification. Usually, when a permit modification is justified, a stricter emission standard is applied.
- Supplemental environmental project. If a facility must upgrade equipment or invest capital expense in order to return to compliance, a portion of that expense may be used to offset the fine amount. However, the amount of offset is generally capped at 50%.

Certainly, circumstances surrounding any violation are considered. However, any violation must be taken as a serious matter and may determine whether any facility continues to operate.

2.9 CURRENT STATUS OF LAND POLLUTION IN THE UNITED STATES

Land regulations, especially CERCLA and RCRA, have had a major impact on improving and preventing primarily land contamination throughout the United States and beyond. CERCLA's principal goal is to evaluate and address unacceptable risks of exposure to environmental contamination at legacy sites. RCRA's principal goal is to address waste when it's generated. RCRA accomplishes this goal by setting standards and procedures to evaluate the potential risks posed by a specific waste so that it can be properly treated, transported, and disposed of safely.

History of Land Pollution

RCRA was essentially created to deal with huge increases in the amount of waste that was being created throughout the United States as a direct result of our industrialization and population increase. The success of RCRA can't be understated. RCRA has changed our behavior in many ways and has involved everyone in participating in taking care of our environment, not just an industry group, for almost all of us now recycle much of our discarded materials. Items that were once considered garbage are now transformed into new products that we use, including used tires, wood and pulp products, plastic, glass, and especially scrap metal, which is now considered a commodity and has grown into an industry of itself.

CERCLA was originally designed as a mechanism to fund cleanup of "orphaned" sites in the United States using the "polluters pay" principle. However, CERCLA has meant much more than just a funding mechanism. CERCLA was also a catalyst in developing investigative techniques and protocols, risk assessment evaluations, remediation technologies, and remediation target values for hundreds of chemical compounds in different media. Over its more than 35-year history, USEPA has recovered over $35 billion to fund cleanup at thousands of CERCLA sites in the United States (USEPA 2021b). In addition, the "polluters pay" principle is used worldwide as the preferred method for protecting human health and the environment at difficult and complex sites of contamination.

2.10 SUMMARY OF LAND POLLUTION REGULATIONS IN THE UNITED STATES

CERCLA and RCRA have been so successful that agriculture is now the biggest polluter in the United States and the world (United Nations 2016). The United Nations estimates that most land pollution is caused by agricultural activities, such as grazing, use of pesticides, fertilizers, irrigation, plowing, and confined animal facilities (United Nations 2016). Agricultural activities release significant amounts of greenhouse gases from plowing, decaying vegetation, and livestock (USEPA 2021e). Fugitive emissions and erosion are also significant and release significant amounts or chemicals into the air and the water. In addition, fertilizers used in agriculture are a significant source of pollution to the oceans of the world (USEPA 2021h). What is utterly amazing at this point is that environmental regulations in the United States largely do not apply to agriculture.

In summary, and since environmental regulations concerned with land have been so successful at reducing the harmful effects from industry in the United States, it is now time to reduce the harmful effects from agriculture and other sources that current environmental regulations have not adequately addressed. Answers and solutions to these questions are addressed in the volume on sustainability.

2.11 REFERENCES

Bates, C. G. and Ciment, J. 2013. *Encyclopedia of Global Social Issues*. M.E. Sharpe Publishers. New York, NY. 1450p.

Carson, R. 1962. *Silent Spring*. Houghton Mifflin. Boston, MA.

Eccleston, Charles S. 2008. *NEPA and Environmental Planning: Tools, Techniques, and Approaches for Practitioners*. CRC Press. New York, NY. 432 pages.

Environmental Education Associates. 2021. The Famous Benjamin Franklin Letter on Lead Poisoning. http://environmentaleducation.com/wp-content/uploads/userfiles/Ben%20Franklin%20Letter%20on%20EEA%281%29.pdf. (Accessed February 15, 2021).

Hahn, R. W. 1994. United States Environmental Policy: Past, Present and Future. *Natural Resource Journal*. John F. Kennedy School of Government, Harvard University. Cambridge, MA. Vol. 34. No. 1. pp. 306–348.

Haynes, W. 1954. *American Chemical Industry—A History*. Vols. I–IV. Van Nostrand Publishers. New York, NY.

Journal of Roman Studies. 2021. *Fresh Water in Roman Law: Rights and Policy*. Vol. 107. www.cambridge.org/core/journals/journal-of-roman-studies/article/fresh-water-in-roman-law-rights-and-policy/548B1C559B3D6ACEDF50C4576DD14603/core-reader#fns01. (Accessed February 15, 2021).

Lazarus, R. 2004. *The Making of Environmental Law*. Cambridge Press. Cambridge, UK. 367 pages.

National Oceanic and Atmospheric Administration (NOAA). 2021. A Brief History of Pollution. https://A Brief History: Pollution Tutorial (noaa.gov). (Accessed June 23, 2021).

Pennsylvania Historical Association. 2021. The Vision of William Penn. http://explorepahistory.com/story.php?storyId=1-9-3&chapter=1. (Accessed February 15, 2021).

Smithsonian. 2021. Air Pollution Goes Back Further Than You Think. www.smithsonianmag.com/science-nature/air-pollution-goes-back-way-further-you-think-180957716/. (Accessed February 15, 2021).

Tarlock, D. A. 2001. *History of Environmental Law. Encyclopedia of Life Support Systems*. University of Chicago Press. Chicago, IL. 1450p.

United Nations. 2016. Human Development Report. United Nations Development Programme. www.hdr.undr.org/sites/2017. (Accessed February 15, 2021).

United Nations. 2021a. The United Nations Environment Programme. www.unenvironment. org/environment-you. (Accessed February 15, 2021).

United States Advisory Council on Historic Preservation. 2021. National History Preservation Act. www.achp.gov. (Accessed February 15, 2021).

United States Census Bureau. 2021. United States Population through Time and World Populations Clock. www.UScensus.gov/USpopluation. (Accessed February 15, 2021).

United States Environmental Protection Agency (USEPA). 2021a. Report on the Environment: LAND. www.epa.gov/report-environment/land. (Accessed June 23, 2021).

United States Environmental Protection Agency. 2021b. National Environmental Protection Act. www.epa.gov/NEPA. (Accessed February 15, 2021).

United States Environmental Protection Agency. 2021c. Summary of Environmental Regulations of the United States. https://epa.gov/summary-environmental-laws. (Accessed February 15, 2021).

United States Environmental Protection Agency. 2021d. Environmental Justice. Environmental Justice | US EPA. (Accessed July 22, 2021).

United States Environmental Protection Agency (USEPA). 2021e. Enforcement Policy, Guidance and Publications. www.epa.gov/enforcement/enforement-policy-guidance-publications. (Accessed February 15, 2021).

United States Environmental Protection Agency (USEPA). 2021f. Enforcement Policy, Guidance and Publications. www.epa.gov/enforcement/criminal-enforement-overview. (Accessed February 15, 2021).

United States Environmental Protection Agency (USEPA). 2021g. Basic Information in Enforcement. www.epa.gov/enforcement/basic-information-enforement. (Accessed February 20, 2021).

United States Environmental Protection Agency. 2021h. Agriculture 101. www.epa.gov/sites/production/files/2015-07/documents/ag_101_agriculture_us_epa_0.pdf. (Accessed February 15, 2021).

3 Land Contamination

3.1 INTRODUCTION

Pollution is everywhere. It's in every breath we inhale, in the water we drink, and on and within the solid earth. Some of these pollutants are naturally occurring; others may be naturally occurring but are emitted, dispersed, or disposed of through a multitude of human activities. Others are not naturally occurring and are solely anthropogenic and originate from a whole host of sources that include farming, industry, power generating plants, automobiles, and residential sources as well.

In this chapter, we begin by first examining the meaning of a pollutant. Second, we will learn how our bodies react to pollutants. Last, we will identify the common types of pollutants and learn about their behavior in nature. This analysis will clarify why some human activities and pollutants are more heavily regulated during subsequent chapters.

3.2 THE NEED FOR REGULATION

There would be no need for environmental laws and regulations or the United States Environmental Protection Agency if there were no contamination. However, it does not just begin and end with chemical contaminants. There are other activities that impact our environment that go beyond what is commonly termed pollution or contamination. Many of these other forms do not introduce chemicals into the natural world. For instance, there are any number of different land uses where humans inflict enormous harm on our environment, including mining, petroleum development, urban land development, wetland destruction, deforestation, forest management techniques, and agriculture. As we are discovering, these activities do not just affect a centralized area where the activity took place. These types of activities often have a ripple effect that negatively impacts surrounding areas and in some cases spreads like a disease that affects the entire planet. As we will discover, this is where sustainability then plays an important and central role. Sustainability goes beyond addressing pollution or contamination and examines how humans, albeit unintentionally, adversely impact our planet as though we ourselves are a form of contamination.

Pollution and **contamination** are used synonymously to mean the introduction into the environment by humans of substances that are harmful or poisonous to human health and ecosystems (van der Perk 2006). Over time, the term pollution evolved to include not only substances but energy wastes such as heat, light, noise, or any human development that impacts the air, land, or water. Pollution now also includes many types of land use that inflict harm on the environment, as stated previously. For our purposes in this chapter, we will concentrate on chemical pollutants. Therefore, we will use the term contaminant, since this term is more often associated with chemical releases. Since we will be discussing chemical contaminants, we will often need to refer to the periodic table of elements for reference and educational

DOI: 10.1201/9781003150107-3

purposes in this chapter. Figure 3.1 presents the periodic table of elements (United States Institute of Health 2021).

A chemical or substance becomes contamination when it is released into the environment either inadvertently or improperly-at the wrong place or in the wrong amounts. For example, milk becomes a contaminant when large quantities are released into a stream. In urban areas, contaminants are everywhere-in the air, soil, water, inside buildings, and in our homes. Most households contain chemicals that would be considered contaminants and hazardous if they were released into the environment or disposed of improperly (see Figure 3.2). These chemicals include (USGS 2006):

- Cleaners
- Pesticides
- Some Paints
- Grease
- Solvents
- Herbicides
- Plastic
- Dirt
- Gasoline
- Lawn Fertilizers
- Oil
- Electronic equipment

In addition, other common household products commonly contain contaminants and can impact the environment if not properly disposed, including (USGS 2006):

- Computer equipment
- Televisions

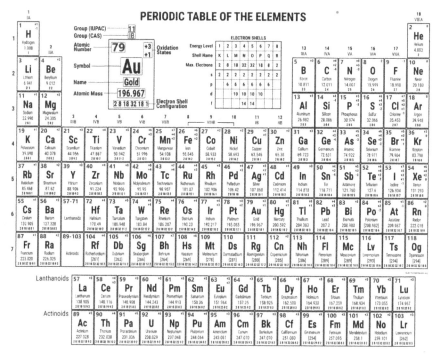

FIGURE 3.1   Periodic table of elements (from United States Institute of Health. *Periodic Table of Elements*. https://pubchem.ncbi.nlm.nih.gov/periodic-table [accessed June 5, 2021] 2021).

Land Contamination

FIGURE 3.2 Household items that can become contaminants (from United States Geological Survey. *Volatile Organic Compounds in the Nation's Ground Water and Drinking-Water Supply Wells*. USGS Circular 1292. Reston, VA, 2006).

- Some electrical equipment
- Most batteries
- Some building products
- Wood with certain applied preservatives or coatings

Thousands of environmental pollutants exist, with the following categories of chemical and organic contaminants commonly present in the environment (Rogers 2020a, 2020b):

- Volatile organic compounds (VOCs)
  - Dense non-aqueous phase liquids
  - Light non-aqueous phase liquids
- Polynuclear aromatic hydrocarbons (PNAs or PAHs)
- Semi-volatile organic compounds (SVOCs)
- Polychlorinated biphenyls
- Pesticides and herbicides
- Heavy metals
- Others, including:
  - Common fertilizers, including nitrates, phosphorus, and potassium
  - Greenhouse gases
  - Chlorofluorocarbons
  - Carbon dioxide

32 Environmental Compliance Handbook

- Carbon monoxide
- Particulates (dust)
- Ozone
- Bacteria such as *E. coli*
- Viruses
- Invasive species
- Pharmaceuticals
- Cyanide
- Asbestos
- Acids
- Bases
- Radioactive compounds
- Dioxins
- Polyfluoroalkyl substances
- Emerging contaminants

Before addressing individual contaminant groups, we must introduce and discuss the general concepts of toxicity by examining the types, differences, and potential effects of exposure to pollutants. The chapter will conclude with an examination of each contaminant category listed previously.

3.3 TOXICITY OF CONTAMINATION

Toxicity or **potency** is the degree to which a chemical or substance is able to inflict damage to an exposed organism (USEPA 1989a). Note that toxicity does not equal risk. The difference between toxicity and risk is based primarily upon the length of exposure to a chemical or substance and whether the dosage received from this exposure is enough to cause harm. Toxic substances surround us and are present at almost all locations on Earth. However, and as stated previously, there is only risk if we are exposed to a substance or chemical long enough and at a high enough dose to cause harm. Toxicity does not assess risk—this determination is reserved for a risk assessment. A **risk assessment** is an evaluation that combines toxicity of a contaminant to possible exposure pathways. A risk assessment is therefore a measure of the nature and magnitude of health risks and ecological receptors from chemical contaminants that may be present in the environment (USEPA 2021a).

There are three basic types of toxic categories:

1. Chemical or substance, including inorganic and organic substances such as acids and bases, flammable liquids, metals, and so on
2. Biological, including bacteria and viruses
3. Physical, including sound and vibration, heat and cold, light, radiation, and so on

Toxicity is measured by the effects on a whole organism, an individual organ, tissue, or even a cell. Populations are most often used to measure toxicity since any one individual typically may have a different level of response to a toxin at a certain dose or concentration. The most common measure of chemical or substance toxicity

Land Contamination

is termed the LD_{50}, defined as the concentration or dose that is lethal to 50% of the population being tested (USEPA 1989a). When direct data are not available, the LD_{50} is estimated by comparing the substance to other similar chemicals and organisms.

Another important measure of toxicity is the **lowest observed adverse effect level** (LOAEL) or **threshold effect value.** The LOAEL is the lowest tested dose of a chemical or substance causing a harmful or adverse health effect (ATSDR 2021a). Factors influencing the LOAEL include:

- The chemical's solubility in body fluids
- The particle size and state of the chemical
- Route of exposure
- Residence time of the chemical in the body
- Individual susceptibility

Units of dose are expressed as the mass of chemical per unit mass of body weight in milligrams per kilogram (mg/kg). These units are employed in order to evaluate the relative toxicities between animals of different species and size. To develop the most accurate human exposure limits, extensive animal studies are initially used to establish extrapolated human dosage limits for a specified chemical, and these estimates are refined by human health studies conducted on individuals known to have been exposed to the same chemical. The Occupational, Safety, and Health Administration has established a list of exposure limits for over 600 individual chemicals and refers to exposure limits as Permissible Exposure Limits (PELs) (ACGIH 2021). Most chemicals or substances have the potential to exhibit some adverse health effect on humans or other organisms given a certain set of circumstances. **Adverse health effect** is defined as a change in body function or cell structure potentially leading to disease or health problems (ATSDR 2021a). The effect that a specific chemical or substance may exhibit on a living organism is dependent upon the following factors (USEPA 1989a, 2005):

- Nature of the chemical or substance
- Concentration
- Route of exposure
- Length of time of exposure
- Individual susceptibility

Chemicals and substances enter the human body through three **routes of exposure:** inhalation, ingestion, and dermal adsorption (USEPA 1989a). When exposure occurs, chemicals may produce one or more of these symptoms:

• Tissue irritation	• Eye irritation	• Rash
• Dizziness	• Bleeding	• Hair loss
• Vision loss	• Hearing loss	• Nausea
• Anxiety	• Narcosis	• Headache
• Vomiting	• Diarrhea	• Pain
• Fever	• Tremors	• Difficulty breathing

• Tissue irritation	• Eye irritation	• Rash
• Psychotic behavior	• Euphoria	• Cancer
• Death	• Burning sensations	• Others

Human response to exposure of a chemical is described as acute or chronic. An **acute** response is generally characterized as a single high dose with rapid onset and disappearance of symptoms. **Chronic** response involves a stimulus lingering for a period of time (ATSDR 2021a). Exposure to a contaminant may not trigger an immediate response. Instead, there is a **latency period**—a duration of time without observable effects. Certain chemicals or substances, such as asbestos and forms of mercury, have a latency period of up to 10 to 20 years (USEPA 2005). Exposure to some chemicals or substances may result in the development of cancer, and any chemical, substance, radionuclide, or radiation contributing to the development or propagation of cancer is termed a carcinogen (United States Department of Health and Human Services [USDHH] 2005). **Cancer** exists when cells in the body become abnormal and grow or multiply out of control (ATSDR 2021a). There is also often some selectivity involved with chemical exposure, as some chemicals may target certain organs, parts of the body, or reproduction. For example, **teragens** may have an adverse effect on a developing fetus, **mutagens** may induce genetic changes that could affect future generations through reproduction, and **heptotoxins** are chemicals posing a risk of liver damage (ATSDR 2021a).

Carcinogens are grouped into five general categories: (ACGIH 2021):

- Group A1: Confirmed human carcinogen
- Group A2: Suspected human carcinogen
- Group A3: Confirmed animal carcinogen with unknown relevance to humans
- Group A4: Not classified as a human carcinogen
- Group A5: Not suspected as a human carcinogen

It is much more difficult to evaluate the toxicity of a mixture of contaminants compared to a single chemical compound or substance. This is because the interaction of a mixture of contaminants may produce enhanced or diminished effects. Common mixtures include gasoline, often containing more that 250 individual chemical compounds; industrial wastes; or a malfunctioning sewage treatment plant, which may result in discharging many chemical and biological contaminants (USEPA 2021b).

The following sections describe the major contaminant categories. For each contaminant category, specific contaminants and their common uses, physical attributes, and toxicities are presented.

3.4   VOLATILE ORGANIC COMPOUNDS

Volatile organic compounds are organic compounds that generally volatilize or evaporate readily under normal atmospheric pressure and temperatures (USGS 2006; USEPA 2021b). Table 3.1 is a list of common VOCs and includes those sought out and analyzed when conducting a subsurface environmental investigation (USGS

Land Contamination

TABLE 3.1
Common uses of VOCs

Compound (Listed Alphabetically)	CAS Registry Number	Common Uses
Acetone [a]		Nail polish remover, paint thinner, laboratory chemical
Benzene[b,c]	67–64–1	Gasoline component, solvent, pharmaceutical, dyes, and plastics
Bromodichloromethane	71–43–2	Trihalomethane (byproduct of water purification)
Bromoform	75–27–4	Trihalomethane (byproduct of water purification)
Bromomethane	75–25–2	Fumigant
Carbon disulfide[d]	75–83–9	Cellulose manufacturing, soil fumigant
Carbon tetrachloride[f]	75–15–0	Solvent, dry cleaning, agriculture, formerly used in fire extinguishers
Chlorobenzene	56–23–5	Solvent, insecticides
Chloroethane	108–90–7	Solvent, pharmaceutical manufacturing
Chloroform	75–00–3	Trihalomethane (byproduct of water purification)
Chloromethane	67–66–3	Solvent, agricultural chemicals and cellulose
Dibromochloromethane	74–87–3	Trihalomethane (byproduct of water purification)
1,1-Dichloroethane	124–48–1	Solvent, manufacture of plastic wrap and adhesives
1,2-Dichloroethane	75–34–3	Solvent, paint ingredient, fumigant, vinyl chloride production
1,1-Dichloroethene	107–06–2	Manufacturing of plastics, adhesives, refrigerants
cis-1,2-Dichloroethene	75–35–4	Fat extraction from meat and fish, refrigerants, pharmaceuticals
trans-1,2-Dichloroethene	156–59–2	Fat extraction from meat and fish, refrigerants, pharmaceuticals
Dichloromethane	156–60–5	Paint and stain remover, aerosol propellant, caffeine extraction
1,2-Dichloropropane	79–09–2	Solvent, stain remover, former soil fumigant
cis-1,3-Dichloropropene	78–87–5	Solvent
trans-1,3-Dichloropropene	10061–01–5	Solvent
Ethyl benzene[g]	10061–02–06	Gasoline component, solvent, styrene manufacturing
2-Hexanone[h]	100–41–4	Paint and paint thinner, used to dissolve oil and waxes
Methyl ethyl ketone (MEK)	591–78–6	Solvent
Methyl isobutyl ketone (MIBK)	78–93–3	Solvent, used in metal extraction
(MIBK)	108–10–1	Former gasoline additive, formerly used to dissolve gallstones
Methyl-tert-butyl ether (MTBE)	1634–04–4	Gasoline component, mothballs, insecticide
(MTBE)	91–20–3	Gasoline component, coating, paint, rubber, adhesives
Naphthalene[i]	100–42–5	Solvent, ingredient in paints and pesticides
Styrene	79–34–5	Solvent, dry cleaning, textile processing
1,1,2,2-Tetrachloroethane	127–18–4	Gasoline component, solvent
Tetrachloroethene (PCE)[e]	108–88–3	Solvent, cosmetic ingredient, aerosol products, textile processes
Toluene[i]	71–55–6	Solvent, manufacturing of plastic wraps
1,1,1-Trichloroethane[e]	79–00–5	Solvent, caffeine extraction, dry cleaning, paint and ink, rubber
1,1,2-Trichloroethane	79–01–6	Polymer
Trichloroethene (TCE)[e]	108–05–4	Rubber, paper, and glass industries, PVC manufacturing
Vinyl acetate	75–01–4	Gasoline component, paint thinner ingredient, plasticizer
Vinyl chloride[j]	1330–20–7	
Xylenes[k]		

Source: United States Geological Survey. 2006. *Volatile Organic Compounds in the Nation's Ground Water and Drinking-Water Supply Wells.* USGS Circular 1292. Reston, VA. 2006: United States Environmental Protection Agency (USEPA). 2021b. EPA's Report on the Environment (ROE). www.epa.gov/report-environment (accessed June 5, 2021).

[a] Levy (2009).
[b] Lide (2008).
[c] ATSDR (2007a).
[d] Holleman and Wiberg (2001).
[e] Doherty, R. E. (2000).
[f] ATSDR (2007b).
[g] ATSDR (1995).
[h] ATSDR (2005a).
[i] ATSDR (2001a).
[j] ATSDR (2006a).
[k] ATSDR (2007c).

2006; USEPA 2021b). Included in Table 3.1 is the Chemical Abstracts Service (CAS) registry number for each chemical compound. The CAS registry identifies each compound with a unique numerical identifier and includes over 143 million substances (American Chemical Society 2021). In general, VOCs have a high vapor pressure (evaporate quickly), low-to-medium solubility, and low molecular weight (USGS 2006). Most VOCs are considered toxic or harmful to humans and other organisms (USEPA 2021c). In addition, many VOCs are flammable and must be handled with extreme care.

The use of VOCs in the United States continues to increase (Storrow 2019; USEPA 2021b). Some VOCs have had and continue to have very heavy usage. An example is gasoline, which contains numerous VOC compounds, and its production and use continue to increase. Current gasoline consumption in the United States is approximately 1.6 billion liters per day (400 million gallons per day) (United States Energy Information Service 2021). Table 3.1 lists common uses of VOCs. As noted, many VOCs are flammable and are components of gasoline and other fuels such as diesel fuel, kerosene, and fuel oil. In addition, VOCs are commonly used as solvents, ingredients in paints, paint thinners, in the manufacturing process of pharmaceuticals, for caffeine extraction, nail polish removers, dry cleaning chemicals, mothballs, pesticides and fumigants, adhesives, refrigerants, and as a byproduct of chlorination for water purification.

For hydrogeological purposes, VOCs are separated into two general categories: dense non-aqueous phase liquids (DNAPLs), and light non-aqueous phase liquids (LNAPLs) (USGS 2009a, 2009b). An additional category of VOCs are the trihalomethanes. Each is discussed separately in the following sections.

3.4.1 Light Nonaqueous Phase Liquids

Light nonaqueous phase liquids are liquids lighter than water and do not mix or dissolve in water readily. Common LNAPL compounds include the VOCs and benzene, ethyl benzene, toluene, and xylenes. These compounds are also often symbolized by the acronym BTEX. BTEX compounds are common components of gasoline and are often used as indicator analytes when evaluating whether there has been a release of gasoline or other similar fuels to the environment (USGS 2009b). LNAPLs also include other compounds lighter than water such as polynuclear aromatic hydrocarbons, described later in this chapter.

BTEX compounds are aromatic hydrocarbons with a **benzene ring** forming the backbone of their molecular structure (Jensen 2009). As depicted within the compound benzene shown in Figure 3.3, the benzene ring consists of a hexagonal arrangement of six carbon atoms located at the vertices. Each atom is bonded to its adjacent atoms by a single covalent bond and by an unusual ring bond of electrons shared by all six carbon atoms. Benzene and ethyl benzene are considered A1 carcinogens (USEPA 2021c).

Health effects from exposure to benzene include headaches, dizziness, drowsiness, rapid heart rate, tremors, and unconsciousness. Exposure to high levels can result in death (ATSDR 2007a). Health effects from exposure to toluene include tiredness, confusion, loss of appetite, memory loss, loss of color vision, and nausea

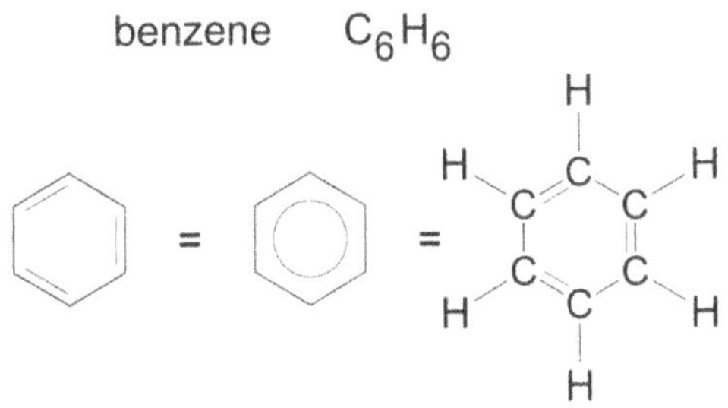

Land Contamination

benzene C_6H_6

FIGURE 3.3 Benzene and the benzene ring.

(ATSDR 2001a), and exposure to ethyl benzene may cause irreversible damage to the inner ear and hearing loss, dizziness, and kidney damage (ATSDR 2007b). Exposure to xylenes can result in headaches, lack of muscle coordination, dizziness, and confusion. Exposure to very high levels of xylenes can cause unconsciousness (ATSDR 2007c).

Methyl-tert-butyl ether is a LNAPL compound that was produced exclusively as a gasoline additive. MTBE belongs to a group of chemicals known as "oxygenates" because they raise the oxygen content of gasoline and thereby raise the octane level. MTBE is a colorless liquid at room temperature and is very volatile and flammable (USEPA 2013). The purpose of adding MTBE to gasoline was to increase the efficiency of combustion in automobiles, enabling them to run more cleanly and emit fewer pollutants. However, the use of MTBE has declined significantly because of health concerns, and MTBE has been detected in many groundwater aquifers used as drinking water sources in the United States (USEPA 2013). The United States has now banned the use of MTBE because of its propensity to contaminate groundwater and the high costs incurred to remove it from groundwater (USEPA 2013). MTBE is also used to dissolve gallstones. Patients treated for gallstones using MTBE have MTBE delivered directly to the gall bladder through surgically inserted tubes. Health effects from exposure to MTBE may include nose and throat irritation, headaches, nausea, dizziness, and mental confusion. Currently, evidence suggesting that MTBE may cause cancer is inadequate (ATSDR 2021b).

3.4.2 Dense Nonaqueous Phase Liquids

Dense nonaqueous phase liquids are liquids denser than water and do not mix or dissolve readily in water (USGS 2006). DNAPL compounds include many common solvents and coal tar (Sutherson and Payne 2005). They are also commonly referred to as chlorinated solvents, halogenated VOCs, or chlorinated VOCs because chlorine is in the atomic structure, and the most common uses of these compounds are for cleaning and degreasing (USGS 2006). Chlorinated VOCs have been in use for nearly 100 years and have been widely used by industry, and many household products contain them. Figure 3.4 shows some household products containing chlorinated solvents (USGS 2006).

FIGURE 3.4 Household products containing chlorinated solvents (from United States Geological Survey. *Volatile Organic Compounds in the Nation's Ground Water and Drinking-Water Supply Wells.* USGS Circular 1292. Reston, VA, 2006).

Halogenated VOCs are a group of organic compounds with a halogen atom as part of their molecular structure. Halogens include the elements fluorine (F), chlorine (Cl), bromine (Br), and iodine (I). Part of the uniqueness of halogenated VOCs is they tend to have a very weak tendency to form hydrogen bonds with water. This lack of affinity for water means halogenated VOCs—especially those with fluorine or chlorine—tend to be hydrophobic and have low solubility (Suthersan and Payne 2005).

Common halogenated VOCs include:

- Tetrachloroethene (PCE)
- cis-1,2-Dichloroethene (DCE)
- Vinyl chloride
- 1,1-Dichloroethene (DCE)
- Carbon tetrachloride
- Chlorobenzene

- Trichloroethene (TCE)
- Trans-1,2-Dichloroethene (DCE)
- 1,1,1-Trichloroethane (TCA)
- Methylene chloride
- Chloroform
- 1,2-Dichlorobenzene

Many VOCs halogenated with chlorine have high electro-negativities and form rather strong bonds with the carbon atoms in their structure. In addition, the substitution of a hydrogen atom by a chlorine atom enhances the inertness of the molecule. This inertness results in many halogenated VOCs being rather persistent when released into the environment. A degradation sequence, however, does exist for a group of the most commonly used halogenated VOCs and is shown in Figure 3.5. This degradation sequence becomes very important when we discuss the fate and

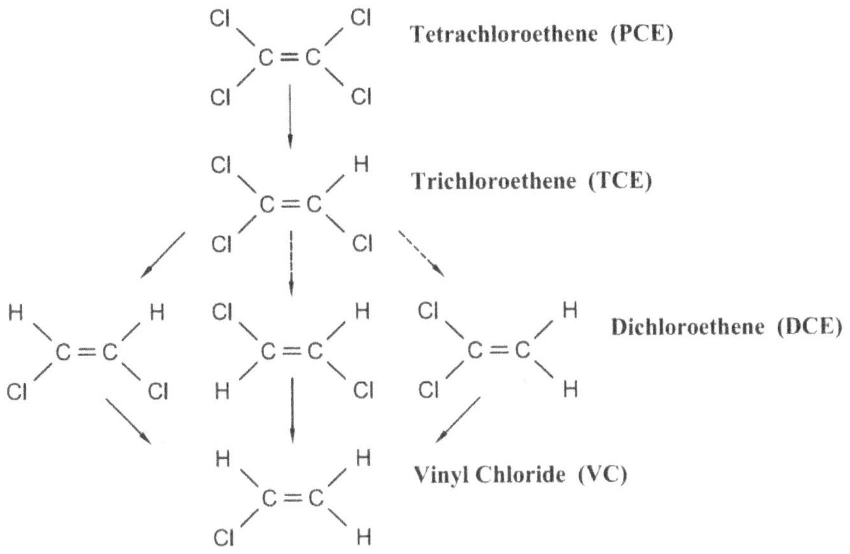

Land Contamination

39

Tetrachloroethene (PCE)

Trichloroethene (TCE)

Dichloroethene (DCE)

Vinyl Chloride (VC)

FIGURE 3.5 Degradation sequence of PCE.

migration of these compounds in the next chapter. Vinyl chloride is an A1 carcinogen, and many of the others are currently under further review (ATSDR 2006a, 2021c; USEPA 2021c). Health effects from overexposure to most chlorinated solvents include lung irritation, difficulty walking and speaking, poor coordination, dizziness, headache, nausea, sleepiness, unconsciousness, and even death (ATSDR 2021c, 2021d, 2021e).

3.4.3 Trihalomethanes

Trihalomethanes are a group of VOCs in which three of the four atoms of methane (CH_4) are replaced by halogen atoms. Many THMs are used as refrigerants and solvents and are also byproducts produced during water purification and chlorination. THMs can form as a byproduct when chlorine and bromine are used to disinfect water for drinking or recreational use; they form from a reaction between chlorine and/or bromine with organic matter in the water being treated (USEPA 2021d). Chloroform has also been used as an anesthetic and is often used in swimming pools as a disinfectant (ATSDR 1997a). Trihalomethanes are currently considered a suspected human carcinogen (Group A2) (USEPA 2021c). According to USEPA (2021d), THMs may cause adverse health effects at high concentrations. Exposure to high concentrations of bromoform may interfere with normal brain function and cause sleepiness (ATSDR 2005b). USEPA has established maximum allowable concentration of THMs in drinking water at 80 parts per billion for the combined concentration of chloroform, bromoform, bromodichloromethane, and dibromochloromethane (USEPA 2021d).

3.5 POLYNUCLEAR AROMATIC HYDROCARBONS

Polynuclear or polycyclic aromatic hydrocarbons are a group of compounds formed synthetically or naturally during the incomplete combustion of coal, oil, tar, gas, wood, and garbage and may also be present in tobacco, medicines, dyes, plastics, pesticides, and charbroiled meat (United States Department of Health and Human Services [USDHH] 1995). Other sources of PAHs include heavy petroleum, including diesel fuels, kerosene, aviation fuels, heavy home-heating oils, oils, used oil, and many lubricants (Missouri Department of Natural Resources 2006). Overall, there are more than 100 distinct PAH compounds. Chemically, PAHs are composed of multiple benzene rings (Figure 3.6), are lighter than water, and are generally not very soluble in water. In addition, PAHs strongly attach themselves or sorb to soil grains, especially organic soils and clay (USEPA 1989b). Because PAHs are not very soluble and are lighter that water, they are considered LNAPLs.

With two rings, naphthalene is the smallest of the PAHs. Naphthalene is a common PAH compound and is characterized by its distinctive odor and common use in mothballs. Anthracene and phenanthrene are composed of three rings. Phenanthrene is an **isomer** of anthracene, and they share many physical properties except boiling point and water solubility (Fetzer 2000). With its five rings, benzo(a)pyrene is the largest PAH molecule and is the only known human carcinogen (Group A1) of the four compounds shown in Figure 3.6 (USEPA 2021c). Common PAH compounds are listed in Table 3.2, along with their CAS registry numbers (USEPA 2021e; ATSDR 2021f).

Animal studies have indicated PAHs can cause harmful effects on the skin, body fluids, and ability to fight disease (ATSDR 2021f). Health effects from exposure to naphthalene may cause damage to red blood cells. Exposure to high levels of naphthalene may cause nausea, vomiting, diarrhea, dizziness, blood in the urine, and a yellow color to the skin (ATSDR 2021f). Currently only benzo(a)pyrene and fluoranthene are considered A1 carcinogens (USEPA 2021c).

3.6 POLYCHLORINATED BIPHENYLS

Polychlorinated biphenyls are a group of synthetically produced compounds (USEPA (2021f). PCBs were produced in the United States from 1929 until 1979, when they

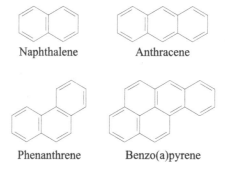

FIGURE 3.6 Molecular structure of the select PAH compounds: naphthalene, anthracene, phenanthrene, and benzo(a)pyrene.

Land Contamination

TABLE 3.2
Common PAH Compounds

Compound (Alphabetically)	CAS Registry Number	Common Uses
Acenaphthene	83–32–9	Dyes, pesticides, pharmaceuticals, oil component
Acenaphthylene	208–96–8	Automobile exhaust, oil and lubricant component
Anthracene	120–12–7	Dyes, stains, wood preservatives, insecticides, oil
Benz[a]anthracene	56–55–3	Automobile exhaust, oil and lubricant component
Benzo[b]fluoranthene	205–99–2	Automobile exhaust, oil and lubricant component
Benzo[k]fluoranthene	92–24–0	Organic semiconductor, oil component
Benzo[ghi]perylene	198–55–0	Photoconductor, pitch, coal tar, tobacco smoke
Benzo[a]pyrene	50–32–8	Pitch, coal tar, automobile exhaust, tobacco smoke
Chrysene	218–01–9	Wood preservative, oil component, dyes
Dibenzo[a,h]anthracene	53–70–3	Wood preservative, insecticides, oil component
Fluoranthene	206–44–0	Automobile exhaust, oil component
Fluorene	86–73–7	Automobile exhaust, dyes, plastics, pesticides
Indeno[1,2,3-cd]pyrene	193–39–5	Pitch, coal tar, oil component
Naphthalene	91–20–3	Mothballs, oil and gasoline component
Phenanthrene	85–01–8	Cigarette smoke, oil and lubricant component
Pyrene	129–00–0	Dyes, coal tar and pitch, oil component

Source: Agency for Toxic Substances and Disease Registry (ATSDR). 2021f. *Factsheet for PAHs*. www.atsdr.cdc.gov/tfacts69.html (accessed June 5, 2021); United States Environmental Protection Agency. *Polycyclic Aromatic Hydrocarbons (PAHs)*. www.epa.gov/sites/production/files/2014-03/documents/pahs_factsheet_cdc_2013.pdf (accessed June 5, 2021).

were banned because of human and environmental health effects (USEPA 2021g). Some of the physical properties of PCBs include (Phillips 1986; Barbalace 2009):

- Very low solubility in water
- High relative solubility in organic solvents, fats, and oil
- Low vapor pressures
- Strongly sorb to soil grains, especially organic-rich soils and clays
- Stable compounds
- Do not readily degrade
- Not flammable
- Resistant to oxidation, reduction, addition, elimination, and electrophilic substitution
- High boiling point

Due to these physical properties, PCBs were widely used in many different industrial and commercial products, including (USEPA 2021g):

- Transformers and capacitors
- Other electrical equipment, including voltage regulators, switches, reclosers, bushings, and electromagnets
- Oil used in motors and hydraulic systems
- Older electrical devices that contain capacitors

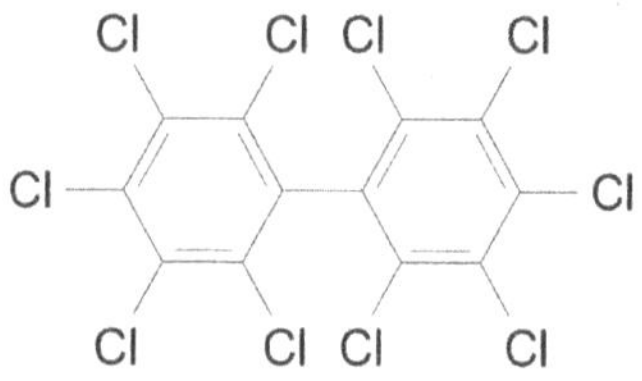

FIGURE 3.7   Basic PCB structure.

- Fluorescent light ballasts
- Cable insulation
- Thermal insulation material, including fiberglass, felt, foam, and cork
- Adhesives and tape
- Oil-based paint
- Caulking
- Plastics
- Carbonless copy paper
- Floor finish

PCBs are mixtures of chlorinated organic compounds called congeners (ATSDR 2021g). A **congener** is a related compound or compounds in a specific chemical family. There are a large number of congeners, with approximately 209 present in PCB mixtures (USEPA 2021f). PCBs were produced synthetically through electrophilic chlorination of a biphenyl molecule with chlorine gas. In the United States, PCBs are commonly known under the trade name aroclors. **Aroclors** are mixtures of PCBs distinguished by a four-digit numbering system. The first two digits refer to the number of carbon atoms in a PCB molecule, and the second two digits indicate the percentage of chlorine by mass in the mixture (USEPA 2021f). For example, PCB aroclor 1254 means the mixture contains 12 carbon atoms and is 54% chlorine by weight (ATSDR 2021g). As the degree of chlorination increases, the melting point increases and the vapor pressure and solubility decrease. Figure 3.7 shows the basic structure of a PCB molecule.

Currently, there is inadequate information to establish whether PCBs are a human carcinogen (USEPA 2021c), but the World Health Organization (2003) suspects they are (Group: A2). Notwithstanding cancer, they pose significant other dangers to human health. PCBs have been shown to cause cancer in laboratory animals and have also been shown to cause a number of serious non-cancer health effects in animals, including the immune system, reproductive system, nervous system, endocrine system, and others (USEPA (2021c).

3.7   SEMI-VOLATILE ORGANIC COMPOUNDS

Semi-volatile organic compounds are a group of organic-based compounds much less volatile than VOCs. A variety of SVOCs are used in clothing and building materials to provide flexibility, water resistance, or stain repellence, as well as to

Land Contamination

inhibit ignition or flame retardants (ATSDR 2002a; USEPA 2010). Common groups of organic compounds associated with SVOCs are phthalates, phenols, amines, and esters.

3.7.1 PHTHALATES

The name phthalates is derived from phthalic acid-itself derived from naphthalene, previously discussed earlier in this chapter. Phthalates exhibit low water solubility, high oil solubility, and low volatility (Fetzer 2000). There are approximately 25 distinctive phthalate compounds, and common ones include:

- Diethyl phthalate (DEP)
- Di-n-butyl phthalate (DBP)
- Bi or Bis(2-ethylhexyl) phthalate (DEHP)

Phthalate compounds are primarily used as plasticizers. Phthalates are added to plastics to increase their flexibility, transparency, durability, and longevity and are also used to soften polyvinyl chloride (PVC)—a common pipe material. More than a billion pounds of phthalates are used in a variety of products every year (ATSDR 2021h). The variety of products using phthalates includes (ATSDR 2002a):

• Coatings	• Plastics	• Pharmaceuticals	• Gelling agents
• Stabilizers	• Binders	• Dispersants	• Eye shadow
• Moisturizer	• Perfume	• Children's toys	• Detergents
• Nail polish	• Hair spray	• Modeling clay	• Packaging
• Waxes	• Printing inks	• Fishing lures	• Caulk
• Paints	• Curtains	• Vinyl upholstery	• Auto interiors
• Tooth paste	• Candles	• Food containers	• Many others

According to ATSDR (2021h) and USEPA (2021g), DEHP is a suspected human carcinogen (Group: A2), while DEP and DBP are not classified as human carcinogens (Group: A4). According to ATSDR, DEHP is not toxic at the low levels usually present in the environment. In animals, DEHP has been found to damage the liver and kidney and has affected the ability to reproduce.

3.7.2 PHENOL

Phenol is a group of organic compounds with a hydroxyl group (-OH) attached to a carbon atom in a benzene ring. Phenol compounds do occur naturally, and their presence in plant foliage discourages herbivores from consuming the plant material (Fetzer 2000; McMurry 2009). Figure 3.8 is a diagram showing the basic structure of a phenol molecule. The simplest phenol compound is carbolic acid (C_6H_5OH), also called phenol.

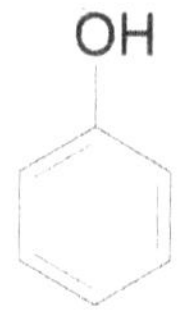

OH

FIGURE 3.8   Basic structure of phenol.

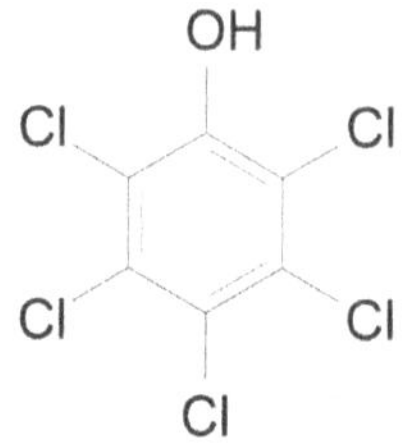

OH

Cl ⟍ ╱ Cl

Cl ⟋ ⟍ Cl

Cl

FIGURE 3.9   Basic molecular structure of pentachlorophenol.

Phenol combined with formaldehyde forms a widely used polymer typically referred to as a phenolic resin. Phenolic resins have a wide range of industrial and commercial applications, including (USEPA 2021h):

- Billiard balls
- Electronic components
- Abrasives
- Foundry applications
- Rubber additives

- Countertops
- Coating
- Adhesives
- Friction products
- Auto brakes

- Plastic
- Composites
- Felt bonding
- Refractory products
- Many others

A compound called pentachlorophenol (PCP) is a phenol compound with the addition of chlorine into its molecular structure (Figure 3.9). PCP has been used as an herbicide, insecticide, fungicide, algaecide, and disinfectant. Other uses include leather, masonry products, and wood preservatives, and utility poles often employ PCP as a wood preservative (USEPA 2006).

Bisphenol A or BPA is an organic phenolic compound with two phenol functioning groups. BPA is a very common compound and is used heavily in the production of plastic products, including bottles for drinking water and numerous other consumer products (USDHH 2008). Some concern about health effects from exposure to BPA have been expressed by the USDHH (2008), as type 7 plastics (polycarbonates) may leach PBA into the liquids they hold. In 2009, Canada banned the use of PBA in polycarbonates in baby bottles (Carwile et al. 2009).

3.7.3 AMINES

Amines are organic compounds that contain nitrogen and are **basic**. Three widely used amine compounds are methylamine, dimethylamine and trimethylamine,

Land Contamination 45

which are prepared by the reaction of ammonia with methanol in the presence of a silicoaluminate catalyst (McMurry 2009). Amines are used in the manufacturing of dyes, plant growth regulators and resins and as a precursor in the manufacturing of tires, pharmaceuticals (ephedrine), pesticides, and surfactants (ATSDR 2021i). Trimethylamine also forms naturally from decomposing plant and animal matter (USEPA 2021i). Amines are often gas at room temperature and pressure and are very soluble in water (USEPA 2021i).

3.7.4 ESTERS

Esters are a group of more than 50 compounds known collectively as acid derivatives. Ester compounds contain a modified carboxylic acid group (-COOH) in which the acidic hydrogen atom has been replaced by a different organic functional group. Common uses of esters include (ATSDR 2021j; USEPA 2021j):

- Flame retardants
- Nail polish removers
- Glue
- Fragrances
- Clothing

Polyester is the most common and most widely produced ester compound. Esters are also naturally occurring and are found in flowers, in fats as triesters derived from glycerol and fatty acids, and in wine (ATSDR 2021j). At high concentrations, ester compounds can irritate the skin and mucous membranes and cause breathing difficulties (USEPA 2021c).

3.8 HEAVY METALS

Heavy metals naturally exist in the environment at amounts termed natural background levels.

When human activities introduce additional heavy metals into the environmental and the background levels are exceeded, they then may be considered contaminants. In terms of their characteristics, heavy metals are generally not soluble in water, except for arsenic and chromium VI at a neutral pH (Murray et al. 2004; Kaufman et al. 2011). Heavy metals are not volatile but are often released into the atmosphere through the combustion of coal, automobile exhaust, metal production, and other methods (Murray et al. 2004). The most common heavy metals in urban areas include (Murray et al. 2004):

• Arsenic	• Barium	• Chromium III	• Chromium VI
• Cadmium	• Copper	• Lead	• Mercury
• Nickel	• Selenium	• Silver	• Zinc

46 Environmental Compliance Handbook

Industrial watersheds are especially prone to heavy metal contamination. In an extensive study of the Rouge River watershed in southeastern Michigan, Murray et al. (2004) discovered elevated levels of barium, cadmium, chromium, copper, nickel, lead, and zinc in surface soil. Table 3.3 lists the common uses for each of the heavy metals listed previously. The toxicity and carcinogenicity of heavy metals varies widely.

A brief summary of their general toxicities follows:

- Arsenic is considered a human carcinogen (USEPA 2021c).
- Barium exposure can cause gastrointestinal disturbances and muscular weakness if it is ingested in a soluble form (ATSDR 2013).
- Cadmium is considered a suspected or probable human carcinogen (ATSDR 2012).

TABLE 3.3
Common Uses for Each of the Select Heavy Metals

Heavy Metal	Atomic Number	Uses
Arsenic	33	Wood preservative, poison, insecticide, pigments, chemical weapons
Barium	56	Superconductors, pigments, fireworks, lubricants, optics
Cadmium	48	Batteries, plastic stabilizer, pigments, metal plating, coatings, alloys
Chromium III	24	Pigments, inks, glass, steel additive, dyes, leather tanning, refractory, alloys
Chromium VI	24	Metal plating, corrosion resistance additive, wood preservative
Copper	29	Wire, building products, piping, jewelry, electromagnetics, brass and other alloys
Lead	82	Batteries, paint, ceramics, firearms, industrial coolant, electrodes, solder, construction materials, alloys, formerly a gasoline additive
Mercury	80	Electrical switches, thermometers, manometers, medical and dental applications, cosmetics, mercury-vapor lamps, formerly used in hat making. Mercury is the only metal evaluated that is a liquid and has a measurable vapor pressure at standard pressure and temperature
Nickel	28	Batteries, steel additive, magnets, coins, alloys
Selenium	34	Photocells, electronics, semiconductors, steel additive, alloys, copier and printing drums, glass manufacturing, pigments
Silver	47	Coins, electronics, circuit boards, alloys, mirrors, decorative items, jewelry, photographic films, batteries
Zinc	30	Metal plating, rust inhibitor (galvanization), brass alloy, batteries, cathodic protection, paint pigment, fire retardant, propellant, photocopying products, medical applications

Source: Pradyot, P. *Handbook of Inorganic Chemical Compounds.* McGraw Hill. New York, NY. 2003: Krebs, R. E. *The History and Use of Earth's Chemical Elements: A Reference Guide.* Greenwood Publishing Group. Oxford, England. 2006.

Land Contamination 47

- Chromium VI is considered a human carcinogen (USEPA 2021c). Chromium III is an essential nutrient that helps the body use sugar, protein, and fat (ATSDR 2021k).
- Copper is essential for good health in small amounts. High levels, however, can be harmful and cause nausea, vomiting, and diarrhea. Very high amounts can damage the liver and kidneys (ATSDR 2018a).
- Lead is considered a suspected or probable human carcinogen (USEPA 2021c). Lead can damage the nervous system and the brain (ATSDR 2021l).
- Mercury exposure may affect the brain and central nervous system (ATSDR 2021m).
- Nickel exposure most often results in an allergic reaction. Nickel may also affect the lungs (ATSDR 2018b).
- Selenium has beneficial and potential harmful effects. Low doses of selenium are necessary to maintain good health. High doses can cause harmful health effects, such as nausea, vomiting, and diarrhea. Prolonged exposure to selenium can cause a disease called selenosis (ATSDR 2018c).
- Silver exposure may result in a condition called arygia if exposure is prolonged. Arygia is a condition that causes a blue-gray discoloration of the skin and other body tissue (ATSDR 2018d).
- Zinc is essential for good health in small amounts. Too little zinc can cause hair and weight loss. Harmful effects do not usually occur until levels of ingestion exceed 10 to 15 times the recommended amount. Effects of overexposure to zinc include stomach cramps, nausea, and vomiting (ATSDR 2018e).

3.9 PESTICIDES AND HERBICIDES

Many of the compounds previously discussed, including several VOCs, SVOCs, and arsenic, are or were ingredients contained in pesticides and herbicides. Simply put, pesticides and herbicides are manufactured to kill things. USEPA (2021k) defines a **pesticide** as preventing, destroying, repelling, or mitigating any pest. An **herbicide** is defined as a substance that is used to kill unwanted plants, commonly referred to as weeds. **Pests** are defined as living organisms occurring where they are not wanted or causing damage to crops or humans or other animals (USEPA 2021k). Examples include:

- Insects
- Mice or other animals
- Unwanted plants (weeds)
- Fungi
- Microorganisms, such as bacteria and viruses
- Invasive species that include many plants and animals

Perhaps the most famous of all pesticides is the now-banned substance called **DDT** (dichlorodiphenyltrichloroethane). It became famous as an environmental contaminant after the book *Silent Spring* written by Rachel Carson was published in 1962. The book catalogued the environmental impacts of the indiscriminant spraying of DDT in the United States and questioned the logic of releasing large amounts of

48 Environmental Compliance Handbook

chemicals into the environment without fully understanding their effects on ecology or human health. DDT was widely used in agriculture and by consumers in the 1940s and 1950s. DDT (CAS Registry Number 50–29–3) is now considered a probable carcinogen (Group: A2) by USEPA (2021c) and is no longer used as a pesticide in the United States after it was banned in 1972. The 2001 United Nations Environmental Program meeting held in Stockholm, Sweden (and put into effect in 2004) permitted its use for "vector control"—organisms that produce pathogens, such as mosquitoes) (USEPA 2021l). The structure of DDT is shown in Figure 3.10.

USEPA banned the use of the pesticide **chlordane** in 1983 because of potential environmental and human health concerns for all applications except termite control (ATSDR 2011; USEPA 2016).

Malathion is an organophosphate insecticide widely used in agriculture, residential landscaping, and public recreation areas. Malathion is used widely to control mosquitoes and the West Nile virus and was used in the 1980s in southern California to combat the Mediterranean Fruit Fly. (ATSDR 2003). Exposure to high amounts of malathion can cause difficulty breathing, chest tightness, vomiting, cramps, diarrhea, blurred vision, sweating, headaches, dizziness, loss of consciousness, and possibly death (ASTDR 2003). If appropriate treatment is provided rapidly, there may be no long-term harmful effects.

Permethrin is a widely used synthetic insecticide and insect repellent used on cotton, wheat, maize, and alfalfa crops. It is also used to kill parasites on chickens and other poultry and as a flea treatment for dogs. Permethrin is considered a neurotoxin and is highly toxic to both freshwater and estuarine aquatic organisms (ATSDR 2005c).

Toxaphene is an insecticide that is composed of a mixture of over 670 chemicals (ATSDR 1997b). Toxaphene was one of the most widely used insecticides in the United States until 1982 when use dropped significantly and then was banned in 1990. It was primarily used in the southern states where it was applied to cotton to control pests. However, it was also used elsewhere to control pests on livestock and to kill unwanted fish in lakes (ATSDR 1997b). Exposure to toxaphene may cause damage to the lungs, nervous system, and kidneys and can even cause death if exposure is extreme (ATSDR 1997b).

Glyphosate and **2,4D** are widely used herbicides in the United States. A common name or trade name for glyphosate is *Roundup*. USEPA (2021c) does not currently classify glyphosate and 2,4D as carcinogenic. However, these two compounds

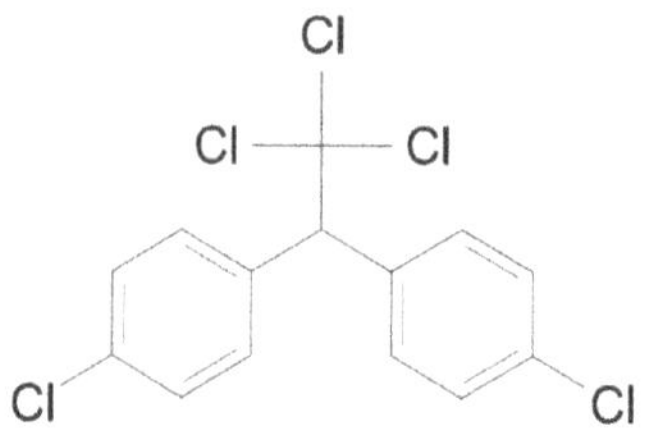

FIGURE 3.10 Basic molecular structure of DDT.

may affect the immune system, kidneys, and liver (USEPA 2021c). Many pesticides and herbicides are **organochlorines**—an organic compound containing at least one covalently bonded chlorine atom. Chlorine released into the environment presents special challenges because: it (1) it is highly toxic, (2) tends to be persistent due to the strength of the chlorine-carbon bonds, and (3) has an affinity for fatty tissues in vertebrates (fish and mammals), thus enabling bioaccumulation, a process where concentrations of a toxin increase at higher trophic levels in a food chain. Numerous studies have noted the presence of trace amounts of organochlorines in human breast milk (Calle et al. 2002).

3.10 DIOXINS

Compounds generally referred to as **dioxins** represent a diverse set of halogenated substances and include other substances called furans. There are 75 dioxin isomers and 135 furan isomers. Dioxins are not intentionally produced and have no known use (ATSDR 2008a); they form unintentionally as a byproduct of many industrial processes involving chlorine, such as waste incineration and combustion. Dioxin compounds may also form as a byproduct during the manufacture of chlorinated compounds and paper bleaching (ATSDR 2008a). A dioxin compound consists of two benzene molecules joined with two oxygen bridges. Figure 3.11 shows the basic structure of dioxin.

According to ATSDR (2008a) the most toxic dioxin compound is 2,3,7,8-tetrachlorodibenzo-p-dioxin, or TCDD (CAS Registry Number 1746–01–6). Exposure to TCDD may lead to a condition known as chloroacne, resulting in severe acne-like skin lesions occurring mainly on the face and upper body. USEPA (2021c) lists many of the dioxin compounds as either known or suspected carcinogens. Dioxin compounds have been shown to bioaccumulate in humans and wildlife, and they behave as teratogens and mutagens (USEPA 2021m).

3.11 FERTILIZERS

Fertilizers are chemical compounds designed to promote plant and fruit growth when applied (USEPA 2021b). The most common fertilizers are nitrogen (N), phosphorus (P), and potassium (K). Nitrogen is found primarily in an organic form in soils but can also occur as nitrate. Because nitrate is very soluble and mobile, it is transported by surface water to rivers, lakes, and streams, where it can promote algal growth. In many cases, the algal growth is extensive. Nitrate can also contaminate drinking water. Phosphorus occurs in soil in organic and inorganic forms but, being more soluble than nitrate, can also be depleted in soil through surface water runoff. Phosphorus can also

FIGURE 3.11 Basic structure of a dioxin molecule.

50   Environmental Compliance Handbook

promote algal growth in rivers, lakes, and streams, because it is a limiting nutrient in fresh water. Potassium (K) in fertilizers is commonly incorporated as potash—an oxide form of potassium that includes the compounds potassium chloride, potassium sulfate, potassium nitrate, and potassium carbonate. The term "potash" comes from the historical practice of extracting potassium carbonate (K_2CO_3) by leaching wood ashes and evaporating the solution in large iron pots. Fertilizers containing potassium generally do not promote algal growth (USEPA 2021b).

The United States Department of Agriculture (USDA) has tracked fertilizer use in the United States since 1960. According to the USDA, approximately 63.5 kilograms (140 pounds) of fertilizer containing N-P-K are applied each year per acre of land farmed, and this amount has increased more than 200% since 1960 (USDA 2021a). In 2018, USDA estimated that 1.41 million square kilometers (350 million acres) were planted for crops which means that approximately 22.2 trillion kilograms (49 trillion pounds) of fertilizers were applied to farm land in the United States (USDA 2021a). In urban areas, fertilizers containing N-P-K are common and routinely applied to residential lawns and golf courses to help maintain green, healthy grass and gardens (USEPA 2021n). According to USEPA (2021c), nitrogen, phosphorus, and potassium fertilizers are not currently known to cause cancer. Exposure to high concentrations may cause nausea and vomiting (USEPA 2021c).

3.12 CYANIDE

Cyanide is any chemical compound containing the cyano group-a carbon atom triple-bonded to a nitrogen atom. Common cyanide compounds include hydrogen cyanide, potassium cyanide, and sodium cyanide. Certain bacteria, fungi, and algae can produce cyanide, and cyanide is present in a number of foods and plants (ATSDR 2006b).

Cyanide compounds occur as gases, liquids, and solids. Inorganic cyanides are commonly salts of the cyano anion CN^-. Organic compounds containing the cyano group are called nitriles. Those compounds that are able to release the cyano group CN^- ion are highly toxic to humans and animals (ATSDR 2006b). Hydrogen cyanide is a colorless gas with a faint, bitter, and almond-like odor. Sodium cyanide and potassium cyanide are both white solids with a bitter, almond-like odor when volatilization occurs. Cyanide and hydrogen cyanide have industrial applications in metal plating, metallurgy, some mining applications for extraction of precious metals, and the manufacturing of plastics (ATSDR 2006b). According to USEPA (2021c), cyanide and related compounds are not classified as carcinogens (Group: A4), but they are highly toxic. According to USEPA (2021c) and ATSDR (2006b), exposure to cyanide can cause rapid breathing, low blood pressure, headaches, coma, and death.

3.13 ASBESTOS

Asbestos (CAS Registry Number 1332–21–4) is the name given to a group of six different fibrous minerals (ATSDR 2001b):

- Amosite
- Chrysotile

Land Contamination

- Crocidolite
- The fibrous forms of tremolite, actinolite, and anthrophyllite

The term asbestos describes a variety of fibrous, nonflammable minerals with flexibility and high tensile strength. Their unique properties were used mostly between the 1940s and 1970s in fireproof insulation, vinyl flooring, pipe insulation, ceiling tiles, brake linings, and roof coatings. Chrysotile, a serpentine mineral, is also known as "white asbestos" and makes up about 95% of asbestos found in buildings in the United States. The other asbestos minerals—crocidolite, amosite, anthophyllite, tremolite, and actinolite—are amphiboles that aren't commonly used in commercial products (USEPA 2021o). Asbestos usually occurs in the form of very thin elongated fibers. If asbestos is disturbed, the fibers may create an airborne dust, which, when inhaled, can penetrate tissues deep within the lungs and cause asbestosis, mesothelioma, and lung cancer. The US Occupational Safety and Health Administration and Environmental Protection Agency make no distinction between the two kinds of asbestos—chrysotile and amphibole. OSHA began regulating workplace asbestos in 1970, around the time when miners and construction workers, who worked with the fibrous material, began reporting serious lung disease. At that time, the United States used about 800,000 metric tons of asbestos per year. OSHA published the Asbestos Standard for the Construction Industry, which outlined four categories of asbestos contamination and specified precautions and disposal techniques for each class. For instance, if asbestos insulation is removed (Class 1), contractors and supervisors trained in asbestos removal must be onsite wearing respirators and protective clothing (USEPA 2021o).

Starting in 1973, USEPA began a process of banning certain uses of asbestos and designated it a Class A1 human carcinogen (USEPA 2021c). USEPA's response to asbestos has changed over the years. In 1983, the agency's asbestos handbook stated that removing the material is always appropriate, while their 1990 handbook acknowledged that asbestos removal may cause more contamination than leaving it in place. USEPA's advice on asbestos is neither to rip it all out in a panic nor to ignore the problem under a false presumption that asbestos is risk free. Asbestos material in buildings should be identified and managed appropriately. The main commercial and industrial value of asbestos lies in its ability to be heat resistant. For this reason, asbestos was widely applied in the United States in manufactured goods and building construction as a heat insulator for these products (USEPA 2021o):

- Roofing, ceiling, and floor tiles
- Window caulking
- Pipe wrap and pipe insulation
- Paper products
- Friction products, such as automobile brakes, clutches, and transmissions
- Heat-resistant fabrics
- Packaging
- Gaskets
- Coatings
- Some vermiculite- and talc-containing products
- Fire-resistant doors

52

Environmental Compliance Handbook

Exposure to asbestos is almost always through the inhalation route. Asbestos affects the lungs and may lead to a condition called asbestosis. USEPA classifies asbestos as a human carcinogen (Group: A1) (USEPA 2021c). According to ATSDR (2001b), we are all exposed to some asbestos in the air we breathe, especially in urban areas. Fortunately, the concentrations in air are generally very low, being on the order of 0.00001 to 0.0001 fibers of asbestos per milliliter of air (USEPA 2021o).

3.14 ACIDS AND BASES

An **acid** increases the concentration of the hydrogen ion H^+ when dissolved in water and lowers the pH (potential hydrogen) of the solution. The bare hydrogen ion, H^+, is short for the hydronium ion, H_3O^+, since a bare H^+ does not exist in a solution. Conversely, a **base** increases the concentration of the hydroxide ion OH- when dissolved in water and raises the pH of the solution (Meyers 2003). Common acids and bases are listed in Table 3.4 (Meyers 2003). We recognize acids and bases by their simple properties, such as taste, and conclude the sour taste of a lemon indicates it must be acidic. Bases tend to taste bitter. On the pH scale, any substance with a pH less than 7 (the neutral point) is acidic, and any substance with a pH greater than 7 is basic. Acids and bases are widely used in industry and are present in many widely consumed foods and drinks (see Table 3.4). Stronger acids and bases are used in household cleaners and detergents, especially those used on glassware and in ovens (Meyers 2003).

Acids or bases are only toxic if they are strong, meaning they are of relatively low or high pH. Exposure to strong acids and bases causes respiratory irritation and

TABLE 3.4
Common Acids and Bases

Common Acids		Common Bases	
Acid	Chemical Formula	Base	Chemical Formula
Acetic Acid	$HC_2H_3O_2$	Ammonia	NH_3
Benzoic Acid	$HC_7H_5O_2$	Aniline	$C_6H_5NH_2$
Boric Acid	H_3BO_3	Dimethylamine	$(CH_3)_2NH$
Carbonic Acid	H_2CO_3	Ethylamine	$C_2H_5NH_2$
Cyanic Acid	HCNO	Hydrazine	H_4N_2
Formic Acid	$HCNO_2$	Hydroxylamine	NH_2OH
Hydrocyanic Acid	HCN	Methylamine	CH_3NH_2
Hydrofluric Acid	HF	Pyridine	C_5H_5N
Hydrogen Sulfide	H_2S	Urea	NH_2CONH_2
Hydrochloric Acid	HCl	Potassium hydroxide	KOH
Nitric acid	HNO_3	Sodium bicarbonate	$NaHCO_3$
Phosphoric Acid	H_3PO_4	Sodium hydroxide	NaOH
Pyuvic Acid	$HC_3H_3O_3$	Calcium hydroxide	$Ca(OH)_2$
Sulfuric Acid	H_2SO_4		

Source: Meyers, R. *The Basics of Chemistry.* Greenwood Press. Westport, CT. 2003.

Land Contamination 53

burning and skin burns. Significant exposure may cause severe burns and even death (USEPA 2021c). Currently, adequate information is not available to evaluate the potential carcinogenic effects of common acids and bases (USEPA 2021c). Ammonia is a common basic chemical widely used as a household cleaning agent and in many industrial applications (ATSDR 2018f). Ammonia is present naturally throughout the environment in air, soil, and water. Exposure to high levels of ammonia may cause lung, skin, and throat irritation. Some people with asthma may react more negatively to the inhalation of ammonia (ATSDR 2018f). Hydrochloric acid (also referred to as hydrogen chloride) is a common acid widely used in industry as a cleaning agent, in the manufacturing of PVC, in making steel, and in making leather. Hydrochloric acid is also present in humans and other organisms as a gastric acid and sometimes exists as an acid mist. This mist may cause skin and lung irritation, and skin burns can occur if you are exposed to a highly concentrated mist for a prolonged period (ATSDR 2002b).

3.15 RADIOACTIVE COMPOUNDS

Radioactive decay occurs when an unstable atomic nucleus spontaneously loses energy by emitting ionizing particles and radiation. This decay, or loss of energy, results in an atom of one type (parent nuclide) transforming into an atom of a different type (daughter nuclide). All elements with an atomic number greater than 80 possess radioactive isotopes, and all isotopes of elements with an atomic number greater than 83 are radioactive (Kathren 1991).

Some radioactive compounds deserving special attention include (Kathren 1991):

* Beryllium
* Cadmium
* Palladium
* Thorium
* Calcium
* Cesium
* Tin
* Uranium
* Carbon
* Iodine
* Radon
* Plutonium
* Potassium
* Strontium
* Radium

Radon (CAS Registry Number 10043–92–2) is the most common radioactive compound present in urban areas, and it has the potential for adverse human health effects (ATSDR 2000). USEPA (2021c) classifies radon as a human carcinogen (Group: A1), and exposure to radon for a long period of time at elevated concentrations may cause cancer. Radon is a decay product of uranium, found naturally in the Earth's crust. It is one of the heaviest substances existing as a gas under normal conditions of pressure and temperature.

The highest average radon concentrations in the United States are found in Iowa, southeastern Pennsylvania, and Appalachian Mountain areas. During the decay process, alpha, beta, and gamma radiation is released. Alpha particles can travel only short distances and cannot travel through your skin. Beta particles can penetrate through your skin but not your whole body. Gamma particles can penetrate your whole body. Radon is normally present at very low levels in outdoor air but may be present at higher levels in indoor air, especially in basements and buildings with poor ventilation and in well water (ATSDR 2000).

3.16 GREENHOUSE GASES

Greenhouse gases are gases in the atmosphere capable of absorbing and emitting radiation within the thermal infrared range (Karl and Trenberth 2003; USEPA 2017, 2018a). Greenhouse gases cause the Earth to warm up, and since the start of the industrial revolution in the early 18th century, levels of greenhouse gases in the Earth's atmosphere have increased (USEPA 2018a). Greenhouse gases are not generally investigated at specific site of environmental contamination. It is important, however, to discuss greenhouse gases because of their potential impacts on climate change. Figure 3.12 shows the increase of carbon dioxide in the atmosphere since 1960 (NOAA 2021). The yearly variation shown in Figure 3.12 is seasonal and is attributed to extraction of carbon dioxide from plant matter during photosynthesis (NOAA 2021). The greenhouse gases include the following compounds (USEPA 2018a):

- Carbon dioxide (CO_2)
- Methane (CH_4)
- Nitrous oxide (N_2O)
- Fluorinated gases, including:
  - Hydrofluorocarbons (HFCs)
  - Perfluorocarbons (PFCs)

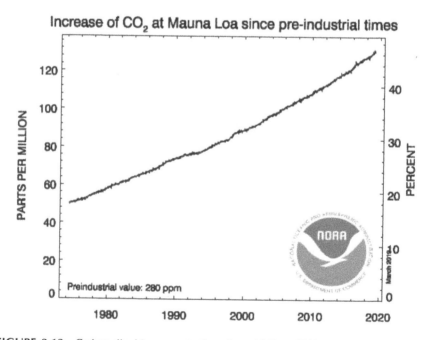

FIGURE 3.12 Carbon dioxide concentrations from 1960 to 2020 at Mauna Loa, Hawaii (from National Oceanic and Atmospheric Administration [NOAA], 2021. *Atmospheric CO_2 at Mauna Loa Observatory*. www.esrl.noaa.gov/gmd/ccgg/trends [accessed June 5, 2021], 2021).

Land Contamination 55

- Sulfur hexafluoride (SF_6)—an industrially produced gas with a global warming potential (GWP) almost 24,000 times that of CO_2. The GWP measures the relative contribution of a gas to global warming based on its ability to absorb infrared radiation, its residence time in the atmosphere, and wavelengths of energy absorbed. For comparison purposes, CO_2 has a GWP = 1.

Greenhouse gases originate from a variety of sources, such as (USEPA 2018a):

- Fossil fuel combustion:
  - Coal
  - Gasoline and diesel fuel in automobiles and trucks
  - Aviation fuels
  - Home heating fuels such as home heating oil and kerosene
- Industrial processes—refineries and cement making. Cement making is an often-overlooked source; it accounts for 5–7% of anthropogenic CO_2
- Waste disposal facilities
- Electrical generation
- Mining
- Residential and commercial sources
- Agriculture

Increasing carbon dioxide levels in the atmosphere have been linked to the burning of fossil fuels (USEPA 2018a), and the evidence is clear. The current total estimated emissions of greenhouse gases worldwide exceed 6,780 million metric tons (USEPA 2018a).

Some negative effects attributed to climate change caused by rising carbon dioxide levels in the atmosphere can be observed in the western portion of North America in Canada and the United States. Millions of acres of forest are being killed by the pine bark beetle moving into areas that were once too cold for the beetle to thrive (United States Forest Service 2021).

3.17 CARBON DIOXIDE

Carbon dioxide is a greenhouse gas that we covered previously, but that is just half the story. We must also discuss the buildup of carbon dioxide in the world's oceans. The oceans were once believed to be so vast and deep that the effects of pollution would never directly threaten them. This has been proven false. The rate of accumulation of pollution is proportional to human population. The more people, the more pollution ends up in the oceans.

Pollution in the oceans does not all originate from direct discharge but is also delivered through the atmosphere from increased concentrations of carbon dioxide in the atmosphere that is partially adsorbed by the oceans and causes acidification of the surface layers. Absorption of carbon dioxide by the oceans has been estimated to be as high as 25%, which represents millions of tons of carbon dioxide absorbed by

56 Environmental Compliance Handbook

the oceans every year. The additional carbon dioxide absorbed by the oceans reacts with ocean water, forms an acid called carbonic acid, and is lowering the pH of the oceans worldwide. In fact, the oceans are acidifying faster than they have in some 300 million years. The effect of the acidification has been linked to bleaching of coral reefs, including the Great Barrier Reef off the northeastern coast of Australia (Natural Resource Defense Council 2021).

3.18 CARBON MONOXIDE

Carbon monoxide (CO) is a colorless, odorless, nonirritating gas that is very toxic to humans and animal life (USDHH 2009). Carbon monoxide is formed naturally and anthropogenically. The human body produces a small amount of CO when red blood cells convert protoporphyrin into bilirubin. Synthetically, most carbon monoxide is created by the incomplete combustion of fossil fuels, with the highest percentage coming from automobile exhaust. Inside homes, significant sources of carbon dioxide emissions can be natural gas and home heating oil furnaces, hot water heaters, appliances, wood-burning stoves, and fireplaces. Carbon monoxide is also created as a byproduct of several industrial processes, including metal melting and chemical synthesis (USDHH 2009).

Carbon monoxide is one of the most common causes of fatal air poisoning (USDHH 2009). When inhaled, carbon monoxide combines with hemoglobin to produce carboxyhemoglobin. A condition known as anoxemia arises because carboxyhemoglobin cannot deliver oxygen to body tissues efficiently. Exposure to high levels of carbon monoxide may cause headache, nausea, vomiting, dizziness, lethargy, seizures, coma, and even death (USDHH 2009). Concentrations of less than 700 parts per million of CO may cause up to 50% of the body's hemoglobin to convert to carboxyhemoglobin. In the United States, OSHA has established a long-term exposure standard for carbon monoxide at 50 parts per million (United States Department of Labor 2021).

3.19 OZONE

Ozone (O_3) is a gas present in Earth's upper atmosphere and at ground level. The ozone occurring at ground level is considered an air pollutant by USEPA (2021p), whereas the ozone occurring in the upper atmosphere is not considered a pollutant, because it is beneficial to life on Earth by acting as a filter of ultraviolet (UV) radiation produced by the sun (USEPA 2021q). Ozone from the ground level to a height of approximately 6 miles in the troposphere is the main constituent of urban smog (USEPA 2021q).

Until the mid 1990s, ozone in the stratosphere had been gradually being destroyed by anthropogenically produced chemicals referred to as ozone-depleting substances (ODSs), including (USEPA 2021q):

- Chlorofluorocarbons (CFCs)
- Hydrochlorofluorocarbons (HCFCs)

Land Contamination

- Halons
- Methyl bromide
- Carbon tetrachloride
- Methyl chloroform

These substances were commonly used in coolants, foaming agents, fire extinguishers, solvents, pesticides, and aerosol propellants. Once released into the environment, they tend to degrade very slowly (USEPA 2021q). It should also be noted that hydrofluorocarbons have been substituted widely for CFCs. While this substitution has slowed the harmful effects brought on by free chlorine in the upper atmosphere, HFCs are potent greenhouse gases. Ground-level ozone is not emitted directly into the air but is created by chemical reactions between oxides of nitrogen (NO_x) and VOCs in the presence of sunlight. Emissions from automobile exhaust, gasoline vapors, chemical solvents, electrical generating facilities, and certain factories are some of the major sources that emit compounds leading to the generation of ozone (USEPA 2021p). Ground-level ozone is of a particular concern in urban regions of the United States. During the summer, strong sunlight and hot weather produce the conditions necessary for producing harmful levels of ozone. Overexposure to ground-level ozone can result in difficulty breathing and other respiratory effects (USEPA 2021p), especially for the elderly, very young, and those with existing respiratory ailments.

3.20 SULFUR DIOXIDE

Sulfur dioxide (SO_2) is released naturally through volcanic eruptions. Anthropogenically, significant sources of sulfur dioxide include automobile exhaust, the burning of coal, and some industrial processes (USEPA 2018b). Sulfur dioxide is a component of smog. Other components of smog include ozone, carbon monoxide, particulate matter, VOCs, and nitrous oxides (USEPA 2018b). When released into the atmosphere, sulfur dioxide often reacts with water vapor and eventually forms sulfuric acid (H_2SO_4). When precipitation occurs with a pH lower than that of natural rain (5.6) it is considered **acid rain** (USEPA 2018b).

3.21 PARTICULATE MATTER

Particulate matter (PM) is also known as particle pollution. PM is a complex mixture of very small particles and liquid droplets. Particle pollution consists of a number of components, including (USEPA 2021r):

- Acids
- Sulfates
- Metals
- Nitrates
- Organic compounds
- Soil or dust particles

The size of the particles is directly related to their potential for causing adverse health effects. Particles of 10 micrometers (μm) or less are small enough to pass through the nose and throat and enter the lings while breathing (USEPA 2021r).

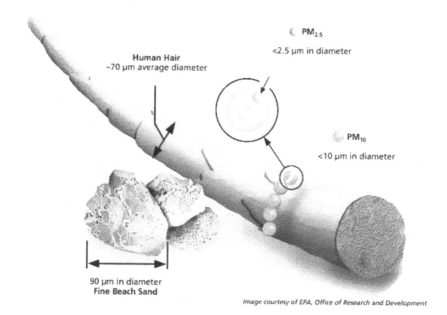

FIGURE 3.13 Particle size representation (from United States Environmental Protection Agency. Particulate Matter (PM) Pollution. www.epa.gov/pm-pollution [accessed June 5, 2021]. 2021r).

Once inhaled, these particles can affect the heart and lungs and cause substantially adverse health affects. Figure 3.13 is a diagram of particle size relative to a human hair (USEPA 2021r).

USEPA groups PM into two categories (USEPA 2021r):

- Particles ranging in size from 2.5 to 10 μm are considered inhalable coarse particles, most likely to be present near roadways and dust-producing industries.
- Particles less than 2.5 μm are considered fine particles and are present in smoke and haze. These particles can be emitted from forest fires or power plants, certain industries, and automobiles.

Adverse health effects caused by inhalation of PM include (USEPA 2021r):

- Increased respiratory irritation, coughing, and difficulty breathing
- Decreased lung function
- Aggravated asthma
- Development of chronic bronchitis
- Irregular heartbeat
- Nonfatal heart attacks
- Premature death in people with heart of lung disease

Land Contamination

An additional concern is the potential exposure to other contaminants sorbed onto PM, which are then ingested or inhaled.

3.22 BACTERIA, PARASITES, AND VIRUSES

Although we do not always think of bacteria, parasites, and viruses as everyday environmental contaminants, they are common and can cause many diseases and adverse health effects. Technically bacteria, parasites, and viruses can contaminate water supplies and are also present in the air, in soil, on and within food, and most other surface surfaces.

Bacteria are organisms made up of just one cell. They are capable of multiplying themselves through a process called "binary fission," whereby a single bacterium grows to approximately twice its normal size and then splits into two daughter cells that are copies of the original bacterium. Bacteria live everywhere, even inside most organisms. Most bacteria are harmless, and some are beneficial by destroying other harmful bacteria within our bodies (Madigan et al. 2008). However, some may cause disease such as tuberculosis. One of the more common harmful bacteria is a group called *E. coli*, which is short for *Escherichia coli*. Most *E. coli* are harmless, but a strain called serotype O157:H7 can cause food poisoning in humans. Three basic shapes of bacteria exist: rounded, rod shaped, and spirals. *E. coli* bacteria is rod shaped. The ability for *E. coli* to survive for a brief period outside the body creates the potential for the bacteria to spread and infect other people. The spread of *E. coli* usually occurs when there is poor sanitation or when untreated sewerage is discharged from municipal wastewater treatment plants, which occasionally occurs during flood events (USEPA 2009). Adverse health effects from exposure to *E. coli* bacteria typically include gastroenteritis, urinary tract infections, skin rashes, and neonatal meningitis (USEPA 2009).

A **virus** is a sub-cellular infectious agent capable of replicating itself inside the cells of another organism (Madigan et al. 2008). They are typically 100 times smaller than a bacterium. Viruses consist of two parts: (1) DNA or RNA molecules carrying genetic information and (2) a protein coat protecting the genes. Some viruses may also have an outside layer of fat surrounding them while they are outside a cell. Viruses spread in many ways, including aerosol routes (coughing and sneezing), infected or contaminated water, and exchange of body fluids. Viruses cause diseases such as the common cold, influenza, chickenpox, mumps, ebola, and HIV (Madigan et al. 2008).

A **parasite** is an organism living on or within a different organism (the host) at the expense of the host organism. Common examples of parasites causing adverse health effects in humans are cryptosporidium and giardia; both may cause severe intestinal disorder (United States Center for Disease Control 2009). Exposure to these two parasites occurs by consuming affected water (United States Center for Disease Control 2009). In humans, giardia creates uncomfortable but curable gastrointestinal symptoms, whereas cryptosporidium can create life-long gastrointestinal symptoms in persons with weak immune systems and may result in death.

60   Environmental Compliance Handbook

3.23 INVASIVE SPECIES

We will consider invasive species as a form of pollution. This is because the definition of invasive species tracks closely with other forms of pollution we have discussed. An **invasive species** is any kind of living organism that is not native to an ecosystem and causes harm that is directly or indirectly introduced by humans or is the result of human activity (United States Department of Agriculture 2021b; USEPA 2018c).

The invasive species mode of invasion is commonly as a stowaway during the transportation of goods. USEPA (2018c) estimated that 30% of invasive species to the Great Lakes were as ballast water in ships carrying cargo. Ballast water is taken into or discharged from a ship as it loads or unloads cargo to accommodate the ship's weight changes. Invasive species are a worldwide problem and have been for hundreds of years. Most wildlife and environmental agencies are losing the battle with invasive species, for one simple and disturbing reason: this type of pollution breeds.

To make this point clear, the United Nations established the Global Invasive Species Program (GISP) in 1997 to develop a global strategy to deal with invasive species (United Nations 2021). Worldwide, there are thousands of invasive species (United Nations 2021). Invasive species are characterized as aggressive when they are transported into areas where they are not naturally occurring usually because they do not have a natural predator or other organisms have not evolved defenses (United Nations 2021). The top 13 invasive species currently causing the most harm in the United States are (USEPA 2018c):

1. Burmese python. Introduced in the Florida Everglades and considered established in 2002; they have reduced deer, raccoon, marsh rabbits, bobcats, possums, and alligators by as much as 99%.
2. Emerald ash borer. Originally from Russia, China, and Japan, they were accidentally introduced in southeastern Michigan in 2002 in infected wood crates. The emerald ash borer has infected and killed hundreds of million of ash trees in the United States.
3. Nutria. Commonly called swamp rats. They were imported from South America into the southern United States for their fur but now have grown out of control and are in the wild. Their endless digging along the banks of rivers and streams kills plants and fish and causes soil erosion.
4. European starling. Since their introduction into the United States in the 1890s, they have become the most successful foraging bird in the United States, with a population now estimated at 200 million.
5. Northern snakehead. The most feared fish in the Chesapeake watershed. A sharp-toothed fish from Korea and China that originally infested the Chesapeake watershed, it has now spread to other parts of Virginia and Delaware.
6. Brown marmorated stink bug. Stink bugs destroy fruits and vegetables to such an extent that they have been know to increase the price of produce. They first showed up in Allentown, Pennsylvania, in 1998 after crawling out of a cargo ship. Originally they are from China.

Land Contamination

7. Feral hogs. Essentially farm pigs gone wild. They are established in 47 states, with massive populations in Texas, Florida, Georgia, Virginia, and North Carolina.
8. Lionfish. Lionfish are native to the Pacific Ocean. They were first spotted off the Miami coast in the mid-1980s. They are characterized as eating everything that they can stuff into their mouths. That's how they obtained their name. They are destroying life on the coral reefs of the Atlantic Ocean.
9. Norway rat. First introduced in the United States in 1775, and now they are everywhere, including Hawaii and Alaska.
10. Tegu. Tegus are brown lizards that are muscular, fast, and love to eat eggs. They are known to harass pets and invade homes where they were released in Florida. Game officials have largely given up on eradicating them from the wild.
11. Asian citrus psyllid. The Asian citrus psyllid carries a bacteria that attacks and kills fruit trees, especially orange trees. Estimates are that 80% of the orange trees in Florida are infected.
12. Brown tree snake. The brown tree snake has infected the island of Guam and has decimated the local bird population. They are so out of control on the island that they are even responsible for power outages.
13. Kudzu. First introduced into the United States in 1877, kudzu is an invasive plant species that grows very fast and smothers other plants in the southeastern United States to as far north as Indiana. Kudzu is now growing and killing other plant species in the United States at a rate of up to 1,000 square kilometers per year.

The numbers of different types of invasive species in North America alone are the following (USDA 2021b):

- Plants—1,911 species
- Insects—6,647 species
- Pathogens—1,278 identified
- Animals—220 species

The number of invasive species and diseases in North America alone is nearly 10,000 (USDA 2021b).

3.24 POLYFLUOROALKYL SUBSTANCES

Per- and polyfluoroalkyl substances are a group of synthetic chemicals that have been manufactured and used in a variety of industries and consumer products worldwide since the 1940s. They are commonly referred to as PFOA or PFAS. These chemicals are very persistent and mobile in the environmental and tend to bioaccumulate. Although these chemicals are no longer manufactured in the United States, they can be present in many consumer goods that include carpet, leather and apparel, textiles, paper, packaging, coatings, rubber, and plastic (USEPA 2018d).

62 Environmental Compliance Handbook

Toxicological studies indicate that these chemicals can cause reproductive and developmental abnormalities, liver and kidney problems, and immunological effects in laboratory animals and cause tumors in animals. The most consistent findings are increased cholesterol levels among exposed populations, with more limited findings related to low infant birth weights, effects on the immune system, cancer, and thyroid hormone disruption (USEPA 2018d). In certain parts of the United States, these chemicals have contaminated water supplies and streams and lakes, prompting health advisories.

3.25 EMERGING CONTAMINANTS

Through research, we are now learning about the presence in the environment of many chemicals and microbes that historically were not considered contaminants (USEPA 2021s). Emerging contaminants originate from urban and agricultural sources and impact soil and groundwater at many urban locations. The potential health risks posed by emerging contaminants are not fully known. Many emerging contaminants enter the environment from residential waste products and agriculture, and this fact has prompted a shift in traditional thinking that most releases of contaminants into the environment were from industrial sources (Barnes et al. 2008).

Emerging contaminants include a wide variety of compounds consisting of (Bell et al. 2019; Barnes et al. 2008; USEPA 2021s):

- Pharmaceuticals and drugs, including:
  - Antibiotics
  - Steroids
  - Antibacterial chemicals
  - Hormones
  - Narcotics
  - Many other legal and illegal drugs

- Insect repellants
- Solvents
- Detergents
- Plasticizers
- Fire retardants
- Veterinary antibiotics
- Pesticides
- Others

Common pharmaceuticals and drugs with the capability to become contaminants if not properly disposed of include (USEPA 2021s):

- Hormones, such as testosterone
- Antibiotics, such as penicillin
- Sildenafil citrate, commonly known as Viagra
- Benzoylmethylecgonine, commonly known as cocaine

Land Contamination 63

Benzoylmethylecgonine or cocaine is a stimulant affecting the central nervous system and also acts as an appetite suppressant. Antibiotics inhibit the growth of bacteria. Sildenafil citrate is an arterial stimulant that was originally intended to treat high blood pressure (Barnes et al. 2008). Testosterone is a male sex hormone, an anabolic steroid, and affects the growth of muscle mass.

Two other emerging contaminants of note include a group of compounds called perchlorates and the compound 1,4-dioxane. **Perchlorates** are colorless and odorless salts. They are a group of compounds including:

- Magnesium perchlorate ($MgClO_4$)
- Potassium perchlorate ($KClO_4$)
- Ammonium perchlorate (NH_4ClO_4)
- Sodium perchlorate ($NaClO_4$)
- Lithium perchlorate ($LiClO_4$)

Perchlorates are very reactive and are commonly used in explosives, fireworks, road flares, and rocket motors (USEPA 2021t; ATSDR 2008b). Perchlorate may also be present in bleach as an impurity. Adverse health affects of exposure to perchlorates include the ability of the thyroid gland to uptake iodine. Iodine is needed to produce hormones that regulate many body functions. USEPA does not currently list any of the perchlorate compounds as human carcinogens (USEPA 2021c).

1,4-dioxane ($C_4H_8O_2$) is a clear liquid that easily dissolves in water. It is one of three isomer varieties of dioxanes and is primarily used as an industrial solvent, with less widespread use in cosmetics, shampoos, and detergents (ATSDR 2007d). The ability of 1,4-dioxane to dissolve so easily in water and its non-biodegradability result in 1,4-dioxane easily contaminating surface water and groundwater. Inadequate information is available for classifying the carcinogenicity of 1,4-dioxane (USEPA 2021c).

1,2,3,-trichloropropane (TCP) is a chlorinated volatile organic compound with high chemical stability and relatively high solubility (Bell et al. 2019 and California Environmental Protection Agency [CalEPA] 2021). TCP is considered a probable human carcinogen (USEPA 2021c). TCP is used as a cleaning solvent and as a pesticide, especially in California. CalEPA established a drinking water notification concentration for TCP of 0.005 micrograms per liter (ug/l). TCP has been detected in groundwater at numerous locations in California, especially within agricultural areas (CalEPA 2021).

3.26 SUMMARY AND CONCLUSION

Now we know there are thousands of different types of pollutants existing everywhere. They are in the air we breathe, the water we drink, the food we eat, and the dirt we play in. They are organic and inorganic. Many naturally occur—many do not. Some we know about, and most we do not. Some are not so toxic, and some are very toxic. Some may cause cancer, and some may not. All of them have the ability to cause some adverse health effects in humans or other organisms or impair or negatively impact the environment if the exposure and dose are just right. Otherwise, they would not be considered pollutants.

64 Environmental Compliance Handbook

This now leads us to the next set of questions we need to explore. How do contaminants behave once they are released into the environment? Do they degrade? Where would we go to find them? How long do they last? Evaluating the behavior of pollutants in the environment is commonly referred to as fate and transport assessment. Analysis of the fate and transport of pollutants once released into the environment is crucial for accurately assessing the risk posed by a specific chemical. Just because a pollutant exists does not mean there will be a risk to human health or the environment. There must be a completed pathway—the pollutant must be transported from its point of release to a place where exposure can occur. Chapter 4 discusses this central concept and describes the fate and transport of many pollutants introduced in this chapter.

3.27 REFERENCES

Agency for Toxic Substances and Disease Registry (ATSDR). 1995. *2-Hexanone*. CAS Registry Number 591–78–6. ATSDR ToxFAQs. Atlanta, GA.

Agency for Toxic Substances and Disease Registry (ATSDR). 1997a. *Chloroform*. CAS Registry Number 127–18–4. ATSDR ToxFAQs. Atlanta, GA.

Agency for Toxic Substances and Disease Registry (ATSDR). 1997b. *Toxaphene*. CAS Registry Number 8001–35–2. ATSDR ToxFAQs. Atlanta, GA.

Agency for Toxic Substances and Disease Registry (ATSDR). 2000. *Radon Toxicity: Who Is At Risk?* ATSDR. Atlanta, GA.

Agency for Toxic Substances and Disease Registry (ATSDR). 2001a. *Toluene*. CAS Registry Number 108–88–3. ATSDR ToxFAQs. Atlanta, GA.

Agency for Toxic Substances and Disease Registry (ATSDR). 2001b. *Asbestos*. CAS Registry Number 1332–21–4. ATSDR ToxFAQs. Atlanta, GA.

Agency for Toxic Substances and Disease Registry (ATSDR). 2002a. *Di(2-ethylhexyl) Phthalate*. CAS Registry Number 117–81–7. ATSDR ToxFAQs. Atlanta, GA.

Agency for Toxic Substances and Disease Registry (ATSDR). 2002b. *Hydrogen Chloride*. CAS Registry Number 7647–01–0. ATSDR ToxFAQs. Atlanta, GA.

Agency for Toxic Substances and Disease Registry (ATSDR). 2003. *Malathion*. CAS Registry Number 121–75–5. ATSDR ToxFAQs. Atlanta, GA.

Agency for Toxic Substances and Disease Registry (ATSDR). 2005a. *Naphthalene*. CAS Registry Number 91–20–3. ATSDR ToxFAQs. Atlanta, GA.

Agency for Toxic Substances and Disease Registry (ATSDR). 2005b. *Bromoform*. CAS Registry Number 75–25–2. ATSDR ToxFAQs. Atlanta, GA.

Agency for Toxic Substances and Disease Registry (ATSDR). 2005c. *Permethrin: Toxicologic Information About Pesticides*. CAS Registry Number 52645–53–1. ATSDR. Atlanta, GA.

Agency for Toxic Substances and Disease Registry (ATSDR). 2006a. *Vinyl Chloride*. CAS Registry Number 75–01–4. ATSDR. Atlanta, GA.

Agency for Toxic Substances and Disease Registry (ATSDR). 2006b. *Cyanide*. CAS Registry Numbers 74–90–8; 143–33–9; 151–50–8; 592–01–8; 544–92–3; 506–61–6; 460–19–5; and 506–77–4 ATSDR. Atlanta, GA.

Agency for Toxic Substances and Disease Registry (ATSDR). 2007a. *Benzene*. CAS Registry Number 71–43–2. ATSDR ToxFAQs. Atlanta, GA.

Agency for Toxic Substances and Disease Registry (ATSDR). 2007b. *Ethybenzene*. CAS Registry Number 100–41–4. ATSDR ToxFAQs. Atlanta, GA.

Agency for Toxic Substances and Disease Registry (ATSDR). 2007c. *Xylene*. CAS Registry Number 1330–20–7. ATSDR ToxFAQs. Atlanta, GA.

Agency for Toxic Substances and Disease Registry (ATSDR). 2007d. *1,4-Dioxane*. CAS Registry Number 123–91–1. ATSDR ToxFAQs. Atlanta, GA.

Land Contamination

Agency for Toxic Substances and Disease Registry (ATSDR). 2008a. *Dioxin*. www.atsdr.cdc. gov/toxprofiles/TP.asp?id=366&tid=63. (Accessed June 5, 2021).

Agency for Toxic Substances and Disease Registry (ATSDR). 2008b. *Perchlorates*. CAS Registry Numbers 10034–81–8, 7778–74–7, 7790–98–9, 7601–89–0, and 7791–03–9. ATSDR ToxFAQs. Atlanta, GA.

Agency for Toxic Substances and Disease Registry (ATSDR). 2011. *Chlordane*. CAS Registry Number 57–74–9. www.atsdr.cdc.gov/substances/toxsubstance.asp?toxid=62. (Accessed June 5, 2021).

Agency for Toxic Substances and Disease Registry (ATSDR). 2012. *Cadmium*. CAS Registry Number 7440–43–9. www.atsdr.cdc.gov/toxprofiles/TP.asp?id=48&tid=15. (Accessed March 24, 2019).

Agency for Toxic Substances and Disease Registry (ATSDR). 2013. *Barium*. CAS Registry Number 7440–39–3. www.atsdr.cdc.gov/toxfaqs/tf.asp?id=326&tid=57. (Accessed March 24, 2019).

Agency for Toxic Substances and Disease Registry (ATSDR). 2018a. *Copper*. Case Registry Number 7440–50–8. www.atsdr.cdc.gov/toxprofiles/TP.asp?id=206&tid=37. (Accessed June 5, 2021).

Agency for Toxic Substances and Disease Registry (ATSDR). 2018b. *Nickel*. Case Registry Number 7440–02–0. www.atsdr.cdc.gov/toxprofiles/tp.asp?id=245&tid=44. (Accessed June 5, 2021).

Agency for Toxic Substances and Disease Registry (ATSDR). 2018c. *Selenium*. Case Registry Number 7782–49–2. www.atsdr.cdc.gov/toxprofiles/tp.asp?id=153&tid=28. (Accessed June 5, 2021).

Agency for Toxic Substances and Disease Registry (ATSDR). 2018d. *Silver*. Case Registry Number 7440–22–4. www.atsdr.cdc.gov/toxprofiles/tp.asp?id=539&tid=97. (Accessed June 5, 2021).

Agency for Toxic Substances and Disease Registry (ATSDR). 2018e. *Zinc*. Case Registry Number 7440–66–6. www.atsdr.cdc.gov/toxprofiles/tp.asp?id=302&tid=54. (Accessed June 5, 2021).

Agency for Toxic Substances and Disease Registry (ATSDR). 2018f. *Ammonia*. Case Registry Number 7664–41–7. www.atsdr.cdc.gov/toxprofiles/tp.asp?id=11&tid=2. (Accessed June 5, 2021).

Agency for Toxic Substances and Disease Registry (ATSDR). 2021a. *ATSDR Glossary of Terms*. Center for Disease Control. http://atsdr.cdc.gov/glossary.html. (Accessed June 5, 2021).

Agency for Toxic Substances and Disease Registry (ATSDR). 2021b. Toxicological Profile for Methyl-tert-butyl-Ether. www.atsdr.cdc.gov/toxprofiles/tp.asp?id=228&tid=41. (Accessed June 5, 2021).

Agency for Toxic Substances and Disease Registry (ATSDR). 2021c. Toxicological Profile for Vinyl Chloride. www.atsdr.cdc.gov/substances/toxsubstance.asp?toxid=51. (Accessed June 5, 2021).

Agency for Toxic Substances and Disease Registry (ATSDR). 2021d. Toxicological Profile for Tetrachloroethylene. www.atsdr.cdc.gov/substances/toxsubstance.asp?toxid=48. (Accessed June 5, 2021).

Agency for Toxic Substances and Disease Registry (ATSDR). 2021e. Toxicological Profile for Trichloroethylene. www.atsdr.cdc.gov/substances/toxsubstance.asp?toxid=30. (Accessed June 5, 2021).

Agency for Toxic Substances and Disease Registry (ATSDR). 2021f. Factsheet for PAHs. www.atsdr.cdc.gov/tfacts69.html. (Accessed June 5, 2021).

Agency for Toxic Substances and Disease Registry (ATSDR). 2021g. Toxicological Profile for Polychlorinated Biphenyls. ATSDR ToxFAQs. Atlanta, GA. www.atsdr.cdc.gov/toxprofiles/tp.asp?id=142&tid=26. (Accessed June 5, 2021).

66   Environmental Compliance Handbook

Agency for Toxic Substances and Disease Registry (ATSDR). 2021h. Phthalates Factsheet. www.cdc.gov/biomonitoring/Phthalates_FactSheet.html. (Accessed June 5, 2021).

Agency for Toxic Substances and Disease Registry (ATSDR). 2021i. Aromatic Amines. www.atsdr.cdc.gov/substances/toxchemicallisting.asp?sysid=36. (Accessed June 5, 2021).

Agency for Toxic Substances and Disease Registry (ATSDR). 2021j. Phosphate Ester Flame Retardants. www.atsdr.cdc.gov/phs/phs.asp?id=1118&tid=239. (Accessed June 5, 2021).

Agency for Toxic Substances and Disease Registry (ATSDR). 2021k. *Chromium*. Case Registry Number 7440–47–3. www.atsdr.cdc.gov/substances/toxsubstance.asp?toxid=17. (Accessed June 5, 2021).

Agency for Toxic Substances and Disease Registry (ATSDR). 2021l. *Lead*. Case Registry Number 7439–92–1. www.atsdr.cdc.gov/substances/toxsubstance.asp?toxid=22. (Accessed June 5, 2021).

Agency for Toxic Substances and Disease Registry (ATSDR). 2021m. *Mercury*. Case Registry Number 7439–97–6. www.atsdr.cdc.gov/substances/toxsubstance.asp?toxid=24. (Accessed June 5, 2021).

American Chemical Society. 2021. CAS Registry Numbers. www.cas.org/support/documentation/chemical-substances. (Accessed March 24, 2019).

American Conference of Governmental Industrial Hygienists (ACGIH). 2021. *Guide to Occupational Exposure Values*. ACGIH. Cincinnati, OH.

Barbalace, R. C. 2009. The Chemistry of Polychlorinated Biphenyls. EnvironmentalChemistry.com. September 2003. www.environmentalchemistry.com/ypgi/chemistry/pcb.html. (Accessed October 26, 2009.

Barnes, K. K., Kolpin, D. W., Furlong, E. T., Zaugg, S. D., Meyer, M. T. and Barber, L. B. 2008. A National Reconnaissance of Pharmaceuticals and Other Organic Wastewater Contaminants in the United States. *Journal of Science in the Total Environment*. Vol. 402. Issues 2–3. pp. 192–200.

Bell, C. A., Gentile, M., Kalve, E., Ross, I., Horst, J. and Suthersan, S. 2019. *Emerging Contaminants*. CRC Press, Boca Raton, FL. 439p.

California Environmental Protection Agency (CalEPA). 2021. 1,2,3,-Trichloropropane (TCP). www.waterboards.ca.gov/drinking_water/certlic/drinkingwater/123TCP.html. (Accessed June 5, 2021).

Calle, E. E., Frumkin, H., Henley, S. J., Savitz, D. A. and Thun, M. J. 2002. Organochlorines and Breast Cancer Risk, CA. *A Cancer Journal for Clinicians*. Vol. 52. p. 301.

Carson, R. 1962. *Silent Spring*. Houghton Mifflin. Boston, MA.

Carwile, J. L., Luu, H. T., Bassett, L. S., Driscoll, D. A., Yuan, C., Chang, J. Y., Ye, X., Calafat, A. M. and Michels, K. B. 2009. Use of Polycarbonate Bottles and Urinary Bisphenol A Concentrations. *Environmental Health Perspectives*, online May 12, 2009. doi:10.1289/ehp.0900604.

Doherty, R. E. 2000. A History of the Production and Use of Carbon Tetrachloride, Tetrachloroethylene, Trichloroethylene, and 1,1,1-Trichloroethane in the United States: Part 1—Historical Background; Carbon Tetrachloride and Tetrachloroethylene. *Journal of Environmental Forensics*. Vol. 1. Issue 2. pp. 69–81.

Fetzer, J. C. 2000. *The Chemistry and Analysis of Large Polycyclic Aromatic Hydrocarbons*. John Wiley & Sons. New York, NY.

Holleman, A. F. and Wiberg, E. 2001. *Inorganic Chemistry*. Academic Press. San Diego, CA.

Jensen, W. B. 2009. The Origin of the Circle Symbol for Aromaticity. *Journal of Chemical Education*. Vol. 86. No. 4. pp. 423–425.

Karl, T. R. and Trenberth, K. E. 2003. Modern Global Climate Change. *Journal of Science*. Vol. 302. pp. 1719–1723.

Kathren, R. 1991. *Radioactivity and the Environment*. Taylor & Francis Publishers, Lieden, The Netherlands.

Kaufman, M. M., Rogers, D. T. and Murray, K. S. 2011. *Urban Watersheds*. CRC Press. Boca Raton, FL. 583 pages.

Land Contamination 67

Krebs, R. E. 2006. *The History and Use of Earth's Chemical Elements: A Reference Guide.* Greenwood Publishing Group. Oxford, England.

Levy, A. B. 2009. Acetone. In: *Ullmann's Encyclopedia of Industrial Chemistry*, 7th Edition. John Wiley & Sons—VCH. New York, NY.

Lide, D. R. 2008. Physical Constants of Organic Compounds. In: *Handbook of Physics and Chemistry*, 89th Edition. CRC Press. Boca Rotan, FL.

Madigan, M. T., Martinko, J. M., Dunlap, P. V. and Clark, D. P. 2008. *Brock Biology of Microorganisms*, 12th Edition. Prentice-Hall, Inc. New York, NY.

McMurry, J. E. 2009. *Fundamentals of Organic Chemistry*. Brook Cole Publishing Company. New York, NY.

Meyers, R. 2003. *The Basics of Chemistry*. Greenwood Press. Westport, CT.

Missouri Department of Natural Resources. 2006. *Missouri Risk-Based Corrective Action for Petroleum Storage Tank Sites Sampling for Polynuclear Aromatic Hydrocarbons*. Hazardous Waste Fact Sheet. Jefferson City, MO.

Murray, K. S., Rogers, D. T. and Kaufman, M. M. 2004. Heavy Metals in an Urban Watershed in Southeastern Michigan. *Journal of Environmental Quality*. Vol. 33. pp. 163–172.

National Oceanic and Atmospheric Administration (NOAA). 2021. Atmospheric CO2 at Mauna Loa Observatory. www.esrl.noaa.gov/gmd/ccgg/trends. (Accessed June 5, 2021).

Natural Resources Defense Council (NRDC). 2021. Ocean Pollution: The Dirty Facts. https://nrdc.org/stories/ocean-pollution-dirty-facts. (Accessed June 5, 2021).

Phillips, D. J. 1986. *PCBs and the Environment: Volume 2*. CRC Press. Boca Raton, FL.

Pradyot, P. 2003. *Handbook of Inorganic Chemical Compounds*. McGraw Hill. New York, NY.

Rogers, D. T. 2020a. *Urban Watersheds: Geology, Contamination, Environmental Regulations, and Sustainability*. CRC Press. Boca Raton, FL. 608p.

Rogers, D. T. 2020b. Geological- and Chemical-Based Environmental Risk Factor Sustainability Model. *European Journal of Sustainable Development*. Vol. 9. No. 4. pp. 303–316.

Storrow, B. 2019. Emissions Growth in the United States and Asia Fueled by Record Carbon Levels in 2018. *Science*. www.sciencemag.org/news/2019/03/emissions-growth-united-states-asia-fueled-record-carbon-levels-2018. (Accessed June 5, 2021).

Suthersan, S. S. and Payne, F. C. 2005. *In Situ Remediation Engineering*. CRC Press. Boca Raton, FL. 430p.

United Nations. 2021. Biodiversity: Invasive Species. United Nations System-Wide Earthwatch. www.un.org/earthwatch/biodiversity/invasivespecies.html. (Accessed June 5, 2021).

United States Center for Disease Control (CDC). 2009. *Cryptosporidium Fact Sheets*. Department of Health and Human Services. CDC. Atlanta, GA.

United States Department of Agriculture (USDA). 2021a. National Invasive Species Information Center (NISIC). www.invasivespeciesinto.gov. (Accessed June 5, 2021).

United States Department of Agriculture (USDA). 2021b. *United States Fertilizer Use and Cost*. USDA. Washington, DC.

United States Department of Health and Human Services (USDHH). 1995. *Toxicological Profile for Polycyclic Aromatic Hydrocarbons*. Agency for Toxic Substances and Disease Registry. Atlanta, GA.

United States Department of Health and Human Services (USDHH). 2005. *Report on Carcinogens*, 11th Edition. Public Health Service. National Toxicology Program. Washington, DC.

United States Department of Health and Human Services (USDHH). 2008. *NTP—CERHR Monograph on the Potential Human Reproductive and Developmental Effects of Bisphenol A*. National Institute of Health Publication No. 08–594. Washington, DC.

United States Department of Health and Human Services (USDHH). 2009. *Draft Toxicological Profile for Carbon Monoxide*. ATSDR. Washington, DC.

United States Department of Labor. 2021. Occupational Safety and Health Administration (OSHA) Permissible Exposure Limit for Carbon Monoxide. www.osha.gov/dsg/annotated-pels/tablez-1.html. (Accessed June 5, 2021).

United States Energy Information Service. 2021. United States Consumption of Gasoline. www.eia.gov/tools/faqs/faq.php?id=23&t=10. (Accessed June 5, 2021).

United States Environmental Protection Agency (USEPA). 1989a. *Risk Assessment Guidance for Superfund. Volume I. Human Health Evaluation Manual (Part A)*. EPA/540/1–89/002. Office of Emergency and Remedial Response. Washington, DC.

United States Environmental Protection Agency (USEPA). 1989b. *Transport and Fate of Contaminants in the Subsurface*. EPA/625/4–89/019. USEPA Center for Environmental Research Information. Cincinnati, OH.

United States Environmental Protection Agency (USEPA). 2005. *Guidelines for Carcinogenic Risk Assessment*. EPA/630/P-03/001F. USEPA. Washington, DC.

United States Environmental Protection Agency (USEPA). 2006. *Pentachlorophenol Consumer Fact Sheet*. USEPA. Washington, DC.

United States Environmental Protection Agency (USEPA). 2009. *E. Coli Bacteria. Total Coliform Rule-Basic Information*. USEPA. Washington, DC.

United States Environmental Protection Agency (USEPA). 2010. Semi-Volatile Organic Compounds. https://cfpub.epa.gov/si/si_public_record_report.cfm?Lab=NRMRL&dirEntryId=226943. (Accessed June 5, 2021).

United States Environmental Protection Agency (USEPA). 2013. Overview of Methyl Tertiary Butyl Ether (MTBE). https://archive.epa.gov/mtbe/web/html/faq.html. (Accessed June 5, 2021).

United States Environmental Protection Agency (USEPA). 2016. Chlordane. www.epa.gov/sites/production/files/2016-09/documents/chlordane.pdf. (Accessed June 5, 2021).

United States Environmental Protection Agency (USEPA). 2017. Overview of Greenhouse Gases. https://19january2017snapshot.epa.gov/ghgemissions/overview-greenhouse-gases_.html. (Accessed June 5, 2021).

United States Environmental Protection Agency (USEPA). 2018a. Global Greenhouse Gas Emissions Data. https://epa.gov/ghgemissions/global-greenhouse-gas-emissions-data. (Accessed June 5, 2021).

United States Environmental Protection Agency (USEPA). 2018b. Sulfur Dioxide (SO2) Pollution. www.epa.gov/so2-pollution. (Accessed June 5, 2021).

United States Environmental Protection Agency (USEPA). 2018c. Invasive Species. www.epa.gov/greatlakes/invasive-species. (Accessed June 5, 2021).

United States Environmental Protection Agency (USEPA). 2018d. Basic Information on PFAS Chemicals. www.epa.gov/pfas/basic-information-pfas. (Accessed June 5, 2021).

United States Environmental Protection Agency (USEPA). 2021a. Risk Assessment: What Is It? www.epa.gov/risk_.html. (Accessed June 5, 2021).

United States Environmental Protection Agency (USEPA). 2021b. EPA's Report on the Environment (ROE). www.epa.gov/report-environment. (Accessed June 5, 2021).

United States Environmental Protection Agency (USEPA). 2021c. Integrated Risk Information System. www.epa.gov/iris. (Accessed June 5, 2021).

United States Environmental Protection Agency (USEPA). 2021d. National Primary Drinking Water Regulations: Trihalomethanes. www.epa.gov/ground-water-and-drinking-water/national-primary-drinking-water-regulations. (Accessed June 5, 2021).

United States Environmental Protection Agency (USEPA). 2021e. Polycyclic Aromatic Hydrocarbons (PAHs). www.epa.gov/sites/production/files/2014-03/documents/pahs_factsheet_cdc_2013.pdf. (Accessed June 2021).

United States Environmental Protection Agency (USEPA). 2021f. Polychlorinated Biphenyls Fact sheet: Basic Information. www.epa.gov/epawaste/hazard/tsd/pcbs/about.htm. (Accessed June 5, 2021).

United States Environmental Protection Agency (USEPA). 2021g. Risk Management for Phthalates. www.epa.gov/assessing-and-managing-chemicals-under-tsca/risk-management-phthalates. (Accessed June 5, 2021).

Land Contamination

69

United States Environmental Protection Agency (USEPA). 2021h. Factsheet for Phenols. www.epa.gov/sites/production/files/2016-09/documents/phenol.pdf. (Accessed June 5, 2021).

United States Environmental Protection Agency (USEPA). 2021i. Factsheet for Triethylamine. www.epa.gov/sites/production/files/2016-09/documents/triethylamine.pdf. (Accessed June 5, 2021).

United States Environmental Protection Agency (USEPA). 2021j. Organic Esters of Phosphoric Acid. https://archive.epa.gov/pesticides/reregistration/web/html/index-198.html. (Accessed June 5, 2021).

United States Environmental Protection Agency (USEPA). 2021k. Pesticides. www.epa.gov/pesticides. (Accessed June 5, 2021).

United States Environmental Protection Agency (USEPA). 2021l. DDT—A Brief History and Status. www.epa.gov/ingredients-used-pesticide-products/ddt-brief-history-and-status. (Accessed June 5, 2021).

United States Environmental Protection Agency (USEPA). 2021m. Dioxin. www.epa.gov/dioxin. (Accessed June 5, 2021).

United States Environmental Protection Agency (USEPA). 2021n. Agriculture Nutrient Management and Fertilizer. www.epa.gov/agriculture/agriculture-nutrient-management-and-fertilizer. (Accessed June 5, 2021).

United States Environmental Protection Agency (USEPA). 2021o. Asbestos. www.epa.gov/asbestos. (Accessed June 5, 2021).

United States Environmental Protection Agency (USEPA). 2021p. Ground Level Ozone. www.epa.gov/ground-level-ozone-pollution. (Accessed June 5, 2021).

United States Environmental Protection Agency (USEPA). 2021q. Upper Atmosphere Ozone. https://www3.epa.gov/region1/eco/uep/ozone.html. (Accessed June 5, 2021).

United States Environmental Protection Agency (USEPA). 2021r. Particulate Matter (PM) Pollution. www.epa.gov/pm-pollution. (Accessed June 5, 2021).

United States Environmental Protection Agency (USEPA). 2021s. Contaminants of Emerging Concern. www.epa.gov/wqc/contaminants-emerging-concern-including-pharmaceuticals-and-personal-care-products. (Accessed June 5, 2021).

United States Environmental Protection Agency (USEPA). 2021t. Technical Fact Sheet on Perchlorate. www.epa.gov/fedfac/technical-fact-sheet-perchlorate. (Accessed June 5, 2021).

United States Geological Survey (USGS). 2006. *Volatile Organic Compounds in the Nation's Ground Water and Drinking-Water Supply Wells*. USGS Circular 1292. Reston, VA.

United States Geological Survey (USGS). 2009a. *Toxic Substance Hydrology Program*. DNAPL. www.usgs.gov/definitions/dnapl_def.html. (Accessed June 5, 2021).

United States Geological Survey (USGS). 2009b. *Toxic Substance Hydrology Program*. BTEX. www.usgs.gov/definitions/btex.html. (Accessed June 5, 2021).

United States Forest Service. 2021. Bark Beetles and Climate Change in the United States. Climate Change Resource Center. www.fs.usda.gov/ccrc/topics/bark-beetles-and-climate-change-united-states. (Accessed June 5, 2021).

United States Institute of Health. 2021. Periodic Table of Elements. https://pubchem.ncbi.nlm.nih.gov/periodic-table. (Accessed October 12, 2019).

Van der Perk, M. 2006. *Soil and Water Contamination: From Molecular to Catchment Scale. Balkema: Proceedings and Monographs in Engineering, Water and Earth Sciences*. Taylor and Francis, London, UK.

World Health Organization (WHO). 2003. *Polychlorinated Biphenyls: Human Health Aspects*. WHO. Geneva, Switzerland.

4 Nature's Response to Land Contamination

4.1 INTRODUCTION

How does nature respond to pollution? That is one of the central questions that will be addressed in this chapter. Nature's response to contaminants is important from a compliance point of view because it influences how they are regulated. In addition, and perhaps most importantly, it provides valuable information on how to manage chemicals to minimize the potential for a release or to prevent a release from occurring. In addition, understanding how contaminants behave in nature provides us with information on how hazardous substances should be regulated.

We now know what contamination is, how and where it originates, and what it can do to our bodies if we are exposed to most forms of contamination. We know that contamination only presents a risk to human health and environment if there is a completed exposure pathway. So, what happens to contaminants once they are released, and how do humans and the environment become exposed to contamination? The answer to these questions is obtained through an understanding of the process called **contaminant fate and transport**, defined as the sequence of anthropogenic and natural events involving a pollution source, its mobilization or transport, and its ultimate fate or resting place, termed a **sink** (Rogers et al. 2007).

It was often thought in the mid-20th century that the Earth would filter contamination if released to the ground and that the Earth would render the contaminant(s) harmless in a relatively short time. We now know that this is just not the case in most instances. Sources of contamination include many human activities performed primarily at the surface, resulting in the release of toxic substances into the environment. These toxic substances may be transported over time or remain relatively close to their source before they are degraded, transformed, or destroyed because (1) contaminants released into the environment are often mixtures, (2) each contaminant is unique chemically, and (3) the geologic environment in which the contaminants are released is also unique. Understanding the physical chemistry of each contaminant, the microbiologic environment, and the geologic environment into which they are released becomes necessary for characterizing their fate and transport in nature (Rogers et al. 2007).

During their transport and before reaching their final sink, certain contaminants may reside at multiple intermediate sinks for different periods of time. Intermediate sinks include surface water, groundwater, and the atmosphere, because the contaminants held within these water-containing sinks will flow and ultimately reach the oceans. Aquifers with a very low hydraulic conductivity are for practical purposes considered final sinks, as are inland bogs and some wetlands. Sediment and soil

DOI: 10.1201/9781003150107-4

72 Environmental Compliance Handbook

can function as intermediate or as final sinks, since erosion may move both of these unconsolidated materials. The oceans are almost always a final sink of contamination, but wave action occasionally brings contaminants onshore. To a large extent, the level of human health and/or environmental risk is a function of two fundamental concepts introduced in this chapter: *mobility* and *persistence*. **Mobility** is a measure of a substance's potential to migrate. **Persistence** is a measure of a substance's ability to remain in the environment before being degraded, transformed, or destroyed (Rogers et al. 2007). In the previous chapter, we noted substances with higher toxicity pose greater health risks, but the risk to humans and the environment grows exponentially if the chemical is both mobile and persistent. For example, a highly toxic but immobile chemical may affect a few people in a warehouse through inhalation, whereas a mobile and persistent chemical of moderate toxicity can contaminate a public water supply or migrate to a different sink where the potential for widespread human exposures and ecosystem damages is much higher.

The following sections discuss how and where contaminant releases occur; how their migration through the soil, groundwater, and atmosphere proceeds; and how they end up in sinks. We conclude with a brief description of the fate and transport for each contaminant group.

4.2 RELEASE OF CONTAMINATION INTO THE ENVIRONMENT

The fate and transport of contamination begin with its release into the environment. Contaminant releases originate from numerous sources and under different circumstances; they vary in degree, concentration, duration, mass, volume, and whether a single contaminant is released or a mixture of contaminants is released. Each of these factors influences their fate and transport. For instance, some releases may be very small and avoid detection. On the other hand, some releases are so large and sudden (e.g., a tanker spill) that they become the leading story of the next newscast. Sudden and large releases increase the probably of severe environmental impairment or destruction, especially if they occur in or near a sensitive ecological area.

Historically, there was no regulation of the disposal of wastes potentially containing contaminants. Until the mid-20th century, the most convenient and least costly method of waste disposal was "up the stack or down the river" (Haynes 1954). We now know the perception of a contaminant leaving your immediate vicinity and being gone forever is not true. Today, many releases of contaminants are permitted but carefully monitored, such as wastewater discharges from industrial and municipal sources and those to the atmosphere from industrial and commercial sources and power plants. The legislation governing such releases (the Clean Water Act of 1972 and the Clean Air Act of 1990) permits a point source (such as a wastewater plant or a smokestack) to release specific compounds at low concentrations to ensure there are no adverse health and environmental effects detectable in the media they are released into (water or air). Some releases, as we shall discover and discuss in detail, are not regulated and include the majority of household wastes and releases of contamination from many agricultural sources. A majority of the other contaminant releases into the environment are unintentional or accidental and occur from

Nature's Response to Land Contamination

numerous sources under a multitude of circumstances, and some of these include (Fetter 1993; Rogers 1996; USGS 2006a):

- Permitted releases of contaminants in wastewater and air emissions
- Spills and leaks from several types of containers or operations, including:
  - Drums of various sizes and shapes
  - Pipelines
  - Aboveground storage tanks
  - Underground storage tanks
  - Tanker trucks and other transport vehicles
  - Tanker ships
  - Railroad tanker cars
  - Aircraft
- Accidents or collisions involving automobiles where gasoline and oil may be released, railroad derailments, and so on
- Septic systems
- Sewer leaks
- Landfills and dumps
- Automobile exhaust
- Sumps and dry wells
- Former disposal lagoons
- Animal feed lots
- Pesticide and herbicide application
- Fertilizer application
- Injection wells
- Application of pesticides, herbicides, and fertilizers
- Application of deicing compounds (i.e., road salts)

Exploring and evaluating the fate and transport of pollutants in the environment is a complex interplay between the environment and the contaminant. And, as Figure 4.1 shows, fate and transport are also influenced by the method and location of contaminant release and the volume, mass, and duration of the release. We will break down and describe these factors influencing the migration and degradation of contaminants and then describe how the fate and transport of each group of substances is influenced by the group's specific chemistry.

4.3 PRINCIPLES OF CONTAMINANT FATE AND TRANSPORT

Interaction with the environment begins immediately after a pollutant has been released. Once released, a pollutant can do three things (Hemond and Fechner-Levy 2000):

- Stay put
- Migrate in soil, water, air, or a combination of media
- Degrade, transform, or get destroyed

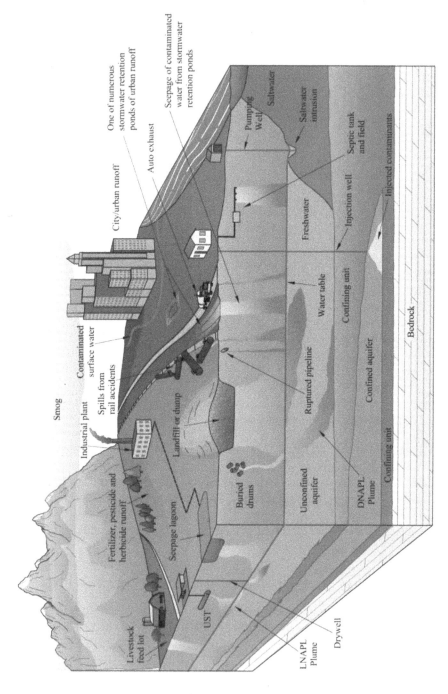

FIGURE 4.1 Sources and locations of many contaminant releases.

Nature's Response to Land Contamination

The factors controlling these three outcomes depend upon the physical chemistry of the contaminant and the characteristics of the receiving environment (USGS 2006a; Rogers et al. 2007). For example, some pollutants may change physical states. An example is many of the more than 250 chemical compounds that make up gasoline that begin to immediately evaporate after exposure to the atmosphere. Other pollutants may degrade in a manner of minutes after being released, while some may last for thousands of years or sometimes longer, as with certain radioactive compounds. Therefore, computing a mass balance should be the first action when assessing any particular release, as this will ensure the mass or volume of contaminant released is accurately measured (Equation 4.1). Conducting a mass balance also serves to validate or refute our current understanding of the behavior of the different environmental elements influencing the pollutant once it has been released.

A simple mass balance is expressed as:

$$\text{Amount released} = \text{amount recovered} + \text{amount lost to the environment (air + water + soil)} \quad (4.1)$$

An additional factor is what media the contaminant is released into first: air, water, or soil. This also influences its behavior and interaction with the environment. Figure 4.2 presents an excellent example of contaminant release into each of the three environmental media (air, water, and land) at the same time. Once in the environment, pollutants can and often move between soil, surface water, groundwater, and the atmosphere, and they can also degrade. Many pollutants degrade quickly if conditions are favorable, yet others can persist and last for years or decades if conditions are favorable (USGS 2006a). Factors degrading contaminants fall into two broad categories: biotic degradation and abiotic degradation. **Biotic** degradation involves microorganisms or fungi and occurs when an organism, such as a bacterium, uses a contaminant as a source of food and either degrades

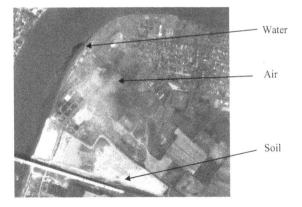

FIGURE 4.2 Aerial photo showing releases to the air, water, and land (from New Jersey Department of Environmental Protection [NJDEP]. Preliminary Assessment Report. NJDEP. Trenton, New Jersey. 2015).

76   Environmental Compliance Handbook

or transforms the contaminant (USGS 2006a). **Abiotic** degradation involves other processes not including microorganisms. Examples of abiotic degradation include photolysis—degradation as a result of exposure to sunlight (USGS 2006a)-and hydrolysis. Hydrolysis involves cleaving a molecule into two parts by the addition of a molecule of water.

When examining pollutant degradation in the environment, it is important not to confuse dilution with degradation. If given enough time, contaminants may become diluted, and this process results in a decrease of the pollutant concentrations per unit volume of the media being measured. However, dilution is not degradation, since it does not involve a chemical transformation of the pollutant. We will discuss degradation in more detail in later sections.

4.3.1 BASIC CONTAMINANT TRANSPORT CONCEPTS

Pollutant transport in the environment is dominated by three physical transport mechanisms (USEPA 1996a; USGS 2006a): advection/convection, molecular diffusion, and dispersion. **Advection** is the horizontal transport of any property of the atmosphere and water. Common examples are the transfer of heat by wind and sediment transport within a flowing stream. **Convection** is the vertical advection of air, water, or other fluid as a result of thermal differences. **Molecular diffusion** is the movement of a chemical from an area of higher concentration to an area of lower concentration due to the random motion of the chemical molecules. **Dispersion** (also referred to as hydrodynamic dispersion) is the tendency for pollutants to spread out from the path of the expected advective flow (USGS 2006a). Occasionally, the effects of diffusion and dispersion are treated together, but for the purposes of this book, we treat them separately.

The rate of advective transport of a pollutant is often expressed in terms of flux density. **Flux density** is the mass of a chemical transported across an imaginary surface of a unit area per unit of time. Equation 4.2 shows this relationship (Hemond and Fechner-Levy 2000), which is independent of the media involved (soil, surface water, groundwater, or the atmosphere).

$$J = CV \tag{4.2}$$

where: J = flux density = (mass/[length $\times$ width] $\times$ Time) or $[M/L^2 \times T]$
C = concentration of the chemical per cubic liter or meter of media $[M/L^3]$
V = velocity [length/time] or $[L/T]$

An example of molecular diffusion is shown in Figure 4.3 (Payne et al. 2008). From the release time to infinity, a pollutant released into a fluid such as air or water will diffuse throughout the fluid at random locations.

Fick's first law of diffusion (Equation 4.3) can be used to predict the diffusive flux of a pollutant (solute) across an imaginary plane as a function of the rate of change in concentration with distance (Hemond and Fechner-Levy 2000).

$$J = -D(\mathrm{d}C/\mathrm{d}x) \qquad \text{(one dimension)} \tag{4.3}$$

Nature's Response to Land Contamination

where: J = the flux density $[M/L^2 \times T]$
D = the Fickian mass transport coefficient $[L^2/T]$
C = chemical concentration $[M/L^3]$
X = the distance over which a concentration change is being considered $[L]$

Note: In simple calculations, the minus sign is often omitted if the direction of Fickian transport is clear.

An example of the effects of subsurface dispersion is presented in Figure 4.4 Hemond and Fechner-Levy 2000).

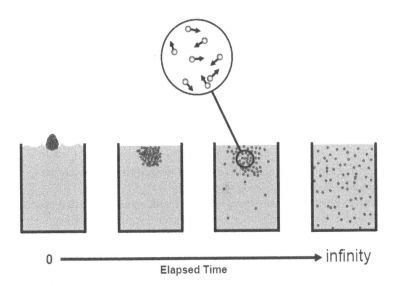

FIGURE 4.3 Molecular diffusion (from Payne, F. C., Quinnan, J. A. and Potter, S. T. *Remediation Hydraulics*. CRC Press. Boca Raton, FL. 2008).

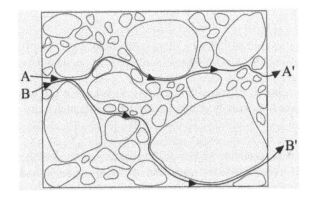

FIGURE 4.4 Effects of dispersion on subsurface migration in a porous media.

78   Environmental Compliance Handbook

General factors that influence or control the rate of migration of a contaminant include (Hornsby 1990):

- Physical and chemical properties of the contaminants themselves
- The geological environment where the release occurs
- Climatological factors
- Vegetation factors

Specific physical properties affecting migration of contaminants in soil include (USGS 2006a; Rogers et al. 2007; Payne et al. 2008):

- Solubility in water. The more soluble a contaminant is in water, the more mobile it will be in the subsurface environment.
- Vapor pressure. As vapor pressure increases, affinity to volatilize increases and the more likely it is that a contaminant will be present in the gas phase.
- Molecular weight. The higher the molecular weight, the greater the energy requirement to transport the contaminant in the horizontal direction. Increased molecular weight may induce the contaminant to migrate downward in areas of steep slopes.
- Chemical stability and persistence. Stable compounds have more time to migrate and may migrate further if conditions are favorable before they are degraded, transformed, or destroyed.
- Sorptive properties. Sorptive processes include adsorption and absorption. The lower the sorptive properties, the higher the migration potential.

Specific geologic factors affecting migration of pollutants in soil include (Rogers 1996; USGS 2006a; Rogers et al. 2007, 2012):

- Composition. Soil type has a significant influence on the migration of contaminants. In general, coarse-grained soils, such as sand and gravel, do not impede the migration of contaminants nearly as much as soils composed of clay.
- Porosity and permeability. Direct evidence of migration potential is hydraulic conductivity. Soils composed of sand and gravel have a hydraulic conductivity 100s to 1,000s of times higher than soils composed of clay. Although soils or sediments composed of clay may have a high relative porosity, they are usually not as permeable because the porosity is generally not interconnected.

In some cases, however, secondary porosity such as root fragments and vertical fractures can make a seemingly impervious clay deposit much more permeable than expected.

- Organic carbon content. As just discussed, an increased total organic content in soil may impede the migration of certain types of contaminants as long as the contaminant mass does not exceed the holding capacity of the soil. Soils or sediments composed of sand and gravel generally have lower organic carbon content compared to soils composed of clay.

Nature's Response to Land Contamination

- Soil chemistry. The pH, redox potential, and other soil chemistry factors influence the contaminant migration of many different types of compounds. Many metals, for instance, are particularly sensitive to pH differences in soil, and these differences—along with the characteristics of each metal-influence their migration patterns.
- Stratigraphy. This is where heterogeneity and the anisotropic nature of the geologic sediments play a significant role at both a micro and macro scale. The geology beneath the surface can and does change dramatically in just a few feet in any direction. The result is differing sediment types and chemical composition, including pH and redox potential, of subsurface layers acting to impede or enhance contaminant migration.
- Unconformities. The presence of unconformities influences the migration of groundwater as well as the migration of contaminants. Hydrogeologically, the presence of an unconformity indicates there is a surface or plane in the subsurface geologic environment. This surface or plane often produces a significant difference in the hydraulic conductivities within the soils or sediments above and below the unconformity, especially if these units are fine-grained sediments such as silts or clays. As a result, contaminants released in this type of location use the unconformity as a sink and migrate much further than expected.

Specific climatological factors affecting migration of pollutants in soil include (Freeze and Cherry 1978; Rogers 2020):

- Freeze-thaw cycles. Freeze-thaw cycles can lead to the development of vertical fractures in the soil to depths approaching 10 meters. These vertical fractures are a type of secondary porosity, and, if present, can greatly increase the migration potential of contaminants vertically through the soil column. In addition, the freezing of near surface soils may trap contaminants at the surface and lead to increased contaminant loading during warmer periods when the ice melts.
- Liquid precipitation. Because water is the universal solvent, geographic locations receiving abundant rainfall play a significant role in enhancing the migration of contaminants, especially if they are soluble. Rainfall also enhances the migration of contamination through the physical transport of particles with sorbed contamination on their surfaces.
- Snowfall. Airborne deposition of contaminants may become temporarily trapped in seasonal snowpack and released when the snowpack melts. Increased contaminant loading to the environment may occur during warmer periods when the snowpack melts (Wania et al. 1998).
- Wind. Many locations within the United States contain significant amounts of wind-blown deposits, especially in the southwest. Contaminants with a high sorption potential may become attached to fine wind-blown sediment grains and transported over long distances. In addition, volatile contaminants released as a gas are routinely transported by wind.
- Water vapor. Humidity plays a significant role in the water cycle by affecting air movement, reacting with contaminants in the gas phase and contaminants

80 Environmental Compliance Handbook

released into the atmosphere attached or sorbed onto particulate matter. Contamination goes along for the ride when precipitation formed around contaminated condensation nuclei is transported from the atmosphere to the lithosphere and then entrained by surface runoff.

- Fog. Fog can be an effective agent for the transfer of acid rain by transferring it to vegetation or other surface materials through direct contact.
- Flood events and hurricanes. Due to their catastrophic nature and magnitudes, floods and hurricanes may not only increase contaminant migration, they can cause significant releases. During the 1993 floods of the Mississippi River, numerous barrels containing hazardous waste were swept away and deposited in the Gulf of Mexico.
- Solar energy. Sunlight breaks down some contaminants through a process called photolysis. In toxic microorganisms, ultraviolet light passes easily through cell walls, cytoplasm, and nuclear membranes and prevents DNA replication.

Specific vegetative factors that affect migration of pollutants in soil include (Nudunuri et al. 1998; Kaufman et al. 2011):

- Roots. These pathways within shallow subsurface geological materials can enhance the migration of contaminants.
- Certain plants can uptake contaminants. Many types of plants and trees have the capability with their root systems to assist in the removal of contaminants from shallow subsurface soil. Contaminants may be stored in plant tissues or are transformed through biologic processes.
- Microorganisms. Microorganisms in the soil often biodegrade many different types of contaminants by using the contaminants themselves as a source of food.

4.3.2 Basic Contaminant Degradation Concepts

The degradation of specific compounds in the environment is expressed in terms their half-life. **Half-life** is the average amount of time required to degrade half or 50% of a specific pollutant population (USEPA 1998a, 1998b). Contaminants degrade through biotic or abiotic processes, and the processes controlling their rate of decay depend upon the following factors (USEPA 1996a, 1996b; USGS 2006a):

- The nature of the release. This group of factors includes:
  - Media receiving the release. Was it into the atmosphere, surface or subsurface soil, surface water, ocean, directly to groundwater, or a combination of media?
  - Amount (volume or mass) of the release.
  - Number of contaminants released. Was it a single contaminant—or a mixture of several contaminants?

Nature's Response to Land Contamination

- Physical state of the release (i.e., liquid, solid, or gas).
- Time duration of the release.
- Geologic environment. The most significant geologic factors include:
  - Soil composition and other physical characteristics such as permeability, porosity, moisture content, composition, extent and distribution, thickness, total organic carbon content, pH, redox potential, dissolved oxygen (if saturated), and other parameters.
  - Depth to bedrock, type, composition, distribution, fractures, permeability, porosity of bedrock and other parameters.
  - Terrain and topography.
  - Potential surface and subsurface migration pathways.
- Climatic factors.
  - Release location. Different climates—deserts, mountains, humid areas, or temperate regions—can influence the type and rate of degradation.
- Surface water features.
  - Distance to surface water bodies and their type. Immature streams, mature streams, rivers, lakes, wetlands, and bogs can differ in pH due to the rock composition of their channels and bottoms and the amount of organic matter they receive from outside, inputs termed allochthonous.
- Weather conditions at the time of the release. Weather conditions are often important and sometimes overlooked as potentially significant. Those conditions affecting degradation and migration include:
  - Temperature
  - Humidity
  - Precipitation
  - Wind speed and direction
- Biologic factors.
  - The type, distribution, and amount of microorganisms will influence the rate and can determine if degradation even occurs.
- Anthropogenic factors. Anthropogenic factors are often overlooked and frequently significant. These include:
  - Physical landscape alteration (i.e., buildings, roads, parking lots, etc.)
  - Surface water drainage modifications, including stormwater control and wetland destruction
  - Alteration of native vegetation
  - Introduction of invasive vegetation
  - Regional contaminant loading, including sources, duration, type, release points, and physical state of contaminants (i.e., solid, liquid, or gas)
  - Developmental history of the area and region

4.3.2.1 Biotic Degradation

Microbes have the ability to oxidize a variety of organic contaminants, including many VOCs, PAHs, and other compounds. This capability arises from their enormous variety, populations, rapid growth, and diversity of environmental niches. Soluble organic compounds with low molecular weights such as alcohols and organic acids are metabolized and degraded rapidly by microbes, perhaps because these compounds also occur naturally and microbes have evolved to degrade them more efficiently (Hemond and Fechner-Levy 2000). Halogenated synthetic or anthropogenic compounds, however, are not easily degraded by microbes (USEPA 1998b), and some contaminants that escape biotic degradation bioaccumulate in the bodies of organisms. The rate of biodegradation by microorganisms generally slows if the organic contaminants posses the following characteristics (Hemond and Fechner-Levy 2000):

- Increasing molecular weight
- Decreasing water solubility
- Presence of benzene or aromatic rings
- A large amount of branching within the molecular structure
- Presence of halogen atoms in the structure (chlorine, fluorine, bromine, or iodine)

4.3.2.2 Abiotic Degradation

Abiotic degradation applies to degradation processes accomplished without microorganisms. Common abiotic degradation processes include:

- Photolysis
- Hydrolysis
- Reduction—oxidation
- Radioactive decay

Photolysis (sometimes referred to as photodegradation or photochemical degradation) occurs in the presence of sunlight. Common examples of this process include the fading of colored and dyed objects and the transformation of plastic object textures from pliable to brittle. Photolysis is most common in the atmosphere, in surface water, and at Earth's surface. The degree of degradation caused by sunlight depends on the wavelength spectrum of the light, intensity of light exposure, and duration (Hemond and Fechner-Levy 2000). If the energy per photon is sufficient to break a specific chemical bond, photolysis can be initiated. Once begun, increased light intensity will result in a faster rate of degradation. Failure to achieve the energy level required to break a bond means degradation will not occur. Ultraviolet light is especially effective at degrading many organic contaminants (USEPA 1996a). Degradation by photolysis occurs within compounds capable of absorbing light energy and is often observed in organic compounds with double bonds between their carbon atoms. Many VOCs, PAHs, and SVOCs characterized by a benzene ring fit this pattern. Under favorable conditions, the degradation of many organic

Nature's Response to Land Contamination

compounds by photolysis may occur in a short period of time—from a few hours to a few days (Lyman et al. 1990).

The process of **hydrolysis** occurs when a water molecule breaks. Contaminant degradation by hydrolysis also destroys a molecule of contaminant. Two types of chemical compounds are susceptible to degradation by hydrolysis (Schwarzenbach et al. 1993):

- Alkyl halides, straight-chained or branch-chained hydrocarbons where one or more hydrogen atoms have been replaced by a chlorine, fluorine, bromine, or iodine atom.
- Esters, compounds containing a modified carboxylic acid group (-COOH), where the acid hydrogen atom has been replaced by a different organic functional group. The process of hydrolysis within this group converts the ester compound into the "parent" organic acid and an alcohol.

In reduction–oxidation degradation reactions (redox), electrons are transferred from one atom to another. **Chemical reduction** is defined as the addition of electrons, and **chemical oxidation** is defined as the loss of electrons (Hemond and Fechner-Levy 2000). In a reaction involving atoms A and B, if atom A gains an electron, it is reduced, and atom B, having donated an electron, is the **reductant**. Because atom B loses an electron, B is oxidized, and atom A is the **oxidant**. Each reaction involving the loss or gain of an electron is termed a **half reaction**. The oxidation of contaminants can occur very rapidly through combustion or incineration. Here, fire transforms the contaminants through oxidation at greatly elevated temperatures and uses the cooking, heating, and transportation applications (Hemond and Fechner-Levy 2000).

4.3.3 Transport and Fate of Contaminants in Soil

Folklore holds that the presence of soil protects groundwater quality by filtering contaminants before they reach and impact groundwater (Hornsby 1990). Soil does have a limited ability to filter contamination; however, it does not do a perfect job of holding, filtering, degrading, transforming, or destroying contaminants. These capabilities also depend upon a number of factors related to the chemistry of the contaminant and the geological environment where the contaminant is released.

Soil is defined as the unconsolidated mineral matter on the immediate surface of the Earth (Soil Science Society of America 1987). Basic to an understanding of soil are the factors affecting its development and ultimate physical structure. The composition, texture, and thickness of soil are influenced by its source material, plant growth, micro- and macro-organisms, climate, topography, process of formation (e.g., alluvial, fluvial, and glacial), and physical and chemical weathering since original formation (Brady and Well 1999). Structurally, soil is composed of three phases: soil gases, soil water, and organic and inorganic solids (Sawhney and Brown 1989). The gas and water phases may constitute 25–50% of the total volume of a surface soil, especially at shallow depths (USEPA 1999). Pollutants released into the soil

84 Environmental Compliance Handbook

can migrate within all three phases. Once a pollutant is resident in soil, these factors determine its migration rate (Schnoor 1996; USEPA 1999; Kaufman et al. 2011):

- Contaminant mass released
- Duration of the release
- Physical chemistry of the contaminant
- Physical chemistry of the soil (e.g., pH, redox potential and mineralogy)
- Amount of water present
- Permeability of the soil
- Retention capacity of the soil
- Distribution of plant matter
- Biological interaction between the contaminant and indigenous microorganisms

All of these factors must be well understood before an accurate assessment of the fate and transport of a contaminant in soil can be evaluated. Contaminants migrate through the soil by two basic processes: diffusion and mass flow. The rates of diffusion and mass flow greatly depend upon the local geology and the physical chemistry of the contaminant. Diffusion of substances through soil and aquifer materials occurs in response to differences in energy from one point to another. These energy gradients may be caused by differences in temperature or chemical concentrations within the contaminated area. In most cases, however, the principal process moving a contaminant through soil is mass flow or advection, because contaminants generally want to move downward through the soil under the force of gravity (USGS 2006a). Contaminant-specific physical and chemical attributes affecting thee migration of contaminants in soil include (USEPA 1996a; Wiedemeier et al. 1999):

- Solubility
- Vapor pressure
- Density
- Chemical stability
- Persistence
- Adsorption potential

In soil, solid-phase contaminants migrate much more slowly than liquid-phase contaminants and tend to remain relatively close to their point of release or deposition (USEPA 1999). Before they can migrate a significant distance, solid-phase contaminants must change phase or undergo a transformation process. For example, heavy metals—a solid-phase contaminant—typically remain at their point of release or deposition. If they undergo oxidation, however, their solubility and other properties enabling migration may increase (Lindsay 1979). And, once a contaminant begins to dissolve in water, it may also be subject to further transformation reactions induced by indigenous bacteria present in the surface soil (Suthersan and Payne 2005). If a source continues to emit contaminant in a solid or liquid form that dissolves in water, the underlying soil will eventually become saturated. The leading edge of contamination will migrate either horizontally or vertically or both as long as the retention capacity of the soil is exceeded (USEPA 1999). When the contaminant release stops,

Nature's Response to Land Contamination

the migration of the liquid contaminant will significantly decrease as the soil regains its retention capacity (Rogers 2020). Since soil is also composed of gas, contaminant migration through the vapor phase is often observed with contaminants having higher relative vapor pressures. VOCs are frequent participants in this type of migration. Capillary forces can also induce the migration of liquid phase contaminants.

4.3.4 TRANSPORT AND FATE OF CONTAMINANTS IN SURFACE WATER

The transport of contaminants in surface water is dominated by turbulent advective flow. Because the rate of flow in a river or stream varies significantly by location and over time, estimating the contaminant flow involves averaging the streamflow variations and contaminant concentrations over a specified time interval. Conducting measurements at multiple locations also provides a more accurate measure of the rates of streamflow and contaminant transport.

Molecular diffusion also influences contaminant migration in surface water. Turbulent flow is characterized by water moving in constantly changing and unpredictable patterns. The swirls resulting from turbulent flow are called **eddies**, and they appear in many sizes, volumes, and velocities. Random mixing of the water within eddies creates turbulent diffusion and also influences mass transport. Wave action can create similar eddying effects in lakes and other non-flowing water bodies. Analytically, the transport rates for chemicals in surface water are expressed in terms of flux density. Flux density is the mass of a chemical transported across an imaginary surface of a unit area per unit of time (Equation 4.2) (Hemond and Fechner-Levy 2000). Fick's first law (see Equation 4.3) is also used to describe the flux density of mass transport by turbulent dispersion (Hemond and Fechner-Levy 2000). Figure 4.5 depicts a municipal or industrial waste water plant discharging to surface water and many of the ensuing contaminant fate and transport processes, including (Rogers 2020):

- Transport of discharged wastewater solute downstream
- Mixing due to turbulent advection and turbulent diffusion
- Photolysis
- Hydrolysis
- Biodegradation
- Transformation
- Volatilization to the atmosphere
- Sorption of contaminants in sediment
- Bioaccumulation of contaminants by animal and plant life
- Dilution

Molecular diffusion and groundwater discharge and recharge are not shown in Figure 4.5.

Since most urban areas of the United States obtain their potable water from surface sources, wastewater discharges are a concern. Treatment costs rise when the source of supply is contaminated, and the risks of biological contamination also increase. Moreover, a majority of urban areas are located along rivers and streams and other surface water bodies, and many of these water bodies have already been degraded—some significantly (USGS 1995a, 1995b; Rogers 2020).

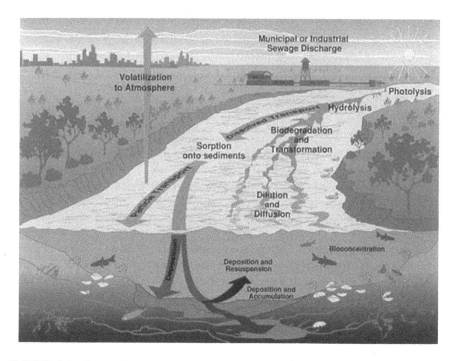

FIGURE 4.5 Fate and transport effects in surface water (from United States Geological Survey, *Contaminants in the Mississippi River*, USGS Circular 1133, Washington, DC, 1995b).

When certain contaminants are released to surface waters through overland flow, stormwater runoff, or wastewater discharge, they may accumulate in sediments. Compounds with a higher likelihood of accumulating have the following physical characteristics (Rogers 2020):

- Low solubility
- High molecular weight
- Low potential to degrade
- High sorption potential

Contaminants with these physical chemistry attributes typically do not sustain themselves in surface water unless the rate of flow is substantial. Even then, they may be carried along the bottom of the stream or river until the carrying capacity of surface water is insufficient and the contaminants settle to the bottom. As shown in Figure 4.5, certain locations in the stream bottom become a sink for these contaminants as they accumulate (Rogers 2020). If the source of contamination persists, greater amounts of the contaminant will be deposited. The accumulation of contaminants in sediments increases the exposure risk to aquatic and terrestrial plant and animal life. If any of the contaminants exhibit bioaccumulation properties, they may proceed up the food chain from bottom-dwelling macro-invertebrates to small fish and eventually to larger fish, predatory birds, and other organisms. Humans are situated at the top of food chain,

Nature's Response to Land Contamination

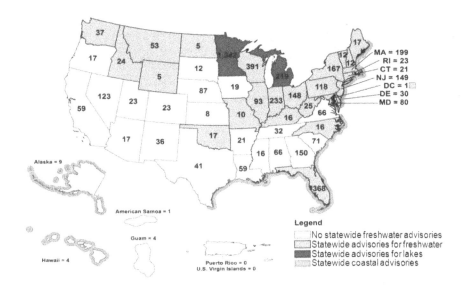

FIGURE 4.6 Health advisories for bioaccumulative contaminants in surface water by state (from United States Environmental Protection Agency. *Advisories and Technical Resources for Fish and Shellfish Consumption.* www.epa.gov/fish-tech [accessed June 10, 2021], 2021b).

and the potential risks to human health must be considered when evaluating the fate and transport of contaminants in surface water (USEPA 2021a).

Contaminants considered bioaccumulative include (USEPA 2021a):

1. Mercury
2. PCBs
3. Chlordane
4. Dioxins
5. DDT

The number of surface water bodies under advisory include (see Figure 4.6) (USEPA 2021a):

- 44% of the nation's total lake acres (excluding the Great Lakes), representing approximately 18 million acres of surface water
- 35% of the nation's total river miles, or approximately 1.3 million miles
- 45% of the nation's contiguous coastal waters
- 100% of the Great Lakes and their connecting waters

4.3.5 TRANSPORT AND FATE OF CONTAMINANTS IN GROUNDWATER

Transport of contaminants in groundwater is dominated by three factors: advection, dispersion, and molecular diffusion. When applied to groundwater, **advection** is the movement of contaminants by the bulk motion of groundwater flow, dispersion is the tendency for contaminants to spread out from the path of the expected advective flow, and diffusion is the action of spreading of molecules from areas of high concentration to areas of low concentration at the molecular level. Groundwater flow lacks turbulent diffusion because velocities are typically much slower. In some instances, however, groundwater does display turbulent dispersion, especially in karst topography, where water flowing beneath the surface flows and behaves much like a stream at the surface. Figure 4.7 shows a spill from an underground storage tank (USGS 2006a). Here, advective transport of contaminants in groundwater is occurring at the water table boundary. Diffusion, biodegradation, volatilization, and recharge from surface precipitation affecting the contaminant migration are also shown. The effects of dispersion and diffusion are represented by the spreading of the contaminant plume as it migrates from a hole or rupture at the bottom of the tank (USGS 1998).

The representation of dispersion in Figure 4.7 is overly simplistic, because the geology of unconsolidated sediments is very complex and typically displays a high degree of heterogeneity and anisotropic distribution patterns. As a result,

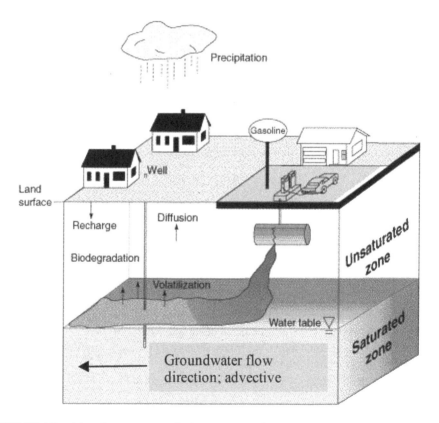

FIGURE 4.7 Advective transport and other processes affecting the migration of contaminants in groundwater (from United States Geological Survey, *Simulating Transport of Volatile Organic Compounds in the Unsaturated Zone Using the Computer Model R-UNSAT*, USGS Fact Sheet 019–08, Washington, DC, 1998).

contaminants migrating in unconsolidated deposits do not migrate uniformly but migrate within the physical parameters of advection dictated by the particular subsurface geology. This concept is represented in Figure 4.8, where a contaminant (solute) is shown migrating in the more permeable layers. More highly permeable layers have a higher hydraulic conductivity and behave as preferred groundwater and contaminant migration pathways. These layers are essentially superhighways for groundwater and contaminant transport, and in some instances, the hydraulic conductivity is from 100 to sometimes 1,000 times greater over distances of just a few centimeters.

Zones with higher permeability move water more quickly and have a higher flux density. If more water moves through these higher-permeability zones, then a potentially greater contaminant mass also moves through. Figure 4.8 shows this relationship at the right of the diagram, down-gradient from the source. As contamination continues to migrate along these flow paths of higher permeability dictated by the subsurface geology, diffusion of contaminants into less permeable zones occurs (Figure 4.9). The top portion of the figure represents the flow paths of contaminants in the early stages of migration, and the later stages of migration are shown in the figure's bottom portion. Over time, the contaminant (solute) has diffused into the less-permeable, lower hydraulic conductivity geologic materials (Payne et al. 2008).

The transport of contaminants in groundwater is also influenced by many of the same factors affecting the migration of contaminants in unsaturated soil or the vadose zone (Rogers 2020):

- Physical chemistry of the contaminants:
  - Solubility
  - Molecular weight
  - Vapor pressure
  - Stability and persistence

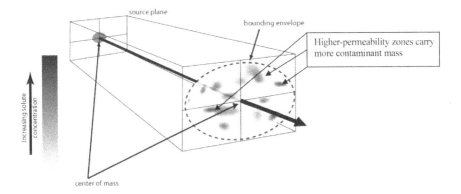

FIGURE 4.8 Contaminant migration in heterogeneous and anisotropic geologic media (adapted from Payne, F. C. et al. *Remediation Hydraulics*, CRC Press, Boca Raton, FL. 2008. With permission).

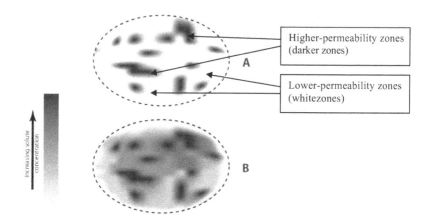

FIGURE 4.9 Effects of diffusion over time within a contaminant plume (adapted from Payne, F. C. et al. *Remediation Hydraulics*, CRC Press, Boca Raton, FL. 2008. With permission).

- Sorption potential
- Type, distribution, and amount of micro-organisms
- Tendency to biodegrade
- Dissolved oxygen content of groundwater
- Geological factors:
  - Stratigraphy (including thickness and distribution of geological units down to micro-stratigraphic scales at the centimeter or even millimeter scale)
  - Presence of unconformities
  - Soil or sediment chemistry
  - Organic carbon content
  - Porosity and permeability
  - Composition
- Climate factors:
  - Freeze and thaw cycles
  - Recharge from surface precipitation
  - Flood events
  - Seasonal climatic variations
- Vegetative factors, including types and distribution of surface vegetation, root networks, and water requirements

Sorption potential has a significant effect on the migration of contaminants in groundwater because it slows the migration of contaminants even as the flow rate of the transporting groundwater remains constant. This effect is termed **retardation** (USGS 2006a), and the degree of retardation present depends upon the specific contaminant's sorptive affinity and the amount of total organic carbon in the aquifer matrix. Many contaminants are captured by pumping wells or migrate to surface water if the travel times and/or distances are short enough before they degrade. Figure 4.10 shows an example of travel times and a capture zone in groundwater beneath urban areas. Any contaminant reaching groundwater within the area marked *capture zone* has the potential to enter the public water supply if the contaminant does not degrade before reaching a public water supply well. Several sources of contamination shown in Figure 4.1 are also shown in Figure 4.10 and include: septic tanks, USTs, landfills, industrial facilities, and power plants. These contaminant sources are typical for urban areas within the United States and pose threats to contaminate a public or private water supply.

In karst topographical settings, the transport of groundwater contamination may behave similarly to surface water (Ford and Williams 2007). Some karst formations may exhibit turbulent advective flow because they have flow rates approaching the velocities observed in surface water flow (Heath 1983). Figure 4.11 shows an example of contaminant flow in a karst aquifer (USGS 1995a).

4.3.6 TRANSPORT AND FATE OF CONTAMINANTS IN THE ATMOSPHERE

Different contaminants affect different portions of the atmosphere. For instance, CFCs affect the protective ozone layer in the upper portions of the atmosphere.

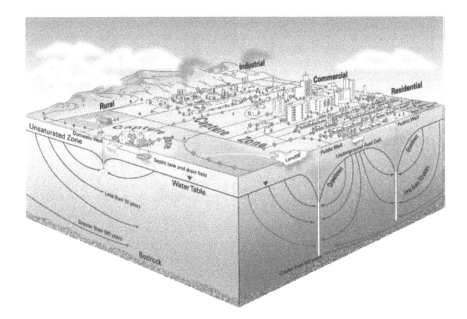

FIGURE 4.10 Potential groundwater travel time and capture zone beneath an urban area (from United States Geological Survey. *Volatile Organic Compounds in Nation's Ground Water and Drinking-Water Supply Wells.* National Water-Quality Assessment Program. USGS Circular 1292. Washington, DC, 2006a).

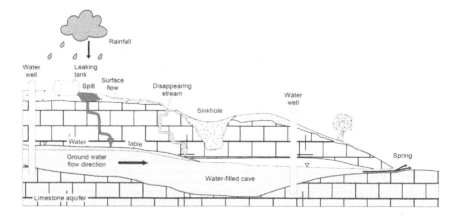

FIGURE 4.11 Contaminant migration in a karst aquifer (from United States Geological Survey 1995a. *Ground-Water Quality Protection*. Open-File Report 95–376. Nashville, Tennessee. www.pubs.usgs.gov/of/1995/ofr-95376 [accessed March 25, 2019], 1995a).

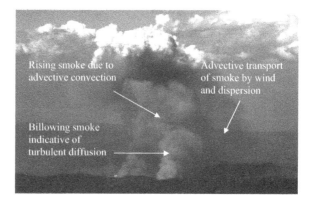

FIGURE 4.12 Smoke rising into the atmosphere from a fire (photograph by Daniel T. Rogers).

Contaminant behavior in the atmosphere is very similar to the behavior of contaminants observed in surface water. Advective transport, turbulent diffusion, and molecular diffusion also influence contaminant migration in the atmosphere (Hemond and Fechner-Levy 2000). Figure 4.12 shows smoke from a fire billowing up (turbulent diffusion) into the atmosphere and the horizontal movement of the smoke by advective transport. We also see a type of advective transport called convection; in this process, air rises due to thermal differences in the atmosphere. Turbulent advective mixing by wind and convection of the atmosphere is most significant within the troposphere (Schlatter 2009). These forces are very effective at transporting contaminants in the gas phase and also do a good job of moving solid particulate matter in the atmosphere (USEPA 2008).

Anthropogenic sources of atmospheric contaminants are significant and present themselves as an array of different contaminants released in high volumes annually

Nature's Response to Land Contamination

(USEPA 1998a). Most contaminants are released into the atmosphere from anthropogenic sources at or near the surface, with most of the impacts occurring in the troposphere and to a lesser degree the stratosphere. Contaminants released near the ground can mix throughout the troposphere in a few weeks, but it can take years or decades for them to reach the stratosphere (Hemond and Fechner-Levy 2000). Records of atmospheric contaminants and their effects can be traced to the 13th century, when King Edward I banned the burning of kiln coal in London due to its impacts on air quality (Wilson and Spengler 1996).

Temperature and *pressure* are two important factors affecting the migration of contaminants in the atmosphere. The reason temperature and pressure play a much more significant role in atmospheric fate and transport than in soil and water is because of the ideal gas laws, which are a combination of the relationships between temperature, density, and pressure. There are two ideal gas laws, which are referred to by the names of the scientists who first proposed each. What became Boyle's law was first proposed by Robert Boyle in 1662; he stated that for a fixed amount of an ideal gas kept at a fixed temperature, pressure and volume are inversely proportional. This means that if the pressure of such a system is doubled, the volume of that system becomes half of its original value. Boyle's law is expressed numerically as follows:

$$PV = K \tag{4.4}$$

where: P = pressure
   V = volume
   K = a constant

Charles' law is named after Jacques Charles, who proposed that in a closed system, the ratio between the temperature and the volume must be a constant. It is numerically expressed as the following:

$$V/T = K \tag{4.5}$$

where: V = volume
   T = temperature in Kelvin scale
   K = a constant

Temperature and pressure typically decrease with increasing altitude in the troposphere. The temperature in the lower portion of the stratosphere is relatively constant and helped give rise to its name, which means "stratified." Table 4.1 lists the standard temperatures and atmospheric pressure within the atmosphere (U. S. Standard Atmosphere 1976).

The composition of the atmosphere is presented in Table 4.2 (U. S. Standard Atmosphere 1976).

Oxygen is a recent addition to the atmosphere in geological terms. The origin of oxygen began with algae production approximately 2.45 billion years ago (Farquhar et al. 2000; Raub and Kirschvink 2008). We do not have a definitive explanation of oxygen's atmospheric origins, but one hypothesis states oxygen levels rose when volcanism providing large amounts of hydrogen to the atmosphere declined. Methane

TABLE 4.1
Standard Atmospheric Temperature and Pressure with Increasing Altitude

Altitude		Pressure	Temperature	
Feet	Meters	(atm)	°F	°C
0	0	1.000	59.0	15.0
2,000	610	0.943	51.9	11.0
4,000	1,219	0.888	44.7	7.0
6,000	1,826	0.836	37.6	3.1
8,000	2,438	0.786	30.5	−0.8
10,000	3,048	0.738	23.3	−5.0
15,000	4,572	0.564	5.5	−14.7
20,000	6,096	0.459	−12	−24.4
30,000	9,144	0.297	−48	−44.4
40,000	13,123	0.185	−67	−55
60,000	18,288	7.1×10^{-2}	−67	−55
80,000	24,384	2.7×10^{-2}	−67	−55
100,000	30,480	1.0×10^{-2}	−67	−55
140,000	42,672	2.0×10^{-3}	74	23.3
180,000	54,864	5.7×10^{-4}	170	76.7
220,000	67.056	1.7×10^{-4}	92	33.3
300,000	91,440	1.5×10^{-5}	27	−2.8
380,000	115,824	7.7×10^{-7}	188	86.7

Source: United States Standard Atmosphere. *The Standard Atmosphere of the United States.* National Oceanic and Atmospheric Administration, National Aeronautics and Space Administration, and the United States Air Force. United States Government Printing Office, Washington, DC. 277p. 1976.

and carbon dioxide were also gradually displaced by oxygen until levels rose rapidly about 2.3 bYBP. The presence of oxygen in the atmosphere plays a significant role and affects contamination in the environment through oxidation reactions and rates of combustion (USEPA 1991). Contaminants initially released into the atmosphere often do not remain in the air; they settle out and contaminate the soil or surface water. Some contaminants, however, remain in the atmosphere for long periods of time, and other contaminants initially released into soil or water may volatilize and contaminate the air. In some cases, air contaminants may settle out of the atmosphere and adsorb onto a soil grain on the land surface only to be picked up later by the wind and sent airborne again. Factors controlling whether a contaminant remains in the atmosphere include:

- Physical and chemical factors, including vapor pressure, molecular weight, solubility, and reactivity
- Geography and local topography
- Climate and weather conditions

TABLE 4.2
Composition of the Atmosphere

Gas	Chemical Symbol	Mean Molecular Weight (m mol^{-1})	Concentration (parts per million by volume [PPMv])
Nitrogen	N_2	28.013	780,840
Oxygen	O_2	31.999	209,460
Argon	Ar	39.948	9,340
Carbon dioxide	CO_2	44.010	384
Neon	Ne	20.180	18.18
Helium	He	4.003	5.24
Methane	CH_4	16.043	1.774
Krypton	Kr	83.798	1.14
Hydrogen	H_2	2.106	0.56
Nitrous oxide	N_2O	44.012	0.32
Xenon	Xe	131.293	0.09
Ozone	O_3	47.998	0.01 to 0.10

Source: United States Standard Atmosphere. *The Standard Atmosphere of the United States.* National Oceanic and Atmospheric Administration, National Aeronautics and Space Administration, and the United States Air Force. United States Government Printing Office, Washington, DC. 277p. 1976.

The average person inhales approximately 20,000 liters of air per day (USEPA 2008). Each year, the World Health Organization (2021) estimates 2.4 million people die from causes directly attributable to air pollution, with the elderly and young children at the most risk. Specific diseases caused from prolonged exposure to air contaminants are chronic and often do not immediately appear after exposure. These diseases include heart disease, lung cancer, and bronchitis. The burning of fossil fuels in power plants and automobile, truck, and bus exhaust account for 90% of all air pollution in the United States (USEPA 2008). Figure 4.13 shows some of the significant sources, methods of transport, and removal of air pollutants in the atmosphere (USEPA 2008).

Deposition of contaminants onto the land from the atmosphere occurs in two different ways:

1. Dry deposition. Dry deposition is typically dust or particulate matter settling out of the air. The amount of dry deposition depends upon the amount of suspended particles, wind speed and duration, and particle size. Contaminants are often present within dry deposition events, especially in urban areas, where they may be sorbed onto the surfaces of particulate matter in the air (USEPA 2008).
2. Wet deposition. Occurs when snow, fog, or a rain droplet forms and then dissolves or carries a contaminant to the surface. Acid rain is a good example of wet deposition (USEPA 2021c).

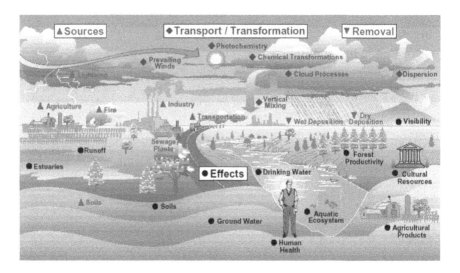

FIGURE 4.13 Sources, transport methods, and removal of air pollution (from United States Environmental Protection Agency. *Taking Toxics Out of the Air*. EPA451/K-98–001. Washington, DC. 1998).

4.4 FATE AND TRANSPORT OF CONTAMINANTS

We now briefly discuss the fate and transport behavior of each contaminant group.

4.4.1 VOCs

VOCs are organic compounds that generally volatilize or evaporate readily under normal atmospheric pressure and temperatures. They usually have a high vapor pressure, low-to-medium solubility, and low molecular weight. As a result of these chemical characteristics, VOCs are common air, soil, and water contaminants (USGS 2006a; USEPA 2008). Automobile exhaust contains VOCs. When combined with other common air pollutants and sunlight, urban smog will form if atmospheric conditions are favorable—that is, there is an ample supply of the combined sources of VOCs and other smog-forming contaminants. This type of smog formation produces **photochemical smog.**

VOCs are also released directly onto the ground surface through leaks or spills at or near the surface. Sources of these leaks include underground storage tanks, service stations, refineries, and pipelines. Because of these surface and shallow subsurface releases, VOCs are common groundwater contaminants and have been detected in the groundwater of numerous aquifers in the United States (USGS 2006a). A study of groundwater in the United States detected VOCs at a concentration of 0.02 ug/L in more than 50% of approximately 3,500 samples collected from 100 different groundwater aquifers across the country (USGS 2006a). The VOCs detected most often included: bromoform, bromodichloromethane, chloroform, chloromethane, 1,1-dichloroethane, dichlorodifluoromethane, methylene chloride,

Nature's Response to Land Contamination

dibromodichloromethane, MTBE, TCE, PCE, 1,1,1-TCA, trans-1,2-dichloroethene, toluene, and trichlorofluoromethane. This same study indicates the vulnerable nature of many aquifers of the United States, and their location corresponds with many urban areas. In fact, of the 28 major urban areas of the United States, 27 have detectable concentrations of VOCs (only Kansas City is missing) and represent a population of 125.4 million—over 41% of the entire US population. Figure 4.14 shows the locations where VOCs were detected (USGS 2006a).

VOCs exist as LNAPLs and DNAPLs. Because LNAPLs are lighter than water, they tend to float on top of groundwater, whereas the heavier-than-water DNAPLs tend to sink through the water column in an aquifer if conditions are favorable, as depicted in Figure 4.15 (USGS 2006a).

Contaminant degradation rates vary widely and depend on many factors, including (1) the nature of the release; (2) the physical chemistry of the contaminants themselves; (3) the geological environment where the contaminants are released; and (4) the presence, type, and distribution of microorganisms (Howard et al. 1997; USEPA 1998b; McKone and Enoch 2002; USGS 2006a).

Degradation rates also vary by media. In general, organic compounds in the atmosphere, including VOCs, degrade more quickly than the same organic compounds released and migrating to subsurface soil and groundwater (USEPA 1998a, 1998b).

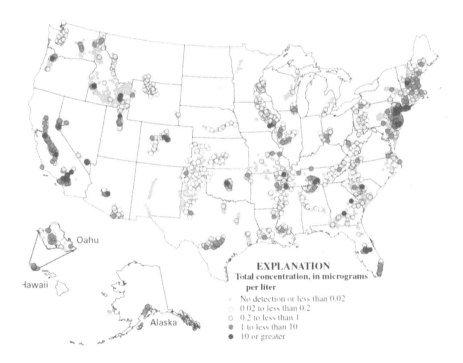

FIGURE 4.14 Occurrence of VOCs in groundwater aquifers of the United States (from United States Geological Survey. *Volatile Organic Compounds in Nation's Ground Water and Drinking-Water Supply Wells*. National Water-Quality Assessment Program. USGS Circular 1292. Washington, DC. 2006a).

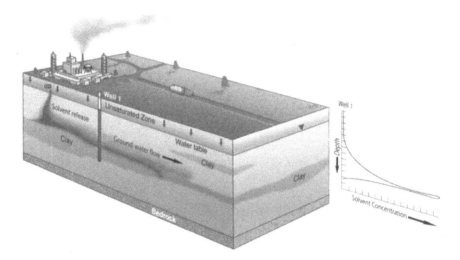

FIGURE 4.15 Migration of DNAPL compounds (from United States Geological Survey. *Volatile Organic Compounds in Nation's Ground Water and Drinking-Water Supply Wells.* National Water-Quality Assessment Program. USGS Circular 1292. Washington, DC. 2006a).

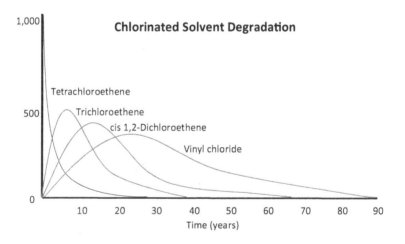

FIGURE 4.16 Chlorinated solvent degradation (modified from Payne, F. C. and Rogers, D. T. Chlorinated Solvent Degradation—A Consecutive, Irreversible Reaction Sequence. *Remediation of Recalcitrant Chlorinated Volatile Organic Compounds in Groundwater.* National Groundwater Association, Dublin, Ohio. Groundwater Management Book 21. P. 97–109.1997).

VOCs that don't degrade very easily include many of the chlorinated solvents, or DNAPL compounds (USEPA 1996a, 1996b). Many chlorinated VOC compounds, including PCE, TCE, 1,1,1-TCA, DCE, and vinyl chloride, are considered very persistent in the environment once released and can remain present for decades (USEPA 1996a; Suthersan et al. 2017). Figure 4.16 shows an example of the degradation of tetrachloroethene.

Nature's Response to Land Contamination

The degradation sequence follows a modified Domenico (1987) analytical solution for mass transport in a consecutive and irreversible fourth-order differential equation with default persistence of 2, 4, 6, and 8 years for each compound (see Figure 4.17). As shown in the figure, the maximum vinyl chloride concentration is achieved approximately 20 years following the release. The example in the figure uses an initial concentration of tetrachloroethene of 1,000 ug/L (Payne and Rogers 1997).

Other VOC compounds, such as benzene, toluene, ethyl benzene, and xylenes (BTEX), are not typically as persistent in the environment and have been known to biodegrade in a few months to years if conditions are favorable (Rogers 1995; USEPA 1996a). The VOC compound MTBE is persistent in the environment. Its relatively high solubility in water and low sorptive properties compared to other common VOC contaminants have resulted in significant MTBE-contaminated groundwater supplies at many urban locations throughout the United States (USGS 2006a). The trihalomethane VOCs include chloroform, bromoform, bromodichloro-methane, and dibromochloromethane. These compounds have been detected in the groundwater of many US aquifers (USGS 2006a). Trihalomethanes have high relative vapor pressures and commonly evaporate quickly when in contact with the atmosphere. Therefore, exposure to trihalomethanes is of special concern during showering

$$\frac{C_x}{C_0} = \exp\left\{\frac{x}{2\alpha_x}\left[1 - \left(1 + \frac{4\lambda\alpha_x}{\upsilon}\right)^{\frac{1}{2}}\right]\right\} erf\left[\frac{Y}{4(\alpha_y x)^{\frac{1}{2}}}\right] erf\left[\frac{Z}{4(\alpha_z x)^{\frac{1}{2}}}\right]$$

where: C_x = contaminant concentration in a downgradient well along the plume centerline at a distance x (mg/L)
C_0 = contaminant concentration in the source well (mg/L)
x = centerline distance between the downgradient well and source well (ft)
α_x, α_y, and α_z = longitudinal, transverse, and vertical dispersivity (ft), respectively

$$D_x = \alpha_x \times v, \; D_y = \alpha_y \times v, \; D_z = \alpha_z \times v$$

λ = degradation rate constant (1/day)
$\lambda = 0.693/t_{1/2}$ (where $t_{1/2}$ is the degradation half-life of the compound)
v = groundwater velocity (ft/day)
Y = source width (ft)
Z = source depth (ft)
erf = error function
exp = exponential function

FIGURE 4.17 Dominico steady-state model for multidimensional transport of decaying contaminant species analytical transport equation (from Domenico, P. A. An Analytical Model for Multidimensional Transport of Decaying Contaminant Species. *Journal of Hydrology.* 91:49–58. 1987).

100 Environmental Compliance Handbook

and washing (ATSDR 1997a, 2005a). They degrade by photolysis when exposed to direct sunlight and can also be degraded by microorganisms (ATSDR 1997a, 2005a).

4.4.2 PAHs

PAHs are LNAPL compounds, and, being lighter than water, they float on surface water and groundwater. They do not readily dissolve in water and have low vapor pressures compared to most VOCs. PAHs are common constituents of automobile exhaust, especially from diesel fuel (USGS 2006a), and are common air contaminants in urban areas. In addition, PAHs have a high sorptive affinity and can therefore attach to particulate matter (USGS 2006a; ATSDR 1996). Many PAH compounds biodegrade under favorable conditions. The half-life of PAHs is shortest in the atmosphere due to photochemical degradation and lasts just a few days or weeks (ATSDR 1996). The half-life of PAH compounds in soil and groundwater is longer and may last for several years or decades. When released to soil, PAHs tend to migrate more slowly than VOCs because of their higher molecular weight and sorption to soils with high organic content.

4.4.3 PCBs

Once in the environment, PCBs do not readily degrade-they remain in the environment for long periods of time and often cycle between air water and soil. PCBs can be carried long distances attached to particulate matter and have been detected in snow and sea water far away from any known point of release (USEPA 1996a). This transport capability is confirmed by their worldwide detection. The lighter the PCB compound (the fewer chlorine atoms in its structure), the farther it can potentially be transported from its release point. PCBs are not very soluble and have high sorptive potential and low vapor pressures. This combination of physical properties largely determines their environmental distribution, as they are only present at high concentrations in water or air when sorbed to particulate matter. Sinks where PCBs are frequently detected include soil near release points and sediments in rivers, streams, and lakes (USEPA 1996a). PCBs bioaccumulate in the leaves and aboveground parts of plants and food crops and in aquatic organisms and fish where PCBs are present in sediments. As a result, humans and other organisms ingesting impacted plant material, food crops, or organisms containing PCBs may bioaccumulate PCBs in their body tissues (ATSDR 2001a; USEPA 2009a).

4.4.4 SVOCs

SVOCs are much less volatile than VOCs, but notable exceptions here are the amine compounds that exist as gases at room temperature and standard pressure (ATSDR 2002a). In terms of solubility, phthalates and phenols do not readily dissolve in water, whereas amines and esters may dissolve, become mobile, and reach groundwater (ATSDR 2002a, 2002b, 2002c, 2002d). When released into the environment,

Nature's Response to Land Contamination 101

SVOCs are commonly detected in soil because they have high sorptive potentials. Many SVOCs, including pentachlorophenol, are degraded by microorganisms under favorable conditions and are also susceptible to photolysis and hydrolysis (ATSDR 2001b, 2002a). Amines and esters degrade in minutes when exposed to direct sunlight (ATSDR 1999a, 2002a, 2002b, 2002c, 2002d).

4.4.5 HEAVY METALS

Heavy metals are released directly to air, water, and soil. In most cases, these contaminants do not remain in the atmosphere for long periods of time because they have high specific gravity and become deposited onto the land surface shortly after being emitted. Mercury, however, has been detected as far away as 50 miles from its source after being released into the atmosphere (USEPA 1997). Lead is also considered a common air pollutant (USEPA 2008). Major sources of lead include metal melting facilities, battery manufacturing, and leaded gasoline and fuels. The good news is there has been a 92% decrease in atmospheric lead concentrations over the period from 1980 through 2018. Contributing to the observed decrease has been the removal of lead from gasoline and fuels, increased efficiency in air pollution control equipment, and the regulation of lead emissions sources (USEPA 2021a).

The solubility of heavy metals in water is very low, except for arsenic and chromium VI at a neutral pH. Due to their inability to form a solution, the preferred sinks for heavy metals are soil and sediments. Many metals undergo some transformation such as oxidation after being released into the environment, but are not destroyed and remain in the environment (ATSDR 1999b, 1999c, 2003, 2004a, 2005b, 2005c, 2007a 2007b, 2007c, 2008a, 2008b). Due to their low solubility and high specific gravities, they tend to remain near release points. This is why increased concentrations of heavy metals are present in the near surface soil of urban areas as a result of releases from anthropogenic sources (Murray et al. 2004). Some heavy metals, such as mercury, accumulate in sediment of lakes, rivers, and streams. Mercury may undergo a transformation through a process known as methylation after being released into the environment, typically when it reaches a surface water body (USEPA 1997). The process of methylation transforms elemental mercury to methyl mercury (CH_3Hg). Methyl mercury is the most toxic form of mercury and has the potential to bioaccumulate in aquatic organisms, including fish (USGS 2000). On land and in sediments, other heavy metals, such as barium, cadmium, chromium, lead, copper, and zinc, can accumulate in plant matter if they are present at sufficient concentrations in soil within the root zone of plants (Nudunuri et al. 1998). Removal of contaminants from near-surface soil is not uncommon and sometimes the preferred alternative. Suitable locations for removal include closed landfill sites requiring a vegetative cap to minimize erosion potential or sites where contamination is shallow, relatively static, and has become a chronic problem (Singer et al. 2003). **Phytoremediation**—the removal of contamination using plants—has been applied to other contaminants, including organic compounds, but with limited success. Murray et al. (2004) evaluated heavy metals in soil in urban soil. This study represents an initial effort to characterize the metal concentrations in surface and near-surface soil in an urban

102 Environmental Compliance Handbook

environment in southeast Michigan that includes the city of Detroit. The study discovered that:

1. Heavy metal concentrations in an urban environment are greatest at the surface.
2. A directional increase of metal concentrations in a west to east trend across the metropolitan area commensurate with an general west to east increase in urbanization and industrial activity.
3. Identified lead as the heavy metal with the highest surface concentration, with mean levels present at more than 16 times background at residential sites and 15.5 times greater than background at industrial sites. More importantly, lead concentrations in surface soil at specific industrial sites and adjacent residential neighborhoods were detected at levels hundreds of times the level that may occur naturally in the soil.
4. Heavy metal concentrations beneath the surface (at depths between 0.5 and 10 m), although elevated, were in general agreement with those of previous studies conducted in Michigan and the United States.
5. Heavy metal concentrations in soils at depths greater than 10 m are in complete agreement with previous studies of heavy metal concentrations in rural areas of Michigan and the United States. This suggests that soil at depths greater than 10 m are generally only affected to a minor extent from anthropogenic heavy metal sources.

The results of the Murray et al. (2004) soil metals study have important implications for land use planning and for future site investigations in any urban area when heavy metal contamination is suspected and include the following:

- Evaluation of human and ecological risk can be achieved by concentrating investigation efforts on heavy metals in surface soils. The results of the study have demonstrated that heavy metal concentrations are highest at the surface and quickly decrease with increasing depth. Therefore, concentrating investigative efforts at the surface will efficiently identify the most elevated heavy metal concentrations derived from anthropogenic sources.
- Innovative regional strategies for the cleanup of sites with metal contamination need to be developed.
- Identification of potential hydrologic and ecological impacts of metals in the near-surface zone is essential for the protection of groundwater. Since this study has identified that the occurrence of heavy metals in an urban environment is highest at the surface, the results can be integrated into evaluations of interflow, base flow, increased leaching from acid rain, infiltration/inflow to public water supplies, runoff to surface water systems, and the potential revegetation capacity at redeveloped sites.
- Mapping of the surface geology in urban areas can pinpoint soils particularly sensitive and prone to contamination.
- Conducting future studies related to the occurrence of heavy metals in other urban areas should consider the effects of historic land use activities and the potential human health issues and risks associated with heavy metals in surface soils.

Nature's Response to Land Contamination

Table 4.3 shows mean heavy metal concentrations in surface soil by land use. Table 4.4 shows mean heavy metal concentrations in subsurface soil by land use.

With respect to dissolved metals in groundwater, results from the 2006 Murray et al. study indicate urbanization and its accompanying industrial activities have

TABLE 4.3
Mean Concentration of Metals in Surface Soil Relative to Land Use

Metal	Commercial			Residential			Industrial		
	N	X	Φ	N	X	Φ	N	X	Φ
As	205	5.1	5.5	77	6.3	4	201	7	9
Ba	151	69	66	71	128	135	118	148	222
Cd	234	2.2	1.5	80	1.1	1.1	151	4.5	13
Cr	282	27	83	67	31	52	197	55	159
Cu	205	32	60	82	30	39	377	113	269
Pb	418	93	300	535	160	250	893	150	380
Hg	167	0.2	0.24	58	0.08	0.07	120	0.2	0.3
Ni	151	16	9.8	35	24	29	132	58	150
Se	164	0.6	0.3	57	0.8	0.8	113	1.9	2.2
Ag	152	0.5	0.2	54	0.8	0.5	114	2.3	2.2
Zn	202	130	310	81	120	124	239	257	534

N = Number of samples; X = Mean (mg kg^{-1}); φ = Standard deviation

Source: Rogers. D. T. *Urban Watersheds: Geology, Contamination, Environmental Regulations, and Sustainability.* CRC Press. Boca Raton, FL. 606p.

TABLE 4.4
Mean Concentration of Metals in Subsurface Soil Relative to Land Use

Metal	Commercial			Residential			Industrial		
	N	X	Φ	N	X	Φ	N	X	Φ
As	95	5.6	4.0	78	2.3	4.2	108	7.1	4.2
Ba	90	55	25	70	61	37	82	76	24
Cd	306	0.5	1.5	86	0.4	0.5	102	1.9	2.5
Cr	292	11.6	14	96	25.4	38	104	27	35
Cu	98	20	14	94	26	57	126	14	17
Pb	906	20	60	220	34	77	185	39	72
Hg	88	0.05	0.01	53	0.1	0.8	85	0.08	0.05
Ni	3	36	3	18	11	7	38	14.5	24
Se	90	0.8	0.6	53	0.35	0.2	82	0.9	0.9
Ag	67	0.6	0.4	48	0.5	1.1	72	2.2	4
Zn	94	67	137	98	66	80	15	60	60

N = Number of samples; X = Mean (mg kg^{-1}); φ = Standard deviation

Source: Rogers. D. T. *Urban Watersheds: Geology, Contamination, Environmental Regulations, and Sustainability.* CRC Press. Boca Raton, FL. 606p.

104   Environmental Compliance Handbook

contributed to heavy metal impacts across all land use categories (industrial, commercial, and residential) and throughout the varied surface geology of the metropolitan area. Chromium is prevalent at industrial sites, and the high levels of Cr in groundwater strongly suggest the chromium is in a hexavalent form—which is much more mobile and soluble than its trivalent form—and has greater potential to damage the environment and impact human health. This is especially significant since hexavalent chromium has a high contaminant risk factor (USEPA 2021d; Rogers 2020). Significantly high contamination ratios were demonstrated by Cr, Cd, and Pb within the study area sand unit, but there were also several metals detected at concentrations exceeding their MCLs in clay-rich soil units, indicating the existence of some mobility within a soil medium previously considered virtually immobile for metals.

4.4.6 PESTICIDES AND HERBICIDES

Pesticides and herbicides are released into the environment for specific purposes and are most commonly detected in near-surface soils and surface water rather than groundwater (USGS 2006b). The highest concentrations of pesticides exist in the nation's streams and sediments within urban areas, where they have been detected in 83% of streams and 70% of sediments (USGS 2006b). The total tonnage of pesticide and herbicide use in the United States has remained constant during the 1980s and early 1990s but has increased by nearly 60% in the last 20 years, mostly for agriculture and home and garden purposes (USEPA 2017).

Permethrin and toxaphene are two other widely used compounds in insecticides and pesticides, respectively. The fate and transport concern with permethrin centers around its use as an insect repellent, application to crops, and flea treatments for pets. All of these activities involve human exposure (ATSDR 2005d), and with a half-life of approximately 28 days, there is often adequate time for human contact. Toxaphene strongly sorbs to soil particles and is not very soluble in water (SRC 2021). Common sinks where toxaphene may be present include sediments and soil, from where it bioaccumulates in fish and mammals. Toxaphene is also in the atmosphere since it evaporates when in a solid form or dissolved (ATSDR 1997b). It is estimated that the half-life of toxaphere is more than 10 years in soil (SRC 2021), so there is a good chance it is still present at appreciable concentrations in the environment.

4.4.7 DIOXINS

Since one method of dioxin formation is through incineration and combustion, dioxin compounds are present in the atmosphere and have been detected around the globe (ATSDR 2006a). When dioxins are released at the surface, they typically sorb to soil particles and are often detected in sediments in lakes, rivers, and streams acting as sinks for dioxins (ATSDR 2006a). Dioxin compounds are considered bioaccumulative contaminants, with the potential to build up in the food chain and yield detectable concentrations in the tissues of many animals (ATSDR 2006a; USEPA 2021d).

Nature's Response to Land Contamination

Due to the presence of chlorine in their atomic structure, dioxins do not readily degrade once they are formed and released into the environment (USEPA 2006).

4.4.8 FERTILIZERS

The most common fertilizers include nitrogen (N), phosphorus (P), and potassium (K). A major sink for fertilizers is surface water because they are applied to the soil surface and are considered soluble in water and mobile-especially in the case of nitrates and phosphorus. Once in surface water, nitrate and phosphorus can promote excessive algal growth. Significant algal growth can deplete the dissolved oxygen in surface water and cause suffocation and death to aquatic organisms. The solubility of some fertilizers, combined with the relationship between surface water and groundwater, may lead to groundwater contamination. The natural process of enrichment of surface waters with plant nutrients is termed **eutrophication**. When anthropogenic activities such as fertilization or sewage discharges accelerate this natural process, **cultural eutrophication** occurs (McGucken 2000).

4.4.9 CYANIDE

In the atmosphere, cyanide is most often present as hydrogen cyanide. When present in surface water, cyanide compounds usually will form hydrogen cyanide and then enter the atmosphere through evaporation. When released to soil, cyanide compounds are considered fairly mobile when the retention capacity of the soil is exceeded and may migrate and contaminate groundwater. Cyanide compounds are degraded by microorganisms when present at low concentrations. When concentrations of cyanide compounds are elevated, they tend to be toxic to microorganisms and resist degradation (ATSDR 2006b). The half-life of cyanide in the atmosphere ranges between 1 and 3 years (ATSDR 2006b). In soil and water, the half-life of cyanide compounds is much more difficult to estimate because the concentration, distribution, and presence of microorganisms available to degrade the cyanide compounds vary.

4.4.10 ASBESTOS

Asbestos fibers do not degrade, evaporate, or dissolve in water and remain virtually unchanged in the environment (ATSDR 2001c). Asbestos originates from naturally occurring minerals and is therefore present in the environment. Average background concentrations of asbestos in air range from 0.00001 to 0.0001 fibers per milliliter of air and are highest in urban areas (ATSDR 2001c).

Small-diameter asbestos fibers can remain suspended in the atmosphere for a long period of time compared to larger fibers (those larger than 10 microns) (ATSDR 2001c). Since asbestos was widely used in building materials, it is most common in urban areas and where natural deposits are present. Asbestos can become airborne through the disturbance of asbestos-containing materials during demolition or remodeling activities.

106   Environmental Compliance Handbook

4.4.11 ACIDS AND BASES

When released into soil, acids and bases neutralize rapidly. They are diluted when they come into contact with water if a difference in pH levels exists. Therefore, if environmental impairment occurs, it must be realized rapidly before the acid or base becomes neutralized. This impairment occurs with the majority of sudden and accidental releases but does not hold true for acid rain which generates effects with slower onsets.

Acids and bases may migrate a significant distance—sometimes more than 1 mile—when released in the atmosphere and may cause significant impairment to living organisms exposed to their vapors (ATSDR 2004b, 2004c).

4.4.12 RADIOACTIVE COMPOUNDS

Radioactive compounds occur naturally, with the most common being radon. Radon is produced from the decay of uranium (ATSDR 2008c) and is present in air, water, and soil. Radon may build up in basements, especially if cracks exist, or other subsurface structures located above natural deposits having higher relative uranium levels. The half-life of radon is approximately 4 days (ATSDR 2008c). Most of the human exposure attributed to other radioactive compounds results from medical devices, diagnostic treatments, testing equipment such as x-ray machines, and cancer therapy (Kathren 1991; ATSDR 2000).

4.4.13 GREENHOUSE GASES

Greenhouse gases decay very slowly and are primarily atmospheric contaminants. Some quantities of these gases are naturally removed from the atmosphere, such as the removal of carbon dioxide during photosynthesis. However, the anthropogenic addition of carbon dioxide and other greenhouse gases into the atmosphere has greatly exceeded the capacity of the natural environmental to remove them (USEPA 2009b). As a result, greenhouse gas concentrations have been increasing (USEPA 2018).

4.4.14 CARBON DIOXIDE

Carbon dioxide is a greenhouse gas, and we discussed its fate in the atmosphere previously, but there is another important method by which carbon dioxide is absorbed, and that is by the oceans. The rate of accumulation of pollution is proportional to the human population. The more people, the more carbon dioxide ends up in the oceans.

Pollution in the oceans does not all originate from direct discharge but is also delivered through the atmosphere from increased concentrations of carbon dioxide in the atmosphere that are partially adsorbed by the oceans and cause acidification of the surface layers. Absorption of carbon dioxide by the oceans has been estimated to be as high as 25%, which represents millions of tons of carbon dioxide absorbed by the oceans every year. The additional carbon dioxide absorbed by the oceans reacts with ocean water, forms an acid called carbonic acid, and is lowering the pH of the

Nature's Response to Land Contamination

oceans worldwide. In fact, the oceans are acidifying faster than they have in some 300 million years. The effect of the acidification has been linked to bleaching of coral reefs, including the Great Barrier Reef off the northeastern coast of Australia (Natural Resource Defense Council 2018).

4.4.15 CARBON MONOXIDE

Carbon monoxide is created when fuel is not burned completely. Carbon monoxide is formed naturally and anthropogenically. The most significant source of carbon monoxide is from automobile exhaust (USEPA 2021a). Inside homes, significant sources of carbon dioxide emissions are natural gas and oil furnaces, hot water heaters, appliances, wood-burning stoves, and fireplaces. Carbon monoxide is also created as a byproduct of several industrial processes, including metal melting and chemical synthesis. From 1980 to 2016, there was a decrease of greater than 80% in carbon dioxide in the United States. This decrease was attributed to improved air pollution control equipment for stationary and mobile sources of air pollution (USEPA 2021a).

4.4.16 OZONE

Ozone is a gas occurring in Earth's upper atmosphere and at ground level. Ozone in the upper atmosphere is greatly beneficial to life on Earth because it filters UV radiation, but ozone occurring at ground level is considered an air pollutant (USEPA 2021a). Ground-level ozone is not emitted directly into the air-it is created by chemical reactions between oxides of nitrogen and VOCs in the presence of sunlight. Emissions from automobile exhaust, gasoline vapors, chemical solvents, electrical generating facilities, and some factories trigger the production of ground-level ozone (USEPA 2009b). This variety of ozone is a concern in urban regions of the United States during the summer, because strong sunlight and hot weather can generate higher levels (USEPA 2009b). Since the Clean Air Act of 1990, atmospheric ozone concentration in the United States has declined 25% (USEPA 2021a). Better control of stationary and mobile sources of air pollution such as automobile exhaust is behind this improvement (USEPA 2021a).

4.4.17 SULFUR DIOXIDE

Sulfur dioxide is a component of smog and also combines with nitrous oxide compounds to eventually form sulfuric acid, which is commonly referred to as acid rain (USEPA 2021a). It is removed from the atmosphere during precipitation and is typically neutralized quickly in soil if the pH of the soil is greater than 7. Some areas of the northeastern United States have soils lacking the ability to effectively neutralize the effects of acid rain, and there have been adverse effects on aquatic life and vegetation in the region. Efforts to reduce sulfur dioxide emissions from stationary and mobile sources have resulted in a decrease of 75% since 1980 in the United States (USEPA 2021a). Nevertheless, the pH of rain in the eastern United States remains acidic.

108 Environmental Compliance Handbook

4.4.18 PARTICULATE MATTER

Urban areas have the highest concentrations of particulate matter, which is a significant distributor of contaminants in the atmosphere. Contaminants such as SVOCs, some VOCs, PCBs, and many pesticides and herbicides may sorb to a soil particle and travel a significant distance through wind action (USEPA 1998a). The size of particulate matter is significant because the largest sizes tend to settle to the ground surface first. Smaller particles can travel around the globe and remain suspended for years if favorable conditions exist (USEPA 2021a). Here is some good news: there was a 20 to 30% decline in atmospheric particulate matter within the urban areas of the United States between 2001 and 2016 (USEPA 2021a).

4.4.19 BACTERIA, PARASITES, AND VIRUSES

Bacteria, parasites, and viruses are present in large numbers everywhere in the environment. They are in and on the food we eat, in and on our bodies, in the air we breathe and the water we drink, in soil, and at depths within the Earth. Many are beneficial, but some have the potential to adversely affect our health and well-being (Madigan et al. 2008).

4.4.20 INVASIVE SPECIES

Invasive species have been around for a few hundred years and now number in the tens of thousands. Invasive species are difficult to impossible to contain once released, so prevention is key. The main reason invasive species are difficult to contain is that the mode of migration is heavily influenced by breeding. With an increasing human population, increased standard of living, a warming climate, and increased human travel, the prospects of increased invasive species issues are almost certain.

4.4.21 POLYFLUOROALKYL SUBSTANCES

PFAS chemicals are persistent, mobile, and toxic. They have been widely produced and used in numerous consumer products for decades and have just recently been identified as a contaminant. Due to their soluble nature, PFAS chemicals are most often detected in surface and groundwater and have also been detected in fish tissues. Reliable estimates on their degradation have not yet been quantified.

4.4.22 EMERGING CONTAMINANTS

Emerging contaminants, including many pharmaceuticals, 1,4-dioxane, 1,2,3-trichloropropane, PFA compounds, and perchlorates, are resistant to degradation and can remain in the environment for long periods of time (Bell et al. 2019). Emerging contaminants have been detected in groundwater, where they can migrate long distances due to their relatively high solubility and resistance to degradation. Emerging contaminants are difficult to investigate because many have entered the environment

Nature's Response to Land Contamination 109

from non-traditional sources such as residential septic systems and agricultural locations, as opposed to industrial sources (Bell et al. 2019).

4.5 SUMMARY AND CONCLUSION

There are tens of thousands of pollutants existing everywhere. After being released into the environment, they migrate in air, soil, and water. Some are persistent, while others are not. Some dissolve in water, and some do not. Some are transported around the globe in the atmosphere, while others are not. Geography, geology, hydrogeology, and atmospheric conditions all play a significant role in affecting the fate and transport of pollutants and determine their final disposition.

Halogenated contaminants (those containing chlorine, fluorine, bromine, or iodine within their structure) generally remain in the environment for long periods of time because, in large part, they are molecularly stable and synthetic compounds, and the natural environment has difficulty in degrading them. Several of these compounds have the ability to accumulate in tissues of living organisms (e.g., PCBs), and through bioaccumulation may expose humans after they work their way up the food chain. Some contaminants can change form, such as a gas to a liquid or a liquid to a gas, and cycle between the soil, air, and water if they last long enough (e.g., some VOCs). Other contaminants not changing form can be found in sinks or areas where they accumulate (e.g., river and lake sediment). Risk only occurs when an exposure pathway exists. Toxicity, therefore, is not the only factor to consider when evaluating risk. Mobility and persistence of a contaminant must also be considered.

4.6 REFERENCES

Agency for Toxic Substances and Disease Registry (ATSDR). 1996. *Polycyclic Aromatic Hydrocarbons*. General Contaminant Class. ATSDR ToxFAQs. Atlanta, GA.

Agency for Toxic Substances and Disease Registry (ATSDR). 1997a. *Chloroform*. CAS Registry Number 127–18–4. ATSDR ToxFAQs. Atlanta, GA.

Agency for Toxic Substances and Disease Registry (ATSDR). 1997b. *Toxaphene*. CAS Registry Number 8001–35–2. ATSDR ToxFAQs. Atlanta, GA.

Agency for Toxic Substances and Disease Registry (ATSDR). 1999a. *Dimethylamine*. CAS Registry Number 124–40–3. ATSDR ToxFAQs. Atlanta, GA.

Agency for Toxic Substances and Disease Registry (ATSDR). 1999b. *Mercury*. CAS Registry Number 7439–97–6. ATSDR ToxFAQs. Atlanta, GA.

Agency for Toxic Substances and Disease Registry (ATSDR). 1999c. *Silver*. CAS Registry Number 7440–22–4. ATSDR ToxFAQs. Atlanta, GA.

Agency for Toxic Substances and Disease Registry (ATSDR). 2000. *Radon Toxicity: Who Is at Risk?* ATSDR. Atlanta, GA.

Agency for Toxic Substances and Disease Registry (ATSDR). 2001a. *Polychlorinated Biphenyls*. ATSDR ToxFAQs. Atlanta, GA.

Agency for Toxic Substances and Disease Registry (ATSDR). 2001b. *Pentachlorophenol*. CAS Registry Number 87–86–5. ATSDR ToxFAQs. Atlanta, GA.

Agency for Toxic Substances and Disease Registry (ATSDR). 2001c. *Asbestos*. CAS Registry Number 1332–21–4. ATSDR ToxFAQs. Atlanta, GA.

Agency for Toxic Substances and Disease Registry (ATSDR). 2002a. *Di(2-ethylhexyl) Phthalate*. CAS Registry Number 117–81–7. ATSDR ToxFAQs. Atlanta, GA.

Agency for Toxic Substances and Disease Registry (ATSDR). 2002b. *Toxicological Profile for Flame Retardant Ester Compounds*. ATSDR ToxFAQs. Atlanta, GA.

Agency for Toxic Substances and Disease Registry (ATSDR). 2002c. *Benzyl Acetate*. CAS Registry Number 140–11–4. ATSDR ToxFAQs. Atlanta, GA.

Agency for Toxic Substances and Disease Registry (ATSDR). 2002d. *Ethyl Acetate*. CAS Registry Number 141–78–6. ATSDR ToxFAQs. Atlanta, GA.

Agency for Toxic Substances and Disease Registry (ATSDR). 2003. *Selenium*. CAS Registry Number 7782–49–2. ATSDR ToxFAQs. Atlanta, GA.

Agency for Toxic Substances and Disease Registry (ATSDR). 2004a. *Copper*. CAS Registry Number 7440–50–8. ATSDR ToxFAQs. Atlanta, GA.

Agency for Toxic Substances and Disease Registry (ATSDR). 2004b. *Ammonia*. CAS Registry Number 7664–41–7. ATSDR ToxFAQs. Atlanta, GA.

Agency for Toxic Substances and Disease Registry (ATSDR). 2004c. *Hydrochloric Acid*. CAS Registry Number 7647–01–0. ATSDR ToxFAQs. Atlanta, GA.

Agency for Toxic Substances and Disease Registry (ATSDR). 2005a. *Bromoform*. CAS Registry Number 75–25–2. ATSDR ToxFAQs. Atlanta, GA.

Agency for Toxic Substances and Disease Registry (ATSDR). 2005b. *Nickel*. CAS Registry Number 7440–02–0. ATSDR ToxFAQs. Atlanta, GA.

Agency for Toxic Substances and Disease Registry (ATSDR). 2005c. *Zinc*. CAS Registry Number 7440–66–6. ATSDR ToxFAQs. Atlanta, GA.

Agency for Toxic Substances and Disease Registry (ATSDR). 2005d. *Permethrin: Toxicologic Information About Pesticides*. CAS Registry Number 52645–53–1. ATSDR. Atlanta, GA.

Agency for Toxic Substances and Disease Registry (ATSDR). 2006a. *Dioxins: Chemical Agent Briefing Sheet*. ATSDR. Atlanta, GA.

Agency for Toxic Substances and Disease Registry (ATSDR). 2006b. *Cyanide*. CAS Registry Number 74–90–8, 143–33–9, 151–50–8, 592–01–8, 544–92–3, 506–61–6, 460–19–5, 506–77–4. ATSDR ToxFAQs. Atlanta, GA.

Agency for Toxic Substances and Disease Registry (ATSDR). 2007a. *Lead*. CAS Registry Number 7439–92–1. ATSDR ToxFAQs. Atlanta, GA.

Agency for Toxic Substances and Disease Registry (ATSDR). 2007b. *Barium*. CAS Registry Number 9440–39–3. ATSDR ToxFAQs. Atlanta, GA.

Agency for Toxic Substances and Disease Registry (ATSDR). 2007c. *Arsenic*. CAS Registry Number 7440–38–2. ATSDR ToxFAQs. Atlanta, GA.

Agency for Toxic Substances and Disease Registry (ATSDR). 2008a. *Cadmium*. CAS Registry Number 7440–43–9. ATSDR ToxFAQs. Atlanta, GA.

Agency for Toxic Substances and Disease Registry (ATSDR). 2008b. *Chromium*. CAS Registry Number 7440–47–3. ATSDR ToxFAQs. Atlanta, GA.

Agency for Toxic Substances and Disease Registry (ATSDR). 2008c. *Radon*. CAS Registry Number 14859–67–7. ATSDR ToxFAQs. Atlanta, GA.

Bell, C. A., Gentile, M., Kalve, E., Ross, I., Horst, J. and Suthersan, S. 2019. *Emerging Contaminants*. CRC Press, Boca Raton, FL. 439p.

Brady, N. C. and Well, R. R. 1999. *The Nature and Properties of Soils*, 12th Edition. Prentice—Hall. Upper Saddle River, NJ.

Domenico, P. A. 1987. An Analytical Model for Multidimensional Transport of Decaying Contaminant Species. *Journal of Hydrology*. Vol. 91. pp. 49–58.

Farquhar, J., Huiming, B. and Thiemens, M. 2000. Atmospheric Influence of Earth's Earliest Sulfur Cycle. *Science*. Vol. 289. No. 5480. pp. 756–758.

Fetter, C. 1993. *Contaminant Hydrogeology*, 2nd Edition. Prentice Hall. Upper Saddle River, NJ.

Ford, D. C. and Williams, P. 2007. *Karst Hydrogeology and Geomorphology*, 2nd Edition. John Wiley & Sons. New York, NY.

Nature's Response to Land Contamination

Freeze, R. A. and Cherry, J. A. 1978. *Groundwater*. Prentice Hall. Upper Saddle River, NJ.

Haynes, W. 1954. *American Chemical Industry—A History*. Vols. I–IV. Van Nostrand Publishers. New York, NY.

Heath, R. C. 1983. *Basic Ground-Water Hydrology*. United States Geological Survey. Water Supply Paper 2220. United States Government Printing Office. Alexandria, VA.

Hemond, H. F. and Fechner-Levy, E. J. 2000. *Chemical Fate and Transport in the Environment*. Academic Press. London, England.

Hornsby, A. G. 1990. *How Contaminants Reach Groundwater*. University of Florida Institute of Food and Agriculture. Gainesville, FL.

Howard, P. H. Michalenko, E. M., Basu, D. K. and Aronson, D. 1997. *Handbook of Environmental Fate and Exposure Data for Organic Chemicals. Volume 5. Solvents 3*. Lewis Publishers. Chelsea, MI.

Kathren, R. 1991. *Radioactivity and the Environment*. Taylor & Francis Publishers, Lieden, The Netherlands.

Kaufman, M. M., Rogers, D. T. and Murray, K. S. 2011. *Urban Watersheds*. CRC Press. Boca Raton, FL. 583 pages.

Lindsay, W. L. 1979. *Chemical Equilibria in Soils*. John Wiley & Sons. New York.

Lyman, W. J. Reehl, W. F. and Rosenblatt, D. M. 1990. *Handbook of Chemical Property Estimation Methods*. American Chemical Society. Washington, DC.

Madigan, M. T., Martinko, J. M., Dunlap, P. V. and Clark, D. P. 2008. *Brock Biology of Microorganisms*, 12th Edition. Prentice-Hall, Inc. New York, NY.

McGucken, W. 2000. *Lake Erie Rehabilitated: Controlling Cultural Eutrophication, 1960s—1990s*. University of Akron Press, Akron, OH.

McKone, T. E. and Enoch, K. G. 2002. *CalTox^{TM}, A Multimedia Total Exposure Model Spreadsheet Users Guide*. Lawrence Berkeley National Laboratory. LBNL 47399. U. of California, Berkeley, CA.

Murray, K. S., Rogers, D. T. and Kaufman, M. M. 2004. Heavy Metals in an Urban Watershed in Michigan. *Journal of Environmental Quality*. Vol. 33. pp. 163–172.

Natural Resources Defense Council (NRDC). 2018. Ocean Pollution: The Dirty Facts. https://nrdc.org/stories/ocean-pollution-dirty-facts. (Accessed November 18, 2018).

New Jersey Department of Environmental Protection (NJDEP). 2015. *Site Investigation Report*. NJDEP. Trenton, NJ. 7,610p.

Nudunuri, K. V., Erickson, L. E. and Govindaraju, R. S. 1998. Modeling the Role of Active Biomass on the Fate and Transport of Heavy Metals in the Presence of Root Exudates. *Journal of Hazardous Waste Research*. Vol. 1. No. 9. pp. 1–25.

Payne, F. C., Quinnan, J. A. and Potter, S. T. 2008. *Remediation Hydraulics*. CRC Press. Boca Raton, FL.

Payne, F. C. and Rogers, D. T. 1997. Chlorinated Solvent Degradation—A Consecutive, Irreversible Reaction Sequence. In: *Remediation of Recalcitrant Chlorinated Volatile Organic Compounds in Groundwater*. National Groundwater Association, Dublin, OH. Groundwater Management Book 21. pp. 97–109.

Raub, T. D. and Kirschvink, J. L. 2008. A Pan-Precambrian Link Between Deglaciation and Environmental Oxidation. In: Cooper, A. K., Barrett, P. J., Stagg, H. Stump, E. and Wise, W. editors. *Antarctica: A Keystone in a Changing World*. The National Academic Press. Washington, DC.

Rogers, D. T. 1995. Intrinsic Bioremediation of Gasoline-Contaminated Groundwater—A Case Study. Air and Waste Management Association. *Annual Meeting*. San Antonio, TX.

Rogers, D. T. 1996. *Environmental Geology of Metropolitan Detroit*. Clayton Environmental Consultants, Novi, MI.

Rogers, D. T. 2020. *Urban Watersheds: Geology, Contamination, Environmental Regulations, and Sustainability*. CRC Press. Boca Raton, FL. 608p.

Rogers, D. T., Murray, K. S. and Kaufman, M. M. 2007. Assessment of Groundwater Contaminant Vulnerability in an Urban Watershed in Southeast Michigan, USA. In: Howard, K. W. F. editor. *Urban Groundwater—Meeting the Challenge*. Taylor & Francis, London, England.

Rogers, D. T., Murray, K. S. and Kaufman, M. M. 2012. Environmental Risk Analysis through Integration of Geologic Vulnerability and Air, Water, and Soil Contaminant Risk Factor Derivation. *International Geological Congress*. Vol. p. 3003. Brisbane, Australia.

Sawhney, B. L. and Brown, K. editors. 1989. *Reactions and Movement of Organic Chemicals in Soils*. Special Publication No 22. Soil Science of America. Madison, WI.

Schlatter, T. W. 2009. *Atmospheric Composition and Vertical Structure*. National Oceanic and Atmospheric Administration (NOAA). Boulder, CO.

Schnoor, J. L. 1996. *Environmental Modeling: Fate and Transport of Pollutants in Water and Soil*. John Wiley & Sons. New York, NY.

Schwarzenbach, R. P., Gschwend, P. M. and Imboden, D. M. 1993. *Environmental Organic Chemistry*. John Wiley & Sons. New York, NY.

Singer, A. C., Crowley, D. E. and Thompson, I. P. 2003. Secondary Plant Metabolites in Phytoremediation and Biotransformation. *Trends in Biotechnology*. Vol. 21. No. 3. pp. 123–130.

Soil Science Society of America. 1987. *Glossary of Soil Science Terms*. Soil Science Society of America. Madison, WI.

SRC. 2021. Environmental Fate Data Base (EFDB). CHEMFATE Chemical Search. http://srcinc.com/what-we-do/efbd.aspx. (Accessed June 10, 2021).

Suthersan, S. S., Horst, J., Schnobrich, M., Welty, N. and McDonough, J. 2017. *Remediation Engineering: Design Concepts*, 2nd Edition. CRC Press. Boca Raton, FL. 603p.

Suthersan, S. S. and Payne, F. C. 2005. *In Situ Remediation Engineering*. CRC Press. Boca Raton, FL. 430p.

United States Environmental Protection Agency (USEPA). 1991. *Air Pollution and Health Risk*. EPA/450/3–90–022. USEPA. Washington, DC.

United States Environmental Protection Agency (USEPA). 1996a. *Transport and Fate of Contaminants in the Subsurface*. EPA/625/4–89/019. USEPA. Washington, DC.

United States Environmental Protection Agency (USEPA). 1996b. *Bioscreen. Natural Attenuation Decision Support System*. Office of Research and Development. Washington, DC.

United States Environmental Protection Agency (USEPA). 1997. *Mercury Study Report to Congress: Volume III: Fate and Transport of Mercury in the Environment*. EPA-454/R-97–005. USEPA. Washington, DC.

United States Environmental Protection Agency (USEPA). 1998a. *Taking Toxics out of the Air*. EPA451/K-98–001. USEPA. Washington, DC.

United States Environmental Protection Agency (USEPA). 1998b. *Chemical Fate Half-Lives for Toxics Release Inventory (TRI) Chemicals*. USEPA. Washington, DC.

United States Environmental Protection Agency (USEPA). 1999. *Fundamentals of Soil Science as Applicable to Management of Hazardous Wastes*. EPA/540/S-98/500. Office of Research and Development. Washington, DC.

United States Environmental Protection Agency (USEPA). 2006. *An Inventory of Sources and Environmental Releases of Dioxin-Like Compounds in the United States for the Years 1987, 1995, and 2000*. EPA/600/P-03/002f. USEPA. Washington, DC.

United States Environmental Protection Agency (USEPA). 2008. *Findings on National Air Quality: Status and Trends Through 2006*. EPA454/R-07–007. USEPA. Research Triangle Park, NC.

United States Environmental Protection Agency (USEPA). 2009a. *Biennial National Listing of Fish Advisors for 2008*. EPA-823-F-09–007. USEPA. Washington, DC.

United States Environmental Protection Agency (USEPA). 2009b. *Ozone—Good Up High Bad Nearby. Air Quality Planning and Standards*. USEPA. Washington, DC.

Nature's Response to Land Contamination

United States Environmental Protection Agency. 2017. *Pesticide Industry Usage. USEPA Biological and Economic Analysis Division*. Office of Pesticide Programs. Washington, DC. 24p.

United States Environmental Protection Agency (USEPA). 2018. Global Greenhouse Gas Emissions Data. https://epa.gov/ghgemissions/global-greenhouse-gas-emissions-data. (Accessed December 31, 2018).

United States Environmental Protection Agency (USEPA). 2021a. EPA's Report on the Environment (ROE). www.epa.gov/report-environment. (Accessed June 10, 2021).

United States Environmental Protection Agency (USEPA). 2021b. Advisories for Fish and Shellfish Consumption. www.epa.gov/fish-tech. (Accessed June 10, 2021).

United States Environmental Protection Agency (USEPA). 2021c. What is Acid Rain. www.epa.gov/acidrain/what/index.html. (Accessed June 10, 2021).

United States Environmental Protection Agency (USEPA). 2021d. Integrate Risk Information System (IRIS). www.epa.gov/ncea/iris/intro.htm. (Accessed June 10, 2021).

United States Geological Survey (USGS). 1995a. *Ground-Water Quality Protection*. Open-File Report 95–376. Nashville, Tennessee. www.pubs.usgs.gov/of/1995/ofr-95376. Accessed December 2009).

United States Geological Survey (USGS). 1995b. *Contaminants in the Mississippi River*. Circular 1133. USGS. Washington, DC.

United States Geological Survey (USGS). 1998. *Simulating Transport of Volatile Organic Compounds in the Unsaturated Zone Using the Computer Model R-UNSAT*. Fact Sheet 019–08. USGS. Washington, DC.

United States Geological Survey (USGS). 2000. *Mercury in the Environment*. Fact Sheet 146–00. USGS. Washington, DC.

United States Geological Survey (USGS). 2006a. *Volatile Organic Compounds in Nation's Ground Water and Drinking-Water Supply Wells. National Water-Quality Assessment Program*. Circular 1292. USGS. Washington, DC.

United States Geological Survey (USGS). 2006b. *Pesticides in the Nation's Streams and Groundwater. 1992–2001—A Summary*. Fact Sheet 2006–3028. USGS. Washington, DC.

United States Standard Atmosphere. 1976. *The Standard Atmosphere of the United States. National Oceanic and Atmospheric Administration, National Aeronautics and Space Administration, and the United States Air Force*. United States Government Printing Office, Washington, DC. 277p.

Wania, F., Hoff, J. T., Jia, C. Q. and Mackey, D. 1998. The Effects of Snow and Ice on the Environmental Behavior of Hydrophobic Organic Chemicals. *Journal of Environmental Pollution*. Vol. 102. pp. 79–95.

Wiedemeier, T. H. Rifai, H. S. Newell, C. J. and Wilson, T. J. 1999. *Natural Attenuation of Fuels and Chlorinated Solvents in the Subsurface*. John Wiley & Sons. New York, NY.

Wilson, R. and Spengler, J. D. editors. 1996. *Particles in Our Air: Concentration and Health Affects*. Harvard University Press. Cambridge, MA.

World Health Organization (WHO). 2021. Health Statistics and Health Information Systems. Mortality Database Tables. www.who.int/healthinfo/morttables/en/index/html. (Accessed June 10, 2021).

5 Resource, Conservation, and Recovery Act, Medical Waste, and Low-Level Radioactive Waste

5.1 INTRODUCTION

The Resource Conservation and Recovery Act was enacted in 1976 and is the principal law in the United States addressing land-based disposal of solid and hazardous waste (USEPA 2021a). It was enacted to address problems that the United States faced from its growing volume of municipal and industrial waste.

In very general terms, RCRA defines a waste as anything that is discarded (USEPA 2021a). RCRA recognizes that a solid waste may not be a solid. Under RCRA, many solid wastes are liquid, semi-solid, or contained gaseous material (USEPA 2021a). RCRA set goals for:

- Protecting human health and the environment from the hazards of waste disposal
- Energy conservation and natural resources
- Reducing the amount of waste generated through source reduction and recycling
- Ensuring the management of waste in an environmentally sound manner

RCRA also established standards for the treatment, storage, and disposal of hazardous waste in the United States. In the United States, RCRA is responsible for (USEPA 2021a):

- Managing approximately 2.5 billion tons of solid, industrial, and hazardous waste
- Overseeing 6,600 facilities with over 20,000 process units in the full permitting universe
- Working to address more than 3,700 existing contaminated facilities in need of cleanup
- Reviewing a possible 2,000 additional facilities that may need cleanup in the future
- Providing nearly $100 million in grant funding to assist states in implementing authorized waste programs
- Providing incentives and opportunities to reduce or avoid greenhouse gas emissions through material and land management practices

DOI: 10.1201/9781003150107-5

It is important to note and highlight some of RCRA's accomplishments nationwide to understand how RCRA is currently viewed and what the future may hold. The major accomplishments of RCRA include (USEPA 2021a):

- Developing a comprehensive nationwide system to manage waste from its point of generation to its final resting place, which is commonly referred to as "cradle to grave"
- Establishing the framework for states to implement effective municipal solid waste and non-hazardous waste management programs
- Preventing contamination from adversely impacting communities and becoming future Superfund sites
- Restoring 18 million acres of contaminated lands, nearly equal to the size of South Carolina
- Creating partnership and award programs to encourage companies to modify manufacturing practices to generate less waste and reuse materials safely
- Enhancing perceptions of wastes as valuable commodities that can be part of new products through USEPA's sustainable materials management efforts
- Creating and enhancing the nation's recycling infrastructure and increasing municipal solid waste recycling/composting rate from less than 7% to nearly 35%

RCRA divides those directly affected by the regulation into three categories (USEPA 2021a):

1. Generators. Generators must determine if they are creating any wastes that would be classified as hazardous. If so, they must obtain a tracking number and manifest, ensure proper storage and labeling of the waste, and keep records.
2. Transporters. Transporters of wastes ensure that they comply with USEPA and USDOT requirements for transportation of hazardous materials (HAZMAT) and must ensure proper packaging, labeling, reporting, and record keeping.
3. Treatment, storage, and disposal facilities (TSDs). Treatment, storage, and disposal Facilities have by far the most complex and stringent requirements. They must obtain permits, which require inspections and monitoring, as well as comply with the manifest system.

There are three main components of RCRA, called subtitles. The first describes and defines non-hazardous waste, the second defines and describes hazardous waste, and the third addresses petroleum underground storage tanks. In addition to the three subtitles, RCRA also includes methods for analytical testing of solid wastes under RCRA called Test Methods for Evaluating Solid Waste (SW-846). This is significant because what the SW-846 provision has created is a consistent set of methods for fully characterizing any solid waste (Cornell University Law 2021; USEPA 2021b).

The three main components of RCRA and SW-846 are described in greater detail in the following sections.

Resource, Conservation, and Recovery Act

5.2 SUBTITLE D—NON-HAZARDOUS WASTE

Regulations under Subtitle D address non-hazardous waste, essentially banned open dumping of waste, and set minimum federal criteria for the operation of municipal and industrial waste landfills, including (USEPA 2021a):

- Design criteria
- Locations restrictions
- Financial assurance
- Corrective action or cleanup
- Closure requirements

States play a lead role in implementing these regulations and may set more stringent requirements but cannot set less stringent requirements. USEPA approves state programs, and if a state does not have an approved program, then the waste facilities must demonstrate that they meet the requirements.

RCRA defines solid waste in broad terms and includes solids, sludges, liquid, semi-solids, or contained gaseous material. In defining waste, USEPA focuses on the actual waste, not materials that were still part of the manufacturing process. USEPA intended to exclude recycling but also wanted to prevent fraudulent recycling efforts. Thus, USEPA uses the following five-factor test to determine the definition of a waste (USEPA 2021a):

1. Whether the material is typically discarded on an industrywide basis
2. Whether the material replaces raw material when it is recycled and the degree to which its composition is similar to that of the raw material
3. The relation of the recovery practice to the principal activity of the facility
4. If the material is handled prior to reclamation in a secure manner that minimizes loss and prevents releases to the environment
5. Other factors, such as the length of time the material is accumulated

Non-hazardous solid wastes include certain hazardous wastes that are exempted from the Subtitle C regulations, such as hazardous wastes from households and from conditionally exempt small quantity generators, which we will define later. Oil and gas exploration and production wastes, such as drilling cuttings, produced water, and drilling fluids, are categorized as "special wastes" and are also exempt from Subtitle C. Subtitle D also includes garbage (e.g., food containers, coffee grounds), non-recycled household appliances, residue from incinerated automobile tires, refuse such as metal scrap, construction materials, and sludge from industrial and municipal waste water facilities and drinking water treatment plants (USEPA 2021a).

5.3 SUBTITLE C—HAZARDOUS WASTE

Regulations under Subtitle C address hazardous waste and were enacted to ensure that hazardous waste is managed safely from the moment it is generated to its final disposal. This is termed "cradle-to-grave." Subtitle C regulations set criteria for hazardous waste generators; transporters; and treatment, storage, and disposal facilities.

118   Environmental Compliance Handbook

This also includes permitting requirements, enforcement, and corrective action or cleanup. An important definition central to RCRA and its enforcement is what defines a hazardous waste. From a universal perspective, a hazardous waste is defined as a waste that could pose a threat to human or public health or the environment if it were to be released. From this basic definition, the potential list of hazardous wastes would be large and diverse, and indeed, this is a true statement. Therefore, USEPA also has the following definition of hazardous waste (USEPA 2021a, 2021c):

> A solid waste, or combination of solid waste, which because of its quantity, concentration, or physical, chemical, or infectious characteristics may (a) cause, or significantly contribute to, an increase in mortality or an increase in serious irreversible, or incapacitating reversible, illness; or (b) pose a substantial present or potential hazard to human health or the environment when improperly treated, stored, transported, or disposed of, or otherwise managed.

This definition is broad as well. However, it does provide a general indication of which wastes USEPA intended to regulate as hazardous, but it obviously does not provide the clear scientific distinctions necessary for waste generators to determine whether their wastes pose a significant threat to warrant regulation or not in most cases. Therefore, USEPA developed more specific criteria for defining hazardous waste and decided on two definitions, a statutory definition and a regulatory definition. The statutory definition is presented previously and served as a general guideline in what would become the regulatory definition, which we discuss in much greater detail in the following. Hazardous wastes can be solids, liquids, and contained gases. They can be byproducts of manufacturing processes; discarded used materials; or discarded unused commercial products, such as residual cleaning fluids (solvents), cleaning products (bleach or ammonia), paints, pigments, pesticides, or a multitude of other items (USEPA 2021a). RCRA divides the regulatory definition of a hazardous waste into three major categories (USEPA 2021a):

1. Characteristic Waste
2. Listed Waste
3. Universal Waste

There are a couple of other categories of hazardous wastes or wastes that are sometimes managed as hazardous, and they include used oil and wastes that are considered hazardous because of what is called the "contained-in rule" or the "derived-from policy." In addition, there are what is called "land disposal restrictions" (LDRs) that must be met for hazardous wastes. We will discuss each in the following sections, along with hazardous waste in its own section.

5.3.1 Characteristic Hazardous Waste

Under RCRA (USEPA 2021a), a characteristic waste is material that is known or tested to exhibit one or more of the following four hazardous traits:

- Ignitibility
- Reactivity

Resource, Conservation, and Recovery Act

- Corrosivity
- Toxicity

Ignitibility characterizes wastes that can create fires under certain conditions, undergo spontaneous combustions, or have a flashpoint of less than 60°C (140°F). Examples include used oil and solvents. Test methods to determine whether a waste is ignitable and therefore a hazardous waste include the (1) Pensky-Martens closed-cup method, (2) the Staflash-closed-cup method, and (3) ignitibility of solids, which are each described in SW-846 Sections 1010, 1020, and 1030, respectively (USEPA 2005, 2021a).

Corrosivity describes materials, including solids, that are acids or bases or that produce acidic and alkaline solutions. Aqueous wastes with a pH of less than or equal to 2.0 or greater than or equal to 12.5 are corrosive and therefore hazardous wastes. A liquid may also be corrosive if it is able to corrode metal containers, such as storage tanks, drums, or barrels. Spent battery acid is an example. Test methods to determine whether a waste exhibits the characteristic of corrosivity are pH electronic measurement and corrosivity towards steel in SW-846 Methods 9040 and 1110, respectively (USEPA 2005, 2021a).

Reactivity describes wastes that are unstable under normal conditions. They can cause explosions or release toxic fumes, gases, or vapors when heated, compressed, or mixed with water. Examples include lithium-sulfur batteries and unused explosives. Wastes are evaluated for reactivity using criteria set forth in the hazardous waste regulations in SW-846 (USEPA 2021d).

Toxicity describes wastes that are harmful or fatal when ingested or absorbed (e.g., wastes containing mercury, lead, DDT, PCBs, etc.). When toxic wastes are disposed, the toxic constituents may leach from the waste and impact groundwater. The characteristic of toxicity contains eight subsections described in the following. A waste is a toxic hazardous waste if it is identified as being toxic by any of the eight subsections (USEPA 2021c):

1. TCLP. Toxicity is defined through application of a laboratory test procedure called toxicity characteristic leaching procedure (TCLP) test. This analytical method is described in USEPA SW-846 Method 1311. The TCLP identifies wastes (as hazardous) that may leach hazardous concentration of toxic substances into the environment. The result of the TCLP test is compared to the regulatory level (RL) (USEPA 2021e). Table 5.1 lists some common chemical compounds with their RLs.

2. Total and WET. Toxicity is defined through application of laboratory test procedures called the "total digestion" and the "waste extract test," commonly referred to as the WET. The results of each of these laboratory tests are compared to their soluble threshold limit concentrations (STLCs) (USEPA 2021e).

3. Acute oral toxicity. This refers to wastes that are toxic because the waste either is an acutely toxic substance or contains an acutely toxic substance if ingested. A waste identified as being toxic would have an acute oral LD50 less than 2,500 mg/kg or simply calculated LD50. We will define and explain LD50 when we discuss health risk assessments in a later chapter.

120 Environmental Compliance Handbook

TABLE 5.1
Select Maximum Concentrations of Contaminants for Toxicity Characteristic

Contaminant	Hazardous Waste Code	Regulatory Level (mg/l)	Contaminant	Hazardous Waste Code	Regulatory Level (mg/l)
Arsenic (As)	D004	5.0	Barium (Ba)	D005	100.0
Benzene	D018	0.5	Cadmium (Cd)	D006	1.0
Carbon Tetrachloride	D019	0.5	Chlordane	D020	0.03
Chlorobenzene	D021	100.0	Chloroform	D022	6.0
Chromium (Cr)	D007	5.0	o-Cresol	D024	200.0
m-Cresol	D024	200.0	p-Cresol	D025	200.0
Cresol (total)	D026	200.0	2,4-D	D016	10.0
1,4-Dichlorobenzene	D027	7.5	1,2-Dichloroethane	D028	0.5
1,-Dichloroethylene	D09	0.7	2,4-Dichlorotoluene	D030	0.13
Endrin	D012	0.02	Heptachlor	D031	0.008
Hexachlorobenzene	D032	0.13	Hexachlorobutadiene	D033	0.5
Hexachloroethane	Do34	3.0	Lead (Pb)	D008	5.0
Lindane	D013	0.4	Mercury (Hg)	D009	0.2
Methoxychlor	D014	10.0	Methyl ethyl ketone	D035	200.0
Nitrobenzene	D036	2.0	Pentachlorophenol	D037	100.0
Pyridine	D038	5.0	Selenium (Se)	D010	1.0
Silver (Ag)	D011	5.0	Tetrachloroethylene	D039	0.7
Toxaphene	D015	0.5	Trichloroethylene	D040	0.5
2,4,5-Trichlorophenol	D041	400.0	2,4,6-Trichlrophenol	D42	2.0
2,4,5-TP (Silvex)	D017	1.0	Vinyl chloride	D043	0.2

mg/l—milligram per liter
Source: United States Environmental Protection Agency. *Resource Conservation and Recovery Act Listed and Characteristic Hazardous Wastes.* Office of Solid Waste. Washington, DC. www.epa.gov/hazardus-wastes (accessed March 24, 2021), 2021e.

4. Acute dermal toxicity. This refers to wastes that are toxic because the waste either is an acutely toxic substance or contains an acutely toxic substance if touched. A waste identified as being toxic would have a dermal LC50 less than 4,300 mg/kg or simply calculated LC50. Again, we will define and explain LC50 when we discuss health risk assessments in a later chapter.

5. Acute inhalation toxicity. This refers to wastes that are toxic because the waste either is an acutely toxic substance or contains an acutely toxic substance if inhaled. A waste is identified as being toxic if it has an inhalation LC50 less than 10,000 mg/kg. Again, we will define and explain LC50 when we discuss health risk assessments in a later chapter. USEPA test method 3810 in SW-846, termed Headspace, is the recommended analytical test to evaluate inhalation toxicity.

6. Acute aquatic toxicity. This refers to wastes that are toxic because the waste is toxic to fish. A waste is aquatically toxic if it produces an LC50 less than

Resource, Conservation, and Recovery Act

500 mg/L when tested using the static acute bioassay procedures for hazardous waste samples (USEPA 2021e).

7. Carcinogenicity. This refers to wastes that are toxic because they contain one or more carcinogenic substances. A waste is identified as being toxic if it contains any of the specified carcinogens at a concentration of greater than or equal to 0.001% by weight.

8. Experience or testing. A waste may be toxic, and therefore a hazardous waste, even if it is not identified as toxic by any of the seven criteria listed previously. At the present time, only wastes containing ethylene glycol (e.g., spent antifreeze) have been identified as toxic in this subsection.

Let's look at a few examples of waste characterization.

Example 1. An auto repair shop uses mineral spirits as a part washer solvent. The solvent does not contain any halogenated or listed solvents, and its flashpoint is greater than 140°F. When the solvent becomes dirty, it is distilled. The distilled solvent is placed back into use. The residual solids left over from the distillation process are the waste that must be characterized. The waste is transferred to an appropriate container and labeled "Parts Washer Waste, Analysis Pending." A representative sample of the waste is collected and sent to a certified laboratory under chain of custody for waste characterization. The laboratory conducts the required tests, and a portion of the results is as follows:

- Lead   0.8 mg/L
- Cadmium   0.5 mg/L
- Chromium   8.0 mg/L
- Selenium   0.5 mg/L
- Silver   0.1 mg/L

Examination of the results shows that lead, cadmium, selenium, and silver do not exceed regulatory levels. However, chromium does exceed the regulatory level. Therefore, the waste would be classified as hazardous for chromium D007.

Example 2. Auto body shop. The exhaust filters in the paint spray booth become saturated with overspray from the painting operation. Since the body shop uses many different types of paint and primers, it's difficult to determine whether the filters are hazardous without laboratory testing. A representative filter is removed and sampled. The remaining filters are placed into a container and are labeled "Waste Filters Pending Analysis." The sample is sent to a certified laboratory under chain of custody for analysis. The laboratory conducts the required tests, and a portion of the results is as follows:

- Lead   9.1 mg/L
- Chromium   0.4 mg/L
- Barium   0.8 mg/L
- Methyl ethyl ketone 10 mg/L

Only lead exceeds the regulatory limit, and the filters are characterized as a hazardous waste. The container label is then immediately changed to a hazardous waste label.

122   Environmental Compliance Handbook

Example 3. General manufacturing facility. The facility receives large steel components, which the facility re-manufactures. The process requires that the components be dismantled, and the surfaces are prepared for re-finishing. The metal components are placed into a sand blasting machine and are cleaned under high-pressure blasting sand. After a few weeks of use, the blasting media (sand) is no longer effective and must be replaced. A representative sample of the sand is sampled, and the used sand from the operation is placed in a metal 55-gallon drum and labeled "Waste Sand from Sand Blasting—Analysis Pending." The sample collected for waste characterization is sent to a certified laboratory under chain of custody for analysis. The laboratory conducts the required tests, and a portion of the results is as follows:

- Arsenic   0.5 mg/L
- Cadmium   3.5 mg/L
- Chromium   25 mg/L
- Lead   40 mg/L

The results of the analysis indicate that the sand waste is indeed a hazardous waste for cadmium, chromium, and lead. The drum is immediately labeled with an appropriate hazardous waste label for D006, D007, and D008 hazardous waste.

5.3.2 LISTED HAZARDOUS WASTE

Under RCRA (USEPA 2021e), a hazardous waste can also be what is called a "listed waste." A listed waste is a waste that appears on one of four RCRA hazardous waste lists (USEPA 2021e):

F-Listed Wastes. F-Listed Wastes identify wastes from many common manufacturing and industrial processes, such as solvents that have been used for cleaning or degreasing. Since the processes producing these wastes occur in many different industry sectors, F-Listed Wastes are known as wastes from non-specific sources. Non-specific means that the waste does not originate from one specific industry or industrial or manufacturing process.

K-Listed Wastes. K-Listed Wastes include certain wastes from specific industries, such as petroleum refining or pesticide manufacturing. Also, certain sludges and wastewaters from treatment and production in these specific industries are examples of source-specific wastes.

P-Listed Wastes. P-Listed Wastes include specific commercial chemical products that have not been used but that will be (or have been) discarded. Industrial chemicals, pesticides, and pharmaceuticals are examples of commercial chemical products that appear on the P-List and become hazardous waste when discarded.

M-Listed Wastes. M-Listed Wastes include certain wastes known to contain mercury, such as fluorescent lamps, mercury switches, the products that house mercury switches, and mercury-containing novelties. These types of wastes may also be included as types of universal wastes, which we will discuss later in this chapter. Therefore, clarification from the applicable regulatory authority may be required.

Resource, Conservation, and Recovery Act

Now that a waste has been characterized as hazardous, it must be disposed of in a timely fashion. Typically, waste containers such as 55-gallon drums that are full and have been characterized as hazardous must be affixed with an appropriate label that identifies them as hazardous. Figure 5.1 is an example of an appropriate label. For some wastes, it may take a few weeks or even longer before enough waste is accumulated to fill a container. The maximum accumulation time allowed ranges between 90 and 270 days and is dependent upon (USEPA 2021e):

- How much waste is accumulated in any month
- The generator status of the facility (i.e., small quantity generator, large quantity generator, etc.)
- Whether the waste is acutely hazardous or extremely hazardous

FIGURE 5.1 Hazardous waste label (from United States Environmental Protection Agency. *Resource Conservation and Recovery Act Listed and Characteristic Hazardous Wastes.* Office of Solid Waste. Washington, DC. www.epa.gov/hazarodus-wastes (accessed March 24, 2021), 2021e.

124 Environmental Compliance Handbook

Typically, the "clock" begins on the first date on which any amount of hazardous waste begins to accumulate during that month (USEPA 2021e).

5.3.3 UNIVERSAL HAZARDOUS WASTE

A universal waste is a category of waste materials designated as hazardous waste but containing materials that are very common. Universal wastes are widely produced by households and businesses alike and typically include the following (USEPA 2021e):

- Televisions
- Computers
- Cell phones
- Other electronic devices such as VCRs, potable DVD players, and so on
- Fluorescent tubes and bulbs, high-intensity discharge lamps, sodium vapor lamps, and electric lamps that contain added mercury, as well as other lamps that exhibit a characteristic of a hazardous waste (e.g., typically lead).
- Pesticides
- Mercury-containing equipment, including thermostats; switches; mercury thermometers; pressure and vacuum gauges; dilators and weighted tubing; mercury gas flow regulators; dental amalgams; counterweights; dampers; and mercury-added novelties such as jewelry, ornaments, and footwear. Of special note for clarification is that mercury-containing equipment may also considered a listed M-Waste. In these cases, clarification from the applicable regulatory authority may be required
- Cathode ray tubes (CRTs), which are typically the glass picture tubes removed from devices such as televisions and computer monitors
- Non-empty aerosol cans

5.3.4 USED OIL

Some states, such as California, regulate used oil as a hazardous waste even if it does not exhibit any characteristics of a hazardous waste. The term **used oil** is a defined term meaning any oil that has been refined from crude oil or any synthetic oil that has been used and, as a result of use, is contaminated with physical or chemical impurities. Other materials that contain or are contaminated with used oil are also commonly considered a hazardous waste as well and are subject to used oil regulations under RCRA 40 CFR Part 279.22 (USEPA 2021f). In general, used oil must be kept in approved containers with secondary containment that does not leak and must be properly labeled as "Used Oil" (USEPA 2021f).

5.3.5 CONTAINED-IN POLICY AND DERIVED-FROM RULE

Materials can be hazardous even if they are not specifically listed or don't exhibit any characteristics of a hazardous waste. This is often confusing, so let's try to make sense of a couple of hazardous waste concepts to understand if a waste is classified

Resource, Conservation, and Recovery Act

as a hazardous waste in a step-by-step approach. For example, let's discuss what is considered "the contained-in rule" (USEPA 2021a).

The contained-in rule or contained-in policy under RCRA gives hazardous waste the power to transform solids that are otherwise not hazardous into a hazardous waste. USEPA separates the contained-in rule from the mixture or derived-from rule. This power to change a solid waste into a hazardous waste stems from USEPA's general policy that "once a hazardous waste, always a hazardous waste." Once a hazardous waste is hazardous, it is presumed forever hazardous regardless of changes in its form or its combination with other substances. This rule provides that any mixture of a listed hazardous waste and a solid waste is itself, in totality, a hazardous waste regardless of the concentration of the hazardous constituents in the waste, even if it no longer exhibits any toxicity (USEPA 2021a).

The derived-from rule differs slightly in concept from the contained-in rule, but the outcome is the same. The way to best explain the derived-from rule is through an example. Let's say an ash is created through incineration of a listed hazardous waste. Since the ash is created from a listed hazardous waste, the ash itself is considered a listed hazardous waste. Thus, the ash produced by burning the listed hazardous waste bears the same hazardous waste code and regulatory status as the original listed waste, regardless of the ash's actual physical and chemical properties (USEPA 2021a).

USEPA's basic position on termination of a material's status as a listed hazardous waste is simple. Once a listed hazardous waste has been classified as a hazardous waste, it generally remains a hazardous waste forever. There are two explicit bases for termination of this status. First, an unlisted waste that is hazardous only because of a characteristic ceases to be a hazardous waste if it no longer exhibits a hazardous waste characteristic. Second, a listed waste, which is a waste containing a listed waste or a waste derived from a listed waste, ceases to be a hazardous waste only if it has been "delisted" from classification. Although a characteristic waste can lose its status as a hazardous waste without action by USEPA, in most cases, a listed waste will remain hazardous until USEPA affirmatively grants a petition to reclassify the material as non-hazardous (USEPA 2021a). Some states, such as California, have a process by which a listed hazardous waste that impacts a media such as soil can be delisted if it is demonstrated that the listed waste is present in insignificant concentrations based on a risk-based evaluation (California Department of Toxic Substance and Control 2021).

By adopting the mixture and derived-from rules, USEPA established incentives for generators to keep waste streams separate and to limit the amount of hazardous waste generated, since the costs for treatment and disposal are much higher for a hazardous waste compared to a non-hazardous waste (2021a).

5.3.6 Land Disposal Restrictions

The 1984 amendments to RCRA called the Hazardous and Solid Waste Amendments (HSWA) required USEPA to established treatment standards termed **land disposal restrictions** for all listed and characteristic hazardous wastes destined for land disposal. For wastes that are restricted, the amendments to RCRA required USEPA to set concentration levels or methods of treatment, both of which are called treatment standards, that substantially diminish the toxicity of the wastes or reduce the likelihood that hazardous constituents contained within the wastes will migrate from the disposal site and potentially impact human health or the environment in an adverse way (USEPA 1991). The major reason for established LDRs was for the prevention of groundwater impacts, usually through the leaching of contaminants from hazardous wastes from landfills and potentially impacting potable water supplies.

LDR provisions focus on eliminating or greatly reducing the probability of impacting groundwater through proper treatment of hazardous or toxic constituents in hazardous waste before land disposal. The processes may include (USEPA 1991):

- Incineration, which destroys organic hazardous compounds.
- Stabilization or immobilization, which binds toxic metals to the mass itself, making it less likely to migrate. An example would be mixing the toxic material with concrete, mortar, a bentonite clay, or other material to render the waste immobile.

There are three main sections of the LDR program (USEPA 1991):

1. Disposal Prohibition. This requires waste-specific treatment standards be met before disposal of the waste on land. A facility can meet such standards by either:
   a. Treating the hazardous chemicals in the waste to meet appropriate levels. Many such methods are approved by USEPA for use, excluding dilution.
   b. Treating hazardous waste with a USEPA-approved technology. Once the waste is appropriately treated, it can then be land disposed.
2. Dilution Prohibition. Ensures that wastes are treated properly and that diluting the waste to meet LDRs is not permitted. This is because of the assumption that dilution does not reduce the overall toxicity of a hazardous waste.
3. Store Prohibition. Hazardous waste cannot be stored indefinitely.

From the moment a hazardous waste is generated, it is subject to LDRs. If a hazardous waste generator produces more than 220 pounds (or 2.2 pounds of acute hazardous waste) of hazardous waste in a calendar month, it must identify the type and nature of the waste and also determine the course of applicable treatment of the waste before land disposal. Generators that produce less than these amounts and are conditionally exempt small quantity generators (CESQGs) of hazardous waste are exempt from LDR requirements, as are waste pesticides and container residues disposed of by farmers on their own land and newly listed wastes for which USEPA has not yet established treatment standards (USEPA 2021e; Ohio EPA 2014).

5.3.7 ANALYTICAL METHODS

Much of what defines a hazardous waste is dependent upon what chemical compounds are contained within the waste. Therefore, for consistency, USEPA developed a manual describing acceptable analytical and sampling methods and quality control for ensuring that wastes are properly characterized. The document USEPA published

Resource, Conservation, and Recovery Act

was entitled *Test Methods for Evaluating Solid Waste SW-846*, or simply SW-846. It was issued in 1980 and has changed through time as technology has advanced in analytical procedures. Currently the SW-846 manual is over 3,500 pages (USEPA 2021d). SW-846 has become one of the most important documents not only for assessing waste but also in conducting investigations at potential hazardous waste sites and environmental investigations in general.

5.3.8 HAZARDOUS WASTE MANIFESTS

RCRA requires that generators of hazardous waste and owners or operators of hazardous waste treatment, storage, or disposal facilities use what is called the Uniform Hazardous Waste Manifest (USEPA Form 8700–22) and, if necessary, the continuation sheet (EPA Form 8700–22A) for both interstate and intrastate transportation. General instructions for completing a manifest are provided at www.epa.gov/hwgenerators/uniform-hazardous-waste-manifest-instructions-sample-form-and-continuation-sheet (USEPA 2021g).

The return receipt of the manifest from the disposal facility must be returned to the generator within 45 days from when the waste was signed and shipped from the generator facility (USEPA 2021a, 2021f).

5.4 SUBTITLE I-PETROLEUM UNDERGROUND STORAGE TANKS

The operation of underground storage tanks (USTs) became subject to RCRA with the enactment of the Hazardous and Solid Waste Amendments of 1984, which amended RCRA by inserting Subtitle I. It was estimated at the time that there were approximately 2 million USTs in the United States. Currently, the estimate is 560,000. The UST regulations under Subtitle I cover tanks storing petroleum or a listed hazardous waste and define the types of tanks permitted. The UST requirements under Subtitle I set standards for (USEPA 2021a):

- Groundwater monitoring
- Double-walled tanks
- Release detection, prevention, and correction
- Spill control
- Overfill control

Up until the middle portion of the 1980s, USTs were largely unregulated, except for those that contained hazardous wastes. It was not until the catastrophic failure of USTs and large-scale groundwater contamination occurred in Provincetown, Massachusetts; Northglenn, Colorado; and Dover-Walpole, Massachusetts, that Congress recognized the need for UST regulation. Subtitle I imposed new obligations and potential liabilities on a large and diverse number of entities and businesses, including (USEPA 2021a):

- Service stations
- Petroleum bulk plants
- Petroleum bulk tanks
- Municipal warehouses

- Shopping centers
- Farms
- Automobile dealerships
- School bus yards
- Factories
- Delivery services

The UST regulations apply to owners and operators of regulated UST systems. An **operator** is defined as any person in control of, or having responsibility for, the daily operation of an UST system. The definition includes current operators only and implies that one must actively manage a tank system to be considered an operator (USEPA 2021a).

An **owner** is defined as any person who owns an UST system, unless that system was taken out of service prior to November 8, 1984. For UST systems no longer in use at that date, the owner is the person who owned the system immediately prior to is discontinuation. A person is defined broadly to include an individual, trust, firm, joint stock company, federal agency, corporation, state, municipality, commission, state political subdivision, any interstate body, or the government of the United States (USEPA 2021a).

The UST regulations impose joint and several liability on UST owners and operators and do not place primary responsibility on one over the other (USEPA 2021a).

When USTs are currently being operated, it is not difficult to identify at least one person responsible for compliance with the UST regulations. By definition, there is no operator of an abandoned or out-of-service UST system. If a UST was in use after November 8, 1984, and the UST remains in the ground, the current property owner may fall within the statutory definition as an owner. If the UST was abandoned before November 8, 1984, by a previous property owner, it may be impossible to locate records identifying the owner of the UST immediately before discontinuation of use because UST registration was not historically required and UST closures were not reported because that was not required (USEPA 2021a).

In amending Subtitle I of RCRA, USEPA required every owner of an UST system to notify the designated state or local agency of the existence of the UST system, age, size, type, locations, product stored, and uses for each UST. Initial notification forms were due by May 8, 1986. In addition to the reporting requirement for operating USTs, owners of USTs taken out of service after January 1974 but still in the ground were required to notify a designated state agency of the existence of the UST by May 8, 1986, as well. This notice also required an UST owner to specify, to the extent known, the date that the UST was taken out of service, the age of the UST when it was taken out of service, size, type, product stored, location, and the type and quantity of substances that remained stored in the UST (USEPA 2021a).

USTs brought into service after December 22, 1988, must meet design and construction requirements. RCRA specified that owners or operators notify the designated state agency within 30 days of the existence of the new UST. The notification form for new USTs requires the owners or operators to certify compliance to the following requirements (USEPA 2021a):

1. That the UST system was properly installed by a state-certified UST contractor
2. That cathodic protection was installed for steel tanks and associated piping

Resource, Conservation, and Recovery Act 129

3. That the financial responsibility requirement was met
4. That release detection was installed and meets regulatory standards
5. That spill and overfill prevention measures were installed and meet regulatory standards

5.5 SUMMARY OF THE RESOURCE CONSERVATION AND RECOVERY ACT

RCRA and the CAA, SWA, and SDWA form the basic fundamentals of environmental protection in the United States, with the exception of cleaning up legacy sites, which we will address next. RCRA was essentially created to deal with huge increases in the amount of wastes that were being created throughout the United States as a direct result of our industrialization and population increase.

The success of RCRA can't be understated. RCRA has changed our behavior in many ways and has involved everyone participating in taking care of our environment, not just industry groups, for almost all of us now recycle. Much of our discarded materials that were once considered garbage is now transformed into new products that we use, including used tires, wood and pulp products, plastic, glass, and especially scrap metal, which is now considered a commodity and has grown into an industry of itself.

Looking back, there are some challenges that RCRA has failed to address proactively, and one of the most significant is municipal or household wastes. We have discussed at length what industry faces in characterizing and disposing of wastes properly in order to protect human health and the environment, but have not addressed with the same vigor our everyday household garbage, which is generally exempted from RCRA but contains many of the same components and chemicals as many hazardous wastes and is not regulated.

As described in previous chapters, there are many common household products that typically become hazardous wastes if they were generated by industry as opposed to households. To many scientists, it does not make sense at all when a chemical which is known to be harmful to human health and the environment and is a hazardous waste but is not fully regulated as such. For household wastes that would be considered hazardous, the United States relies on the honor system. The Earth does not care how or by whom the material is released into the environment; it's toxic to the Earth just the same.

Next, we will discuss medical waste and low-level radioactive waste before moving on to the next chapter, which is dedicated to the Comprehensive Environmental Response Compensation and Liability Act, which focuses on addressing cleaning up contaminated sites.

5.6 MEDICAL WASTE

Medical waste is a subset of waste primarily generated at health care facilities, such as hospitals, physicians' offices and clinics, blood banks, veterinary hospitals and clinics, medical research centers, and laboratories. **Medical waste** is defined as health care waste that may contain or is contaminated with blood, body fluids,

130 Environmental Compliance Handbook

or other potentially infectious materials (USEPA 2021h). Concern surrounding the potential health hazards of medical waste grew suddenly after medical wastes washed up along beaches and the coastline of the eastern United States in the 1980s (USEPA 2021h).

USEPA was the agency responsible for regulating medical waste from 1988 when medical waste first became regulated under the Medical Waste Tracking Act of 1988. However, the Act expired in 1991. Currently, medical waste is regulated by each state.

Historically, treatment and disposal of medicals wastes typically involved incineration. However, because of concerns over lowered air quality through the incineration of medical wastes, continued incineration has been discouraged (USEPA 2021h). Alternatives for treatment and disposal of medical wastes include (USEPA 2021h):

- Thermal treatment, such as microwaves
- Steam sterilization, such as an autoclave
- Electropyrolysis
- Chemical systems

In general, each state requires that medical waste be carefully collected in appropriately labeled containers, be treated, and then be disposed of under a manifest as a regulated medical waste (USEPA 2021h). There are several different types of containers, depending on the type of medical waste.

5.7 LOW-LEVEL RADIOACTIVE WASTE

Low-level radioactive wastes are often associated with medical facilities such as x-ray and other machines at medical and dental facilities. Low-level radioactive wastes are regulated under the Low-Level Radioactive Waste Policy Act (LLRWPA) of 1980, as amended in 1985. Under the act, each state is responsible for management and disposal requirements of low-level radioactive wastes (USEPA 2021i). The LLRWPA does not regulate high-level radioactive wastes, which is the responsibility of the Department of Energy. Typically, high-level radioactive wastes are associated with nuclear power plants and those generated by or associated with the United States Department of Defense (Nuclear Regulatory Commission [NRC] 2021).

Each facility planning on installing a low-level radioactive source must obtain a license from the NRC or state. The facility must be designed, constructed, and operated to meet safety standards, depending on the source (NRC 2021). Disposal of low-level radioactive sources is strictly regulated. Disposal options are limited, and licensed transporters are generally few. Therefore, advance planning is key. Manifests are also complex and comprehensive. An example manifest is available for review at https://nrc.gov/cdn/legacy/waste/llw-disposal.ml20178a433-nureg-br-0204-rev3-form-541-example-filled.jpg (3542×2729) (nrc.gov).

5.8 REFERENCES

California Department of Toxic Substance and Control (DTSC). 2021. Hazardous Waste. www.dtsc.ccelern.csus.edu/wasteclass. (Accessed March 24, 2021).

Resource, Conservation, and Recovery Act

Cornell University Law. 2021. Resource Conservation and Recovery Act (RCRA) Overview. www.law.cornell.edu/wex/rcra. (Accessed March 24, 2021).

Nuclear Regulatory Commission (NRC). 2021. Low-Level Waste Disposal. Low-Level Waste Disposal | NRC.gov. (Accessed November 9, 2021).

Ohio Environmental Protection Agency (Ohio EPA). 2014. *Land Disposal Restrictions (An Overview)*. Division of Materials and Waste Management. Ohio EPA, Columbus, Ohio. 3 pages.

United States Environmental Protection Agency (USEPA). 1991. *Land Disposal Restrictions: Summary of Requirements*. USEPA Office of Solid Waste and Emergency Response. Washington, DC. 86 pages.

United States Environmental Protection Agency (USEPA). 2005. *Introduction to Hazardous Waste Identification (40CFR Parts 261)*. EPA530-K-05–012. USEPA Office of Solid Waste and Emergency Response (5305 W). Washington, DC. 30 pages.

United States Environmental Protection Agency. 2021a. Summary of the Resource Conservation and Recovery Act. www.epa.gov/rcra/summary. (Accessed March 24, 2021).

United States Environmental Protection Agency (USEPA). 2021b. Criteria for the Definition of Solid Waste and Solid and Hazardous Waste Exclusions. https://.www.epa.gov/criteria-definition-solid-waste. (Accessed March 24, 2021).

United States Environmental Protection Agency (USEPA). 2021c. Defining Hazardous Waste. www.epa.gov/hw/defining-hazardous-waste-listed-characteristic-and-mixed-radiological-wastes. (Accessed March 24, 2021).

United States Environmental Protection Agency (USEPA). 2021d. *USEPA Office of Solid Waste. SW-846 Test Methods Manual*. USEPA. Washington, DC. 1,357 pages www. USEPA.gov/sw-846 (Accessed March 24, 2021).

United States Environmental Protection Agency (USEPA). 2021e. *Resource Conservation and Recovery Act Listed and Characteristic Hazardous Wastes*. Office of Solid Waste. Washington, DC. www.epa.gov/hazarodus-wastes. (Accessed March 24, 2021).

United States Environmental Protection Agency (USEPA). 2021f. *Used Oil Regulations Under the Resource Conservation and Recovery Act*. Office of Solid Waste. Washington, DC. www.epa.gov.used-oil. (Accessed March 24, 2021).

United States Environmental Protection Agency (USEPA). 2021g. Hazardous Waste Manifests. www.epa.gov/hwgenerators/uniform-hazardous-waste-manifest-instructions-sample-form-and-continuation-sheet. (Accessed March 24, 2021).

United States Environmental Protection Agency (USEPA). 2021h. Medical Waste. www.epa. gov/medicalwaste. (Accessed November 9, 2021).

United States Environmental Protection Agency (USEPA). 2021i. Radiation Regulations and Laws. www.epa.gov/radiationregulationsandlaws. (Accessed November 9, 2021).

6 Comprehensive Environmental Response, Compensation, and Liability Act

6.1 INTRODUCTION

The Comprehensive Environmental Response, Compensation, and Liability Act of 1980 is commonly referred to as Superfund or Polluters Pay Law. CERCLA authorized USEPA to recover natural resource damages caused by hazardous substances released into the environment. Created out of CERCLA was the Toxic Substance and Disease Registry (ATSDR) (USEPA 2021a).

In simple terms, CERCLA deals with our past because its main focus is to investigate and clean up legacy sites.

6.2 POTENTIAL RESPONSIBLE PARTIES

USEPA may identify responsible parties for contamination from releases of hazardous substances to the environment (polluters) and either compel them to investigate and clean up sites, or it may undertake the investigation and cleanup itself using the Superfund (a trust fund) and recover all response costs and even more in some instances from polluters by referring to the Department of Justice for cost recovery proceedings. As of 2016, there were 1,328 sites listed, 391 sites delisted, and 55 new sites proposed for a total of 1,774 Superfund sites. Figure 6.1 is a map showing the distribution of CERCLA sites in the United States (USEPA 2021a).[

Approximately 70% of CERCLA investigation and cleanup activities have been paid for by responsible parties (PRPs). The exceptions occur when the responsible party cannot be found or is unable to pay for the cleanup. The majority of the funding came from a tax on the petroleum and chemical industries, but since 2001, most of the funding now comes from taxpayers (USEPA 2021a).

USEPA's enforcement program under CERCLA is based on the **"polluters pay"** principle, which stipulates that the party responsible for the pollution pays for cleaning up the pollution. Over the past 35 years, USEPA has recovered over $35 billion from potential responsible parties for cleanup at CERCLA sites. This principle is one of many highlights of USEPA environmental protection that has been copied worldwide (USEPA 2021c).

DOI: 10.1201/9781003150107-6

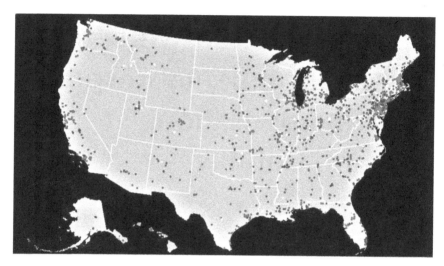

FIGURE 6.1 CERCLA or Superfund sites in the United States (from United States Environmental Protection Agency. *CERCLA Sites in the United States.* www.epa.gov/CERCLA-sites [accessed March 24, 2021], 2021b.)

6.3 CERCLA HISTORY

CERCLA was enacted in 1980 in response to threats from sites of environmental contamination, including three sites in particular: Times Beach in Missouri, Love Canal in New York, and the Valley of the Drums in Kentucky. As part of CERCLA, USEPA established a hazardous ranking system (HRS) to prioritize each Superfund site related to its threat to human health and the environment. In 1982, USEPA published its first list of Superfund sites and titled it the National Priority List (NPL) (USEPA 2021a).

Other prominent sites that have not only shaped Superfund but made significant contributions to environmental science include:

- Toms Rivers and the Reich Farm Superfund site in New Jersey
- Wells G and H Superfund site in Woburn, Massachusetts
- PG&E Hinckley Site in California

A brief discussion of each of these sites is given in the following sections.

6.3.1 TIMES BEACH

Times Beach is now a ghost town located approximately 17 miles southwest of St. Louis, Missouri. Times Beach was a town of more than 2,000 people before all of them were evacuated in 1983 because of dioxin contamination. During the 1960s, a pharmaceutical and chemical company was located in Times Beach and produced some thick oily wastes that contained dioxin. Approximately 18,500

CERCLA 135

gallons of the dioxin-containing oily waste ended up with a waste oil contractor who mixed the material with other sources of waste oil and sprayed the material at three local horse arenas and 23 miles of roads in Times Beach as a dust suppressant (USEPA 2021d).

The Centers for Disease Control (CDC) initially became involved in August 1971 when local citizens reported the deaths of numerous horses that used one or more of the horse arenas sprayed with the dust suppressant. The CDC discovered an old tank containing approximately 4,300 gallons of oily waste that contained dioxin at a concentration of 340 ppm. In 1974, the CDC collected soil samples from the area and detected trichlorophenol, PCBs, and dioxin. USEPA did not become heavily involved until 1979, when it discovered 90 drums of oily waste at the site with dioxin concentrations as high as 2,000 ppm. In May and June 1982, USEPA collected soil samples from the area and detected dioxin as high as 0.3 ppm in soil. The CDC-recommended cleanup level for dioxin at the time was as low as 0.001 ppm. On December 23, 1982, the CDC publicly recommended that Times Beach not be inhabited. Subsequently, 800 residential properties and 30 businesses were abandoned. In 1997, the cleanup was completed at a cost of nearly $200 million (Sun 1983; USEPA 2021d).

6.3.2 Love Canal

Love Canal is a neighborhood within Niagara Falls, New York. The site that became the epicenter of contamination was a 29-acre parcel that was used as a landfill. Originally planned as a canal, the site was abandoned at the turn of the 20th century and was used as a dump by the city of Niagara Falls. In 1942, the Niagara Power and Development Company granted permission to Hooker Chemical to dump wastes in the canal. In 1948, the City of Niagara ended disposal of refuse into the canal, and the Hooker Chemical Company became the sole user and owner of the site. Hooker Chemical ceased dumping in 1952 and capped the dump with clay. During its lifespan of 5 years, an estimated total of 21,800 tons of chemicals, mostly consisting of sludges, caustics, alkalines, chlorinated hydrocarbons, and volatile organic hydrocarbons, were disposed in the landfill, many in 55-gallon drums (Beck 1979; USEPA 2021e).

By the early 1950s, the City of Niagara Falls was in the midst of a population increase and purchased the site from Hooker Chemical to build a school. In 1955, 99th Street School opened with 400 children enrolled. The area surrounding the dump continued to develop in the 1950s, with a total of 800 private residences built in the surrounding area. Soon after development, residents began to complain of odors and black substances in their yards and public playgrounds (USEPA 2021e).

In 1977, the New York State Department of Health and Environmental Conservation conducted a large scientific study of the air, groundwater, and soil in the area and detected a number of organic compounds in the basements of several residences adjacent to the former dump. Compounds detected included benzene, chlorinated hydrocarbons, PCBs, and dioxin (Beck 1979; USEPA 2021f). By 1978, Love Canal had become a national media event, and then on August 7, 1978, then-President Jimmy Carter announced that Love Canal was a federal health emergency. In total, 950 families were evacuated. This was the first time that emergency funds were used for a situation that was not a natural disaster. When CERCLA was passed, Love Canal was placed first on the list of Superfund sites. USEPA finished the cleanup of Love Canal in 2004 at an estimated cost of $400 million (USEPA 2004, 2021e).

6.3.3 Valley of the Drums

The Valley of the Drums site is a 23-acre site in northern Kentucky, near Louisville. The site became a collection site for wastes dating back to the 1960s. It initially caught the attention of local governmental officials when some drums that were onsite caught fire and took firefighters more than a week to extinguish. However, no action was taken at the time since no laws existed to take any further action. In 1978, a Kentucky Department of Natural Resources and Environmental Protection (KDNREP) investigation of the property estimated that over 100,000 drums of waste were located onsite, an estimated 27,000 drums were buried, and the remaining containers were discharged into onsite pits and trenches (USEPA 2021g). Figure 6.2 shows some of the drums discovered.

The KDNREP investigation revealed that many of the drums onsite were deteriorated and their contents spilled onto the ground and washed into a nearby creek. Analysis of collected samples detected the presence of several heavy metals, PCBs,

FIGURE 6.2 Photograph of Valley of the Drums Superfund site, circa 1979 (from United States Environmental Protection Agency. *Superfund Site Profile: Valley of Drums, Kentucky.* www.epa.gov/valley-of-drums-profile [accessed March 24, 2021], 2021g).

CERCLA

and 140 other chemical substances. USEPA initiated an emergency cleanup in 1979 but quickly realized that the magnitude of the project was beyond what it was prepared for, and this was used by members of Congress as an additional reason CERCLA was needed (USEPA 2021g).

The Valley of the Drums site was placed on the Superfund list in 1983 when a large-scale cleanup of the site began, which ended in 1990. In 1996, the site was removed from the Superfund list. However, more buried drums were discovered during an inspection of the site in 2008 outside the cleanup area. Currently, USEPA is monitoring the site and has authorized an additional $400,000 in funds to remove and properly dispose of more of the discovered buried drums (USEPA 2021g).

6.3.4 WELLS G AND H SUPERFUND SITE IN WOBURN, MASSACHUSETTS

Groundwater extraction Wells G and H were developed by the City of Woburn, Massachusetts, in 1964 and 1967, respectively. The wells were capable of pumping 2 million gallons of water per day and were initially intended to supplement an existing water system that supplied water to the residents of Woburn. In 1979, the Massachusetts Department of Environmental Protection (MDEP) conducted analytical testing of water from the two wells and discovered the presence of several chlorinated volatile organic compounds in the samples that included (USEPA 1989a):

- 1,1,1-trichloroethane (TCA)
- Trans-1,2-dichloroethene (trans DCE)
- Tetrachloroethene (PCE)
- Trichloroethene (TCE)

Concentrations ranged from 1 to 400 ppb. MDEP immediately shut the wells down. In 1981, USEPA conducted a hydrogeologic investigation of a 10-square-mile area surrounding the wells to evaluate the source of the contamination and define the extent of groundwater contamination. Five facilities were identified as sources of contamination, and the site was placed on the Superfund list in December 1982 (USEPA 1989a). Remedial response actions were first initiated by USEPA in 1983 and are still being conducted more than 35 years later. USEPA has been maintaining close community relations since 1986. Community involvement has been significant and was even the subject of a movie titled *A Civil Action*.

6.3.5 CIBA-GEIGY AND REICH FARM SITE IN TOMS RIVER, NEW JERSEY

The Ciba-Geigy Superfund site is located in Toms River, New Jersey. The site is approximately 1,400 acres in size, of which 320 acres were developed and were formerly used to manufacture synthetic organic pigments. From 1952 to 1966, the facility discharged treated effluent into Toms River, which then discharged directly into the Atlantic Ocean. Beginning in 1966, the facility discharged effluent directly into the Atlantic Ocean through a pipeline. In addition, the facility disposed of solid wastes onsite at several locations (USEPA 1989b).

138 Environmental Compliance Handbook

During the 1970s and early 1980s, the facility conducted several investigations at the direction of the New Jersey Department of Environmental Protection (NJDEP). USEPA eventually listed the site as a Superfund site in 1982. In 1984, NJDEP discovered that the facility was illegally disposing of drums containing hazardous waste onsite, which led to a criminal investigation. As a result of the criminal investigation, several company officials were charged with illegal disposal of chemical waste, filing false documents, making false and misleading statements, and illegal disposal of hazardous waste (1989b).

In 1985, USEPA began conducting a remedial investigation of the site to fully define the extent of contamination in soil and groundwater both onsite and offsite. The investigation was completed in 1988 (USEPA 1989b). Remedial activities included excavation of drums, contaminated soil, and groundwater treatment that continues to this day more than 30 years later.

The Reich Farm Superfund Site in Toms River, New Jersey, is in close proximity to the Ciba-Geigy Superfund Site and was designated a Superfund site in 1983. According to USEPA (1988), a 3-acre portion of the farm was leased for temporary storage of used 55-gallon drums in 1971. However, it was subsequently discovered that wastes from the drums may have been dumped into trenches and that this activity contaminated groundwater (USEPA 1988).

Compounds detect in groundwater included (USEPA 1988):

- Styrene-acrylonitrile trimer
- Tetrachloroethene (PCE)
- Trichloroethene (TCE)
- 1,1,1-trichloroethane (TCA)
- Chlorobenzene
- 1,2-dichloroethene (DCE)
- Other volatile organic compounds

These contaminants listed and many others not listed contaminated groundwater, which residents of Toms River used as a source of drinking water. This led to a significant toxicological risk evaluation that is still ongoing (USEPA 2021h).

6.3.6 PG&E Hinkley Site in California

The PG&E site was located 2 miles from the town of Hinkley, California, and 12 miles west of Barstow, California. Between 1952 and 1966, PG&E used hexavalent chromium as corrosion protection in its cooling tower water. The wastewater from this process was discharged to unlined ponds onsite, which subsequently contaminated groundwater. The resulting groundwater contaminant plume of hexavalent chromium migrated from the site, was estimated to be 8 miles long and 2 miles wide, and impacted drinking water in Hinkley (California Regional Water Quality Control Board 2021). In 1993, citizens of Hinkley filed suit against PG&E and settled in 1996 for $333 million (San Francisco Chronicle 1996). This site was also the subject of a movie titled *Erin Brockovich*. Remediation at the site is ongoing.

CERCLA 139

Although this site is not actually a federal Superfund site, it does follow a similar profile as the five other sites briefly described previously. Similarities include a significant toxicological risk study, large groundwater contaminant plume, and high remedial cost.

6.3.7 Summary of CERCLA History

The sites highlighted previously have often been referred to as the sites that influenced the formation and passage of CERCLA or have demonstrated a lack of knowledge of how contaminants behave in the subsurface geological environment. In addition, four of the sites involve agriculture or agricultural land. These examples have provided scientific information that has highlighted:

- The need for groundwater protection zones
- The need, development, and procedures for conducting toxicological risk evaluations
- The development of groundwater modeling
- The difficulty in remediating contaminated groundwater
- The enormous cost of remediating contaminated groundwater
- The toxicity, mobility, and persistence of chlorinated VOCs
- The toxicity, mobility, and persistence of hexavalent chromium

6.4 SITE SCORING

USEPA developed the Hazard Ranking System to score potential sites of environmental contamination. A score of 28.5 was usually high enough to be placed on the NPL or Superfund list. The HRS is a numerical process that assigns integers to factors that relate to risk based on site conditions. The factors are grouped into three categories (USEPA 2021a):

1. Likelihood that a site has released or has the potential to release hazardous substances into the environment
2. Characteristics of the waste, including toxicity, quantity, and so on
3. Potential human exposure pathways

There are four pathways for the potential human exposure category (USEPA 2021a):

1. Surface water migration (i.e., drinking water sources, sensitive habitat such as a wetland, human food chain, etc.)
2. Soil exposure (i.e., dermal contact, dust, etc.)
3. Groundwater migration (i.e., drinking water)
4. Air migration

After scores are calculated for one or all the pathways, they are combined using a root-mean-square equation to determine overall site score. It should be noted that some sites score high for just one pathway if the potential exposure is severe enough

140 Environmental Compliance Handbook

and the toxicity of the contaminant or contaminants of concern are also high. The identification of a site on the NPL is intended primarily to guide USEPA in the following (USEPA 2021a):

- Determining which sites warrant further investigation to assess the nature and extent of the risks to human health and the environment
- Identifying what CERCLA-financed remedial actions may be appropriate
- Notifying the public of site which USEPA believes warrant further investigation
- Notifying PRPs that USEPA may initiate a CERCLA-financed remedial action

A potential responsible party is a possible polluter who may eventually be liable under CERCLA for contamination or misuse of a particular property or resource. Four classes of PRPs may be liable for contamination at a Superfund site (USEPA 2021a):

1. The current owner or operator of the site
2. The owner or operator of a site at the time that disposal of a hazardous substance, pollutant, or contaminant occurred
3. A person who arranged for the disposal of a hazardous substance, pollutant, or contaminant at a site
4. A person who transported a hazardous substance, pollutant, or contaminant to a site, who also selected that site for the disposal of the hazardous substances, pollutants, or contaminants

Inclusion of a site on the Superfund list does not itself require PRPs to initiate action to clean up a site, nor does it assign liability to any one entity or person. The Superfund list serves primarily informational purposes, notifying the government and the public of those sites or releases that appear to warrant a remedial action. Under Superfund, the key difference in the authority to address hazardous substances and pollutants or contaminants is that the cleanup of pollutants or contaminants, which are not hazardous substances, cannot be compelled by unilateral administrative order (USEPA 2021a).

An important provision of CERCLA is that if a PRP spends money to clean up a CERCLA site, that party may sue other PRPs in a contribution action. CERCLA liability has generally been judicially established as joint and several among PRPs to the government for cleanup costs in that each PRP is hypothetically responsible for all costs, but CERCLA liability is allocable among PRPs in contribution based on comparative fault. An "orphan share" is the share of costs at a Superfund site that is attributed to a PRP that is unidentifiable or insolvent. USEPA, as a matter of long-standing policy, attempts to treat all PRPs equitably and fairly (USEPA 2021a).

6.5 REMEDIATION CRITERIA

Most states have passed their own versions of CERCLA so that sites that do not qualify to become Superfund sites but have had a release of hazardous substances

CERCLA 141

above cleanup levels get addressed. Cleanup levels established at the state level vary from state to state depending on various factors. However, much is based on USEPA exposure scenarios and toxicological data. States have established cleanup levels for different types of land use, typically residential and non-residential (industrial or commercial), and by media or exposure pathways, including:

- Groundwater or drinking water
- Soil
- Groundwater–surface water interface
- Inhalation to indoor air
- Inhalation to outdoor or ambient air

Table 6.1 lists the generic residential cleanup levels for select chemical compounds for the state of Michigan (MDEQ 2016). The exposure pathways listed in the table are residential direct contact, commercial-industrial direct contact, and construction worker direct contact. The complete list can be viewed at www.michigan.gov/documents/deq/deq-rrd-chem-CleanupCriteriaTSD_527410_7.pdf (MDEQ 2016). The values listed under the groundwater–surface water interface in Table 6.1 are values within a zone where groundwater becomes surface water before mixing with surface water or exposure to the atmosphere. The values listed for air are presented in micrograms per kilogram in soil immediately beneath a building for indoor air and immediately beneath the groundwater surface if an obstruction is not present (MDEQ 2016). The values listed for indoor and ambient air should be considered screening values since they are not actual air samples collected from a typical breathing zone. In Michigan, groundwater is defined as any water encountered beneath the

TABLE 6.1
Generic Residential Cleanup Levels for Select Compounds

Contaminant	CAS No.	Soil (ug/kg)	Groundwater (ug/l)	Surface–Water Interface (ug/l)	Indoor Air (ug/kg)	Ambient Air (ug/kg)
Volatile Organic Compounds						
Benzene	71432	100	5	4,000	1,600	13,000
Bromobenzene	108861	550	18	NA	3.10 E+5	4.50 E+5
Bromochloromethane	74755	1,600	80	ID	1,200	9,100
Bromodichloromethane	75274	1,600	80	ID	1,200	9,100
Bromoform	75252	1,600	80	ID	1.50 E+5	9.0 E+5
Bromomethane	74839	200	10	35	860	11,000
n-Butylbenzene	104518	2.60 E+5	80	44,000	5.4 E+7	2.90 E+7
sec-Butylbenzene	135988	1,600	80	ID	1.0 E+5	1.0 E+5
tert-Butylbenzene	98066	78,000	80	ID	3.1 E+8	9.70 E+9
Carbon tetrachloride	56235	100	5	900	190	3,500
Chlorobenzene	108907	2,000	100	25	1.3 E+5	7.70 E+5

(Continued)

142 Environmental Compliance Handbook

TABLE 6.1 *(Continued)*
Generic Residential Cleanup Levels for Select Compounds

Contaminant	CAS No.	Soil (ug/kg)	Ground-water (ug/l)	Surface–Water Interface (ug/l)	Indoor Air (ug/kg)	Ambient Air (ug/kg)
Chloroethane	75003	8,600	430	1,100	2.9 E +6	3.0 E+7
Chloroform	67663	1,600	80	350	7,200	45,000
Chloromethane	74873	5,200	260	ID	2,300	40,000
2-Chlorotoluene	95498	3,300	150	ID	2.70 E+5	1.20 E+6
4-Chlorotoluene	106434	900	150	360	4.3 E+5	9.60 E+6
1,2-Dibromo-3-chloropropane	96128	100	.2	ID	260	260
1,2-Dibromoethane	106934	20	80	ID	1.1 E+5	2.60 E+6
Dibromomethane	74953	1,600	80	ID	7,000	7,000
1,2-Dichlorobenzene	95501	14,000	600	13	1.1 E+7	3.90 E+7
1,3-Dichlorobenzene	541731	170	6.6	28	26,000	79,000
1,4-Dichlorobenzene	106467	1,700	75	17	19,000	77,000
Dichlorodifluoromethane	75718	95,000	80	ID	5.3 E+7	5.50 E+8
1,1-Dichloroethane	75353	18,000	880	740	2.30 E+5	2.10 E+6
1,2-Dichloroethane	10762	100	5	360	2,100	6,200
1,1-Dichloroethene	75354	140	7	130	62	1,100
cis-1,2-Dichloroethene	156592	1,400	70	620	22,000	1.80 E+5
trans-1,2-Dichloroethene	156592	2,000	100	1,500	23,000	2.80 E+5
1,2-Dichloroporopane	78875	100	5	230	4,000	25,000
1,4-Dioxane	123911	1,700	85	2,800	NLV	NLV
Ethylbenzene	100414	1,500	74	18	87,000	7.20 E+5
Hexachlorobutadiene	87673	26,000	50	ID	1.30 E+5	1.30 E+5
Isopropylbenzene	98828	91,000	800	28	4.0 E+5	1.70 E+6
Methylene chloride	75092	100	5	1,500	45,000	2.10 E+5
Methyl-tert butyl ether	163404	800	40	7,100	9.9 E+6	2.50 E+7
1,1,1,2-Tetrachloroethane	630206	1,500	8.5	78	36,000	54,000
1,1,2,2-Tetrachloroethane	79345	170	8.5	35	4,300	10,000
Tetrachloroethene	127184	100	5	60	11,000	1.70 E+5
Toluene	108883	16,000	790	270	3.3 E+5	2.80 E+6
1,2,4-Trichlorobenzene	120821	4,200	70	99	9.6 E+6	2.80 E+7
1,2,3,-Trichlorobenzene	87616	4,200	70	99	9.6 E+6	2.80 E+7
1,1,1-Trichloroethane	75556	4,000	220	89	2.50 E+5	3.80 E+6
1,1,2-Trichloroethane	79005	100	5.0	330	4,600	17,000

ug/kg = micrograms per kilogram
ug/L = micrograms per liter
NA = Not available
ID = Insufficient data
NLV = Not likely to volatilize

Source: Michigan Department of Environmental Quality. *Cleanup Criteria and Screening Levels Development and Application.* www.michigan.gov/documents/deq/deq-rrd-chem-CleanupCriteriaTSD_527410_7.pdf (accessed March 24, 2021), 2016.

CERCLA   **143**

surface of the ground regardless of quantity or if it occurs in a naturally occurring formation.

As a comparison, Table 6.2 presents soil cleanup criteria for New Jersey (New Jersey Department of Environmental Protection 2008, 2015, 2017), Illinois (Illinois Environmental Protection Agency [IEPA] 2017), California Department of Toxic Substances Control Chemical Look-Up Table (California DTSC 2013; California Environmental Protection Agency 2005), and USEPA regional soil screening levels (RSLs) (USEPA 2021i).

TABLE 6.2
Soil Cleanup Levels for New Jersey, Illinois, New York, California, and USEPA

Contaminant	CAS ID No	New Jersey[1] (mg/kg)	Illinois[2] (mg/kg)	California[3] (mg/kg)	USEPA[4] (mg/kg)
Acetone	67641	1,000	25	20	2.9
Benzene	71432	3	0.03	5	0.0023
Bromobenzene	108861	NL	NL	NL	0.042
Bromochloromethane	74755	11	NL	NL	0.021
Bromodichloromethane	75274	NL	0.6	NL	0.00036
Bromoform	75252	86	0.8	NL	0.00087
Bromomethane	74839	79	NL	NL	0.0019
Methyl ethyl ketone	789833	1,000	17	NL	1.2
tert-Butylbenzene	98066	NL	NL	NL	1.6
Carbon tetrachloride	56235	2	0.07	NL	0.00018
Chlorobenzene	108907	37	1	NL	0.053
Chloroethane	75003	NL	NL	NL	0.081
Chloroform	67663	19	0.6	NL	0.000061
Chloromethane	74873	520	NL	NL	0.049
2-Chlorotoluene	95498	NL	NL	NL	0.23
4-Chlorotoluene	106434	NL	NL	NL	0.24
1,2-Dibromo-3-chloropropane	96128	NL	NL	NL	0.081
1,2-Dibromoethane	106934	NL	NL	NL	0.0000021
Dibromomethane	74953	NL	NL	NL	0.0021
1,2-Dichlorobenzene	95501	5,100	17	NL	0.03
1,3-Dichlorobenzene	541731	5,100	NL	NL	0.00082
1,4-Dichlorobenzene	106467	570	2	NL	0.00046
Dichlorodifluoromethane	75718	NL	NL	NL	0.03
1,1-Dichloroethane	75353	8	0.06	NL	0.00078
1,2-Dichloroethane	10762	6	0.02	NL	0.000048
1,1-Dichloroethene	75354	8	0.06	5	0.00078
cis-1,2-Dichloroethene	156592	4	0.4	5	0.011
trans-1,2-Dichloroethene	156592	4	0.7	NL	0.11
1,2-Dichloroporopane	78875	10	0.03	NL	0.00015
2,2-Dichloropropane	594207	NL	NL	NL	0.013
1,3-Dichloropropane	142289	NL	2NL	NL	0.13
1,4-Dioxane	123911	NL	NL	10	0.000094

(Continued)

TABLE 6.2 *(Continued)*

Soil Cleanup Levels for New Jersey, Illinois, New York, California, and USEPA

Contaminant	CAS ID No	New Jersey[1] (mg/kg)	Illinois[2] (mg/kg)	California[3] (mg/kg)	USEPA[4] (mg/kg)
Ethylbenzene	100414	1,000	13	5	0.0017
Hexachlorobutadiene	87673	1	NL	5	0.00027
Isopropylbenzene	98828	NL	NL	NL	0.084
Methylene chloride	75092	49	0.02	10	0.0029
Methyl-tert butyl ether	1634044	NL	0.32	NL	0.0032
1,1,1,2-Tetrachloroethane	630206	170	NL	NL	0.00022
1,1,2,2-Tetrachloroethane	79345	34	NL	NL	0.00003
Tetrachloroethene	127184	4	0.06	5	0.0051
Toluene	108883	1,000	12	5	0.76
1,2,4-Trichlorobenzene	120821	68	5	NL	0.0034
1,2,3,-Trichlorobenzene	87616	NL	NL	NL	0.021
1,1,1-Trichloroethane	75556	210	2	NL	2.8
Trichloroethene	75694	23	0.06	5	0.00018
Trichlorofluoromethane	96184	NL	NL	NL	3.3

mg/kg = milligrams per kilogram

NL = Not listed

Source: [1] New Jersey Department of Environmental Protection. *Introduction to Site-Specific Impact to Ground Water Soil Remediation Standards Guidance Document.* Trenton, New Jersey. 10 pages, www.nj.gov/dep/srp/guidance/rs/igw (accessed March 24, 2021), 2008; New Jersey Department of Environmental Protection. *Site Remediation Program: Remediation Standards,* www.nj.gov/dep/srp/guidance (accessed March 24, 2021), 2015; and New Jersey Department of Environmental Protection. *Soil Cleanup Criteria,* www.nj.gov.srp/guidance/scc (accessed May 5, 2017), 2017.

[2] Illinois Environmental Protection Agency. 2017. *Tiered Approach to Corrective Action Objectives (TACO).* Springfield, Illinois. 284 pages.

[3] California Department of Toxic Substances and Control. 2013. *Chemical Look-Up Table. Technical Memorandum.* Sacramento, California. 5 pages; and California Environmental Protection Agency (CalEPA). 2005. *Use of Human Health Screening Levels (CHHSLs) in Evaluation of Contaminated Properties.* Sacramento, California. 65 pages.

[4] United States Environmental Protection Agency. *Regional Screening Levels.* www.epa.gov/RSLs (accessed March 24, 2021), 2021i.

Criteria presented in Table 6.2 generally correspond to a carcinogenic risk of 1 in 100,000 and a hazard quotient of 1 for residential properties. The values presented for New Jersey are not for protection of groundwater. Soil remediation values for New Jersey protective of groundwater are calculated by NJDEP on a site-by-site basis. The lessons from examining Tables 6.1 and 6.2 are the following:

- Cleanup or remediation criteria vary substantially by medium, including soil, groundwater, surface water, indoor air, and outdoor air
- Cleanup or remediation criteria very substantially based on land use, which typically includes residential, commercial, industrial, construction worker, or recreational
- Cleanup or remediation criteria vary between each state and with USEPA

6.6 SUMMARY OF COMPREHENSIVE ENVIRONMENTAL RESPONSE, COMPENSATION, AND LIABILITY ACT

CERCLA was originally designed as a mechanism to fund cleanup of "orphaned" sites in the United States using the "polluters pay" principle. However, CERCLA has meant much more than just a funding mechanism. CERCLA was also a catalyst in developing investigative techniques and protocols, risk assessment evaluations, remediation technologies, and remediation target values for hundreds of chemical compounds in different media. Over its more than 35-year history, USEPA has recovered over $35 billion to fund cleanup at thousands of CERCLA sites in the United States (USEPA 2021j). In addition, the "polluters pay" principle is used worldwide as the preferred method for protecting human health and the environment at difficult and complex sites of contamination.

6.7 REFERENCES

Beck, E. C. 1979. *The Love Canal Tragedy*. Journal of the Environmental Protection Agency. Washington, DC. www.epa.gov/epa/aboutepa/love-canal-tragedy.html. (Accessed March 24, 2021).

California Department of Toxic Substances and Control (DTSC). 2013. *Chemical Look-Up Table*. Technical Memorandum. Sacramento, CA. 5 pages.

California Environmental Protection Agency (CalEPA). 2005. *Use of Human Health Screening Levels (CHHSLs) in Evaluation of Contaminated Properties*. Sacramento, CA. 65 pages.

California Regional Water Quality Control Board. 2021. PG & E Hinkley Chromium Cleanup. www.waterboards.co.gov/lahontan/water_issues/projects/pge/. (Accessed March 24, 2021).

Illinois Environmental Protection Agency (IEPA). 2017. *Tiered Approach to Corrective Action Objectives (TACO)*. IEPA. Springfield, IL. 284 pages.

Michigan Department of Environmental Quality. 2016. Cleanup Criteria and Screening Levels Development and Application. www.michigan.gov/documents/deq/deq-rrd-chem-CleanupCriteriaTSD_527410_7.pdf. (Accessed March 24, 2021).

New Jersey Department of Environmental Protection (NJDEP). 2008. *Introduction to Site-Specific Impact to Ground Water Soil Remediation Standards Guidance Document*. Trenton, NJ. 10 pages. www.nj.gov/dep/srp/guidance/rs/igw. (Accessed March 24, 2021).

New Jersey Department of Environmental Protection (NJDEP). 2015. Site Remediation Program: Remediation Standards. www.nj.gov/dep/srp/guidance. (Accessed March 24, 2021).

New Jersey Department of Environmental Protection (NJDEP). 2017. Soil Cleanup Criteria. www.nj.gov.srp/guidance/scc. (Accessed March 24, 2021).

San Francisco Chronicle. 1996. PG&E to Pay 333 Million in Pollution Suit. www.sfgate.com/article/PG-E-to-Pay-333-million-in-pollution-suit-3303933.php. (Accessed March 24, 2021).

Sun, M. 1983. Missouri's Costly Dioxin Lesson. *Science*. Vol. 219. pp. 367–369.

United States Environmental Protection Agency (USEPA). 1988. *Superfund Record of Decision: Reich Farms, Toms River, New Jersey*. EPA/ROD/R-02–88/070. Office of Emergency and Remedial Response. Washington, DC. 108 pages.

United States Environmental Protection Agency (USEPA). 1989a. *Superfund Record of Decision: Wells G & H, Woburn, Massachusetts*. EPA/ROD/R-01–89/036. Office of Emergency and Remedial Response. Washington, DC. 85 pages.

United States Environmental Protection Agency (USEPA). 1989b. *Superfund Record of Decision: Ciba-Geigy, Toms River, New Jersey*. EPA/ROD/R-02–89/076. Office of Emergency and Remedial Response. Washington, DC. 120 pages.

146 Environmental Compliance Handbook

United States Environmental Protection Agency (USEPA). 2004. EPA Removes Love Canal from Superfund List. www.epa.gov/admpress/love-canal. (Accessed March 24, 2021).

United States Environmental Protections Agency (USEPA). 2021a. CERCLA. www.epa.gov. superfund. (Accessed March 24, 2021).

United States Environmental Protection Agency (USEPA). 2021b. CERCLA Sites in the United States. www.epa.gov/CERCLA-sites. (Accessed March 24, 2021).

United States Environmental Protection Agency. (USEPA). 2021c. Laws and Regulations. www.epa.gov/laws-regulations/regulations. (Accessed March 24, 2021).

United States Environmental Protection Agency (USEPA). 2021d. Times Beach, Missouri Archive. www.epa.gove/times-beach. (Accessed March 24, 2021).

United States Environmental Protection Agency (USEPA). 2021e. Love Canal Tragedy. www. epa.gov/history/love-canal. (Accessed March 24, 2021).

United States Environmental Protection Agency (USEPA). 2021f. *Resource Conservation and Recovery Act Listed and Characteristic Hazardous Wastes*. Office of Solid Waste. Washington, DC. www.epa.gov/hazarodus-wastes. (Accessed March 24, 2021).

United States Environmental Protection Agency (USEPA). 2021g. Superfund Site Profile: Valley of Drums, Kentucky. www.epa.gov/valley-of-drums-profile. (Accessed March 24, 2021).

United States Environmental Protection Agency (USEPA). 2021h. EPA to Update Community on Toms River, New Jersey Superfund Site. https://archive.epa.gov/epa/newsreleases/epa-update-community-toms-river-nj-superfund. (Accessed March 24, 2021).

United States Environmental Protection Agency (USEPA). 2021i. Regional Screening Levels. www.epa.gov/RSLs. (Accessed March 24, 2021).

United States Environmental Protection Agency. 2021j. Superfund Enforcement: 35 Years of Protecting Communities and the Environment. 7p. www.epa.gov/enforcement/superfund-enforcement-35-years-protecting-communities-and-environment. (Accessed March 24, 2021).

7 Hazardous Materials Transportation Act, Superfund Amendments and Reauthorization Act, Emergency Planning and Community Right to Know Act, Toxic Substance Control Act, Pollution Prevention Act, Brownfield Revitalization Act

7.1 INTRODUCTION

Once RCRA and CERCLA were in effect and USEPA and the regulated community gained experience in applying these new environmental regulations, USEPA and others noticed that some improvements were necessary. Hence, the Hazardous Material Transportation Act, Toxic Substance Control Act, Superfund Amendments and Reauthorization Act, Emergency Planning and Community Right to Know Act (EPCRA), Brownfield Revitalization Act, and Pollution Prevention Act were promulgated to improve the effectiveness of the intent of RCRA and CERCLA and also to address areas that RCRA and CERCLA did not specifically cover.

7.2 HAZARDOUS MATERIALS TRANSPORTATION ACT

In the 1970s, numerous landfills throughout the United States began to refuse acceptance of wastes considered hazardous because of liability concerns from what would

DOI: 10.1201/9781003150107-7

147

148 Environmental Compliance Handbook

later become known as Superfund liability, which greatly increased the disposal costs of wastes at those waste sites that did accept hazardous wastes for disposal. This led to increased illegal dumping of wastes in vacant lots, along highways, and in many isolated and wooded areas and farms. At the same time, there were many incidents and accidents involving hazardous materials during transportation, causing property damage and endangering the public (USEPA 2021a). During this time, the United States Department of Transportation estimated that 75% of hazardous waste shipments within the United States violated the regulations in effect at the time because of the following (USEPA 2021a):

- Inconsistencies in the regulations between individual states
- Lack of inspectors to enforce those regulations in effect at the time
- Lack of proper training of those inspectors who were present at the time
- Poor coordination between those federal and applicable state agencies responsible for regulating transportation routes and methods, including the United States Coast Guard, Federal Aviation Authority, Federal Highway Administration, and Federal Railroad Administration

As a result of these issues, Congress passed the Hazardous Materials Transportation Act in 1975. The HMTA is the principal law governing the transportation of hazardous materials. Its purpose is to protect against the risks to life, property, and the environment that are inherent in the transportation of hazardous materials within the United States and beyond. The responsibility for enforcing the act has been delegated to the United States Department of Transportation (USDOT), not the USEPA (USEPA 2021a).

USDOT estimates that currently, there are 500,000 shipments of hazardous materials in the United States every day, and more than 90% of these shipments are transported by truck. According to USDOT, approximately 50% of those materials that are shipped daily contain corrosive or flammable petroleum products, while the remaining shipments represent any of the 2,700 other chemicals considered hazardous by USDOT. The act was passed to improve the uniformity of existing regulations for transporting hazardous materials and to prevent spills and illegal dumping or disposal that endangered the public and the environment, which was a problem that was sometimes made worse by fragmented regulations (USEPA 2021a).

The definition of hazardous materials includes those materials designated by the secretary of the United States Department of Transportation as posing an unreasonable threat to the public and the environment. The term "hazardous material" includes all of the following (USEPA 2021a):

- Hazardous substance
- Hazardous waste
- Marine pollutants
- Elevated temperature material
- Materials identified in Part 172.101 of the Act
- Materials meeting the definitions contained in Part 173 of the act

Hazardous Materials Transportation Act 149

The act has four main elements (USEPA 2021a):

- Procedures and policies
- Material designations and labeling
- Packaging requirements
- Operational rules

HMTA regulations apply when an individual or company satisfies one or more of the following (USEPA 2021a):

- When hazardous materials are transported in commerce
- When a company or individual designs, manufacturers, fabricates, inspects, marks, maintains, reconditions, repairs, or tests a package, container, or packaging component that is represented, marked, certified, or sold as qualified for use in transporting hazardous material in commerce
- When a company or individual prepares or accepts hazardous material for transportation in commerce
- When a company or individual is responsible for the safety of transporting hazardous material in commerce
- When a company or individual certifies compliance with any requirement under the act
- When a company or individual misrepresents whether such a person or company is engaged in any activity under the previously listed requirements

Essentially, even though every person involved in the preparation of the transportation of hazardous materials shares some responsibility to ensure its safety, the primary burden of liability falls on the shipper of the material. Carriers are only responsible to ensure that required information accompanying hazardous materials packaging is immediately available to personnel who would respond to an incident or conduct a hazardous materials investigation. The act is implemented through several agencies and is based on the type of transportation and type of hazardous material being shipped (USEPA 2021a).

Procedures of proper handling and preparation for handling hazardous materials, as well as finding instructional information about implementing the act correctly, include (USEPA 2021a):

- Training requirements. The ACT describes what an employer who ships hazardous material must do to train employees on the safe loading, unloading, handling, storing, and transporting of hazardous material and emergency preparedness for responding to an accident or incident involving the transportation of hazardous material.
- Beginning and completing training. A hazardous material employer shall begin training employees no later than 6 months after employment begins.
- Certification of training. After training is complete, each employer shall certify, with documentation, that the employees have received training and have been tested on appropriate areas of responsibility.

A record of current training is required by each HAZMAT employer for each HAZMAT employee for as long as the employee is employed by that employer as a HAZMAT employee and for 90 days thereafter. The record is required to include (USEPA 2021a):

- The HAZMAT employee's name
- The most recent training completion date of the HAZMAT employee's training
- A description, copy, or the location of the training materials used to meet the requirements
- The name and address of the person who provided the training
- Certification that the HAZMAT employee has been trained and tested

7.2.1 MATERIAL DESIGNATION AND LABELING

Each package, freight container, and transport vehicle carrying a hazardous material must have markings that are (USEPA 2021a):

- Durable
- In English
- Printed or affixed on the surface of the shipping package or on a label, tag, or sign on the package
- Displayed on a background of sharply contrasting color
- Not obscured by other labels or attachments
- Located away from any other marking that could reduce its effectiveness

Each non-bulk package, container, or small tank must be labeled with a code corresponding to the hazard class of the hazardous material being transported and must also follow design and placement requirements. Each piece of bulk packaging, freight container, unit load device, transport vehicle, or rail car containing any quantity of a hazardous material must be placarded corresponding to the hazard class of the hazardous material being transported and must follow design and placement requirements (USEPA 2021a).

There are nine hazardous classes (see Figure 7.1) (United States Department of Transportation 2013):

1. Class 1: Explosives
2. Class 2: Gases
3. Class 3: Flammable Liquid and Combustible Liquid
4. Class 4: Flammable Solid, Spontaneously Combustible, and Dangerous When Wet
5. Class 5: Oxidizer and Organic Peroxide
6. Class 6: Poison (Toxic) and Poison Inhalation Hazard
7. Class 7: Radioactive
8. Class 8: Corrosive
9. Class 9: Miscellaneous

Hazardous Materials Transportation Act

The contents of any package containing a hazardous material and the material of construction of the package itself must be resistant to significant chemical or galvanic reactions that can potentially compromise the integrity of the package. In addition, hazardous materials may not be mixed together with other materials regardless of whether they are hazardous if they potentially can create a reaction causing the following (USDOT 2021):

- Combustion or generating heat
- Flammable gas
- Poisonous gas
- Asphyxiant gas
- Formation of unstable or corrosive materials

It is the responsibility of the shipper of a hazardous material to determine that the compatibility between the hazardous material and the packaging is sufficient for safe transportation (USDOT 2021).

Regulations providing for immediate emergency response information in an incident, as well as requirements for the development and implementation of security plans, must be adhered to by any person who offers transportation in commerce or transports in commerce hazardous materials. Placarding on vehicles is contained in 49 CFR Subpart F Part 172 and is required on all four sides of a vehicle carrying a hazardous material. An example placard is presented in Figure 7.2. Placards are composed of four squares that form a diamond-shaped placard (USDOT 2021).

The upper portion of a placard contains information concerning the fire hazard of the material on a scale of zero (0), meaning the material will not burn, to four (4), meaning the material has a flashpoint of 73°F or less, and this indication is typically

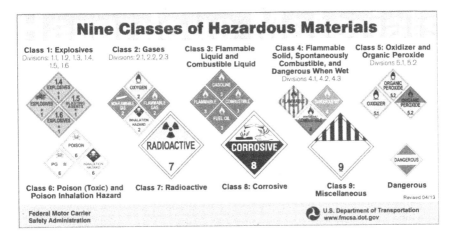

FIGURE 7.1 USDOT hazardous classes (from United States Department of Transportation. *Federal Motor Carrier Safety Administration [FMCSA] Placard Requirements.* www.fmcsa.dot.gov [accessed March 24, 2021], 2013).

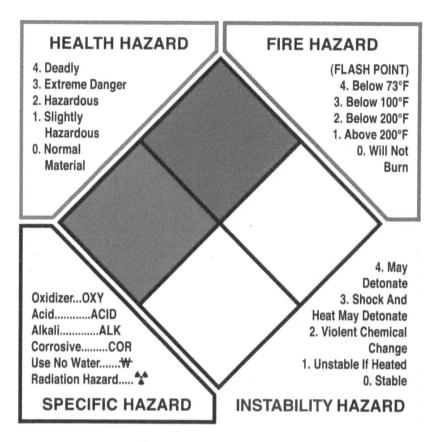

FIGURE 7.2 USDOT hazardous material classifications and example placard (from United States Department of Transportation. *Summary of the Federal Hazardous Material Transportation Act.* Office of Hazardous Materials Safety. www.USDOT.gov/HMTA [accessed March 24, 2021], 2021).

colored red. The left portion of the placard contains information concerning health hazard information on a scale of zero (0), meaning normal materials, to four (4), meaning the material is considered deadly. The right portion of the placard contains information on the potential reactivity of the hazardous material on a scale of zero (0), meaning that the material is considered stable, to four (4), meaning the material is not stable and may detonate under normal conditions. The final portion of the placard is at the bottom and contains contamination information about the specific hazard of the material (USDOT 2021).

Examples of specific hazards that may be present include information like the following (USDOT 2021):

- Acid
- Alkali
- Corrosive

Hazardous Materials Transportation Act

- Oxidizer
- Use no water or water reactive
- Radiation hazard
- Explosive
- Flammable

Immediate notification of a hazardous material incident by the carrier is required at the earliest practical moment for incidents that occur during the course of transportation (including loading, unloading, and temporary storage) in which, as a direct result of the hazardous materials, any one of the following occurs (USDOT 2021):

- A person is killed
- A person receives an injury requiring admittance to a hospital
- The general public is evacuated for 1 hour or more
- A major transportation artery or facility is closed or shut down for 1 hour or more
- Fire, breakage, spillage, or suspected contamination occurs involving an infectious substance other than a diagnostic specimen or regulated medical waste
- A release of a marine pollutant occurs in a quantity exceeding 450 L (119 gallons) for a liquid or 400 kg (882 pounds) for a solid
- A situation exists of such a nature (e.g., a continuing danger of life exists at the scene of the incident) that, in the judgment of the person in possession of the hazardous material, it should be reported to the National Response Center even though it does not meet the other criteria listed

Each notice shall be given via telephone to the National Response Center at 800.424.8802. Incidents involving etiologic agents should be also made to the Centers for Disease Control at 800.232.1024. A written report is also required to be submitted to USDOT and others as directed using DOT Form F 5800.1 for all incidents involving the transportation of hazardous materials unless specifically directed by USDOT (USDOT 2021).

7.3 TOXIC SUBSTANCE CONTROL ACT

The Toxic Substance Control Act authorizes USEPA to screen existing and new chemicals used in manufacturing and commerce to identify dangerous products or uses that should be subject to federal control. TSCA was first passed in 1976 and was amended in 1986, 1988, 1990 (twice), 1992, 2007, 2008, 2010, and 2016. Both naturally occurring and synthetic chemicals are subject to TSCA, with the exception of chemicals already regulated under other federal laws that include food, drugs, cosmetics, firearms, ammunition, pesticides, herbicides, and tobacco (Schierow 2013).

The purpose of TSCA was to protect the public from unreasonable risk of injury to health or the environment by regulating the manufacture and sale of chemicals. TSCA does not address wastes produced as byproducts of manufacturing. Instead, TSCA attempts to exert direct government control over which types of chemicals can and cannot be used in actual use and production. For example, the use of chlorofluorocarbons in manufacturing is now strictly prohibited in all manufacturing processes in the United States, even if no chlorofluorocarbons are released into the atmosphere. The types of chemicals regulated under TSCA are in two broad groups: new and existing. New chemicals are defined as any chemical substance which is not included in the chemical substance list compiled and published under TSCA, Section 8(b). This list includes all of the chemical substances manufactured or imported into the United States prior to 1979 that were grandfathered into TSCA and were considered safe and included over 62,000 chemicals (USEPA 2021b).

When first passed, TSCA limited the manufacture, processing, commercial distribution, use, and disposal of chemical substances including polychlorinated biphenyls, asbestos, radon, and lead-based paint. At first USEPA could only limit the commercial use of new chemicals and even then only immediately after the chemicals were introduced in commerce. Once a chemical substance established itself in the commercial world, USEPA could only restrict new uses of the substance. Thus, USEPA's influence was limited under the 1976 version of TSCA. This created much criticism of TSCA from environmental groups through the years that focused on the "after the fact focus" of TSCA in that it failed to protect individuals before substances were available in products. In addition, TSCA was difficult to implement because the approximately 62,000 substances that were grandfathered in remained on the market. Last, fully assessing the risks posed by a new chemical was expensive and time consuming (Markell 2014).

7.3.1 Restricting Chemicals

To put things into perspective, as of 2014, the list of chemicals in use in the United States is greater than 84,000, and the list of chemicals USEPA considers fully evaluated numbers only 250 (USEPA 2014). Even though TSCA gives the authority to USEPA to test existing chemicals, USEPA has had difficulty in obtaining the data needed from industry to determine their risks, and the cost to conduct the tests themselves has been evaluated as too costly (Markell 2014).

USEPA has been successful in restricting only nine chemicals in its history of over 40 years since promulgated in 1976 and none since 1984. Those substances that are currently restricted are (USEPA 2021b):

- Polychlorinated biphenyls
- Chlorofluorocarbons
- Dioxin
- Asbestos
- Hexavalent chromium
- Four nitrite compounds that are either:

  - Mixed mono- and diamides of an organic acid
  - Triethanolanime salt of a substituted organic acid
  - Triethanolanime salt of tricarboxylic acid
  - Tricarboxylic acid

Hazardous Materials Transportation Act

Even though many critics seem to believe that TSCA has failed in many respects, it still represents a potentially powerful tool for pollution prevention and can and should play a vital role in our society (Markell 2014).

In June 2016, TSCA was amended significantly and gave USEPA the authority and responsibility to proactively evaluate the risks of all chemical substances, which will be conducted in a multi-step process that is outlined as the following (National Law Review 2016):

- Developing a screening process to identify high-priority chemical substances
- Designating chemical substances as high or low priority
- Evaluating the risks of high-priority chemicals
- Determining whether any chemical presents unreasonable risk
- Issuing rules to restrict the use of substances that present an unreasonable risk

The 2016 amendments define high-priority substances as those chemicals that may present an unreasonable risk of injury to health or the environment because of a potential hazard and a potential route of exposure, including an unreasonable risk to a potentially exposed or susceptible subpopulation that includes but is not limited to any of the following (National Law Review 2016):

- Infants
- Workers
- Children
- Elderly
- Pregnant women

Accordingly, USEPA has stated that its intent is to accord preference to those substances with persistence and bioaccumulative tendencies and known human carcinogens with high acute and chronic toxicity (USEPA 2021b). In the near term, USEPA has identified ten chemicals as high priority and is currently conducting risk evaluations. These are the following (USEPA 2021b):

- 1,4-Dioxane
- Asbestos
- Cyclic Aliphatic Bromide Cluster
- N-methylpyrolidone
- Tetrachloroethylene
- 1-Bromopropane
- Carbon Tetrachloride
- Methylene Chloride
- Pigment Violet 29
- Trichloroethylene

Analysis of the list shows four of the substances belong to the same chemical group called chlorinated or halogenated volatile organic compounds (CVOCs or HVOCs, respectively) and are also known as chlorinated solvents. Those are carbon tetrachloride, methylene chloride, tetrachloroethylene, and trichloroethylene. In addition, 1,4-dioxane is associated with the breakdown of tetrachloroethylene and trichlorethylene. Therefore, of the ten chemicals USEPA has initially placed on its high-priority list, half belong to or are associated with the same group of compounds, CVOCs or chlorinated solvents. We shall discuss this in much greater detail in the upcoming chapters.

7.3.2 Regulating PCBs under TSCA

PCBs are a group of synthetic organic chemical compounds consisting of carbon, hydrogen, and chlorine atoms. According to USEPA (2021b), PCBs are still released into the environment from sources that include:

- Poorly maintained hazardous wastes sites that contain PCBs
- Illegal or improper dumping of PCB wastes
- Leaks or releases from electrical transformers containing PCBs
- Disposal of PCB-containing consumer products into municipal or other landfills not designed to handle hazardous waste
- Burning some wastes in municipal and industrial incinerators

According to USEPA (2021b), PCBs are not defined as a hazardous waste under RCRA. PCBs are considered TSCA wastes and are regulated under Title 40 CFR Part 761 Section 261.8.

Cleanup of PCBs that have been released into the environment varies widely depending on where the PCBs were detected or reside in the environment. For instance, if PCBs are detected in river or lake sediments, which is often the case, cleanup requirements may be very low because PCBs tend to bioaccumulate and do not readily degrade and potentially enter the food chain. In general, if the source of PCBs detected originated at a concentration greater than 50 parts per million (ppm), it would be considered a PCB waste, and TSCA regulations apply. Cleaning up PCBs that originate from a source greater than 50 ppm requires approval from USEPA (USEPA 2021b).

7.4 SUPERFUND AMENDMENTS AND REAUTHORIZATION ACT AND EMERGENCY PLANNING AND COMMUNITY RIGHT-TO-KNOW ACT

Superfund Amendments and Reauthorization Act of 1986 revised CERCLA, largely in response to a 1984 tragedy in Bhopal, India, where thousands of residents were killed or injured by exposure to a chemical, methyl isocyanate (MIC), gas that leaked from a Union Carbide plant (USEPA 2021c). In response to continuing community concerns regarding hazardous materials and the Bhopal, India, tragedy, Title III of SARA was developed and is known at the Emergency Planning and Community Right-to-Know Act.

During the early morning hours of December 3, 1984, a Union Carbide plant in a village just south of Bhopal, India, released approximately 40 tons of MIC into the air. MIC is used in manufacturing pesticides and is considered a lethal chemical. The gas quickly and silently diffused near the ground surface and in the end killed thousands of people and injured tens of thousands of residents (Eckerman 2005). Title III of SARA, commonly called EPCRA, was a direct result of the tragedy in India.

There are many examples of incidents in the United States that involve hazardous substances, and the purpose of EPCRA was to prevent or minimize the potential for

Hazardous Materials Transportation Act

FIGURE 7.3 PCB guide (from United States Environmental Protection Agency. *Summary of the Toxic Substance Control Act.* https://epa.gov/laws-regulations/summary-toxic-substances-control-act [accessed March 24, 2021]), 2021b.

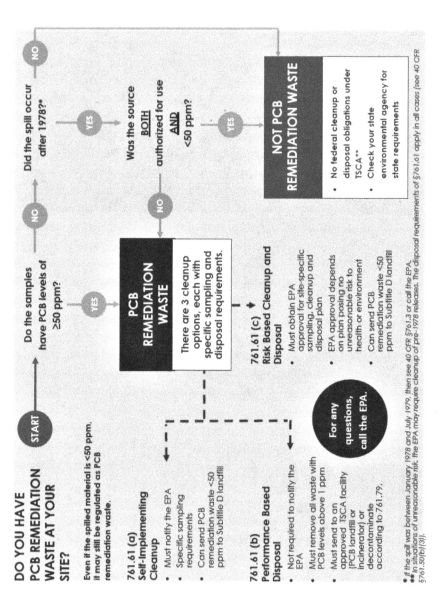

FIGURE 7.4 Flowchart for PCB wastes (from United States Environmental Protection Agency. *Summary of the Toxic Substance Control Act.* https://epa.gov/laws-regulations/summary-toxic-substances-control-act [accessed March 24, 2021] 2021b).

Hazardous Materials Transportation Act

an incident like what occurred in India from ever happening in the United States (USEPA 2021c). The purpose of EPCRA includes:

1. To encourage and support emergency planning for responding to chemical accidents
2. To provide local governments and the public with information about chemical hazards in their communities

To facilitate cooperation between industry, interested citizens, environmental and other public-interest groups and organizations, and the government at all levels, EPCRA establishes an ongoing forum at the local level called the local emergency planning committee (LEPC). LEPCs are governed by the state emergency response commission (SERC) in each state. Under EPCRA, SERCs and LEPCs are responsible for (USEPA 2021d):

1. Preparing emergency plans to protect the public from chemical accidents
2. Establishing procedures to warn and, if necessary, evacuate the public in case of an emergency
3. Providing the public citizens and local governments with information about hazardous chemicals and accidental releases of chemical in their communities
4. Assisting in the preparation of public reports on annual release to toxic chemicals into the air, water, and soil

EPRCRA does not place limits on which chemicals can be stored, used, released, disposed of, or transferred at a facility. It only requires a facility to document, notify, and report information (USEPA 2021d). Emergency planning under EPCRA is found in Sections 301–303 and is to ensure that state and local communities are prepared to respond to a potential chemical accident. As a first step, each state had to establish a SERC. In turn, the SERC designated local emergency planning districts. For each district, the SERC appoints supervisors and coordinates the activities of the LEPCs. The LEPC, in turn, must develop an emergency response plan for its district and review it annually. The membership of the LEPC includes representatives from public and private organizations as well as a representative from every facility subject to EPCRA emergency planning requirements, which we will discuss later in this section (USEPA 2021d).

The plan developed by the LEPC must have the following information (USEPA 2021d):

- Identify affected facilities and transportation routes
- Describe emergency notification and response procedures
- Designate community and facility emergency coordinators
- Describe methods to determine the occurrence and extent of a release
- Identify available response equipment and personnel
- Outline evacuation plans

160 Environmental Compliance Handbook

- Describe training and practice programs and schedules
- Contain methods and schedules for exercising the plan

USEPA has established the requirements of when a facility is subject to EPCRA emergency planning reporting. USEPA has published a list of what it terms "extremely hazardous substances (EHS)." For each EHS, the list includes its name, the Chemical Abstract Service number of the substance, and a number that is called the threshold planning quantity (TPQ) and is expressed in pounds. If a facility has within its boundaries an amount of an EHS equal to or greater than the TPQ, the facility is subject to EPCRA and must notify both the SERC and the LEPC. The facility must also appoint an emergency response coordinator to work with the LEPC in developing and implementing the local emergency plan at the facility (USEPA 2021d).

Emergency release notification requirements are described in Section 304 of EPCRA and define if a facility is subject to release reporting requirements even if it is not subject to the provisions described in Sections 301–303. Section 304 applies to a facility which stores, produces, or uses a **hazardous chemical**, defined as any chemical which is a physical hazard or health hazard, and releases a reportable quantity of a substance contained in either the list of extremely hazardous substances or the list of CERCLA hazardous substances (USEPA 2021d).

Under EPCRA, the RQ is the determining factor in whether a release must be reported. This is a number that is expressed in pounds that is assigned to each chemical in USEPA's list of extremely hazardous substances or the CERCLA list of hazardous substances. If the amount of a chemical released to the environment exceeds its corresponding RQ, the facility must immediately report the release to the appropriate LEPC and SERC and provide written follow-up as soon as practicable. The written follow-up report must include an update to the original notification; any additional information in the response action conducted; known or anticipated health risks incurred; and any information on medical care needed by any exposed person or population (USEPA 2021d).

Community right-to-know reporting requirements are described in Sections 311 and 312. The purpose of these requirements is to increase community awareness of chemical hazards and to assist in emergency planning, when and if ever needed. Sections 311 and 312 apply to any facility that is required by OSHA under its hazard communication standard to prepare or have available a safety data sheet (SDS) for a hazardous chemical that is onsite, for any one day in a calendar year, and at an amount equal to or greater than the following threshold limits established by USEPA (2021d):

- 10,000 pounds (4,500 kg) for hazardous chemicals
- The lower of 500 pounds (230 kg) or the threshold planning quantity for extremely hazardous substances

If a facility is subject to reporting under these sections, it must submit information to the SERC and LEPC and the local fire department with jurisdiction over the facility under two categories, SDS reporting and inventory reporting. SDS reporting requirements specifically provide information to the local community about mixtures and

Hazardous Materials Transportation Act

chemicals present at a facility and their associated hazards. For all substances whose onsite quantities exceed the threshold values, the facility must submit the following (USEPA 2021d):

- A copy of the SDS for each above-threshold chemical onsite or a list of the chemicals grouped into categories
- Submit any change within 3 months as an SDS or list of additional chemicals that meet the reporting criteria

Inventory reporting is designed to provide information on the amounts, location, and storage condition of hazardous chemicals and mixtures containing hazardous chemicals present at any facility. The inventory has two forms. The first is called Tier I. The Tier I form contains aggregate information for applicable hazard categories and must be submitted yearly by March 1. The Tier II form contains more detailed information, including specific names of each chemical. This form is submitted upon request of the agencies authorized to receive the Tier I form. It can also be submitted yearly in lieu of the Tier I form (USEPA 2021d). Toxic chemical release inventory reporting is described in Section 313 of EPCRA. Under this section, the USEPA is required to establish a toxics release inventory (TRI), which is an inventory of routine toxic chemical emission from certain facilities. The original requirements under this section have been expanded by the passage of the Pollution Prevention Act of 1990 and will be described further in a later section. The TRI must now include information on source reduction, recycling, and treatment. To obtain this data, EPCRA requires each required facility to submit a toxic chemical release inventory form (Form R) to the USEPA and designated state officials each year on July 1 (USEPA 2021d).

A facility must file a Form R if the facility (USEPA 2021d):

- Has more than ten full-time employees
- Is in a specified Standard Industrial Classification Code
- Manufactures more than 25,000 lb/year of a listed toxic chemical
- Processes more than 25,000 lb/year of a listed toxic chemical
- Otherwise uses more than 10,000 lb/year of a listed toxic chemical
- Manufactures, processes, or otherwise uses a listed persistent bio-accumulative toxic chemical above the respective PBT reporting threshold. PBT reporting thresholds can vary anywhere from 0.1 grams for dioxin compounds to 100 pounds (45 kg) for lead

USEPA updated some aspects of the Form R rules in 1999, 2001, and 2006, which related to adding chemicals or adjusting thresholds for chemicals on either the list of extremely hazardous substances or the list of CERCLA hazardous substances. Therefore, companies should review the lists routinely for changes and modifications (USEPA 2021d). Table 7.1 is the partial list of extremely hazardous substances defined in Section 302 of EPCRA. The full list can be viewed at www.epa.gov/epcra/final-rule-extremely-hazardous-substance-list-and-threshold-planning-quantities-emergency (USEPA 2015, 2021e).

162 Environmental Compliance Handbook

TABLE 7.1
Partial List of Extremely Hazardous Substances

Compound	Compound	Compound
Acetone cyanohydrin	Acetone thiosemicarbazide	Acrolein
Acrylamide	Acrylonitrile	Acryloyl chloride
Adiponitrile	Aldicarb	Aldrin
Allyl alcohol	Allylamine	Aluminum phosphide
Aminopterin	Amiton	Amiton oxalate
Ammonia	Amphetamine	Aniline
Aniline, 2,4,6-trimethyl	Antimony pentafluoride	Antimycin A
ANTU (Alpha-Naphthlthiourea)	Arsenic pentoxide	Arsenous oxide
Arsenous trichloride	Arsine	Azidoazide azide
Azinphos-ethyl	Azinphos-methyl	Benzal chloride
Benzenamine, 3-(trifluoromethyl)-	Benzenearsonic acid	Benzimidazole, 4,5-dichloro-2-(trifluoromethyl)-
Benzothrochloride	Benzyl chloride	Benzyl cyanide
Bicyclo(2.2.1)heptane-2-carbonitrile	Bis(chloromethyl) ketone	Bitoscanate
Boron trichloride	Boron trifluoride	Boron trifluoride w/methyl ether
Bromadiolone	Bromine	Cadmium oxide
Cadmium stearate	Calcium arsenate	Camphechlor
Cantharidin	Carbachol chloride	Carbamic acid
Carbofuran	Carbon disulfide	Carbophenothion
Chlordane	Chlorfenvinfos	Chlorine
Chlorine trifluoride	Chlormethos	Chlormequat chloride
Chloracetic acid	2-chlorothanol	Chloroethyl chloroformate
Chloroform	Chloromethyl methyl ether	Chlorophacinone
Chloroxuron	Chlorthiophos	Chromin chloride
Cobalt carbonyl	Cobalt, (2,2-(1,2-ethanediylbis)	Colchicine
Coumaphos	Cresol, -o	Crimidine
Crotonaldehyde	Crotonaldehyde, (E)	Cyanide
Cyanogen bromide	Cyanigen iodide	Cyanophos
Cyanuric fluoride	Cycloheximide	Cyclohexylamine
Decaborane	Demeton	Demeton-S-methyl
Dialifor	Diborane	Dichloroethyl ether
Dichloromethylphenylsilane	Dichlorvos	Dicrotophos
Diepoxybutane	Diethyl chlorophosphate	Digitoxin
Diglycidyl ether	Digoxin	Dimefox
Dimethoate	Dimethyl mercury	Dimethyl phosphorochloridothioate
Dimethyl-p-phenylenediamine	Dimethyl-p-phenylebediamine	Dimethylcadmium
Dimethyldichlorosilane	Dimethylhydrazine	Dimetilan
Dinitrocresol	Dinoseb	Dineterb
Dioxathion	Diphacinone	Disulfoton
Dithiazanine iodide	Dithobiuret	Endosulfan
Endoothion	Endrin	Epichlorohydrin
EPN	Ergocalciferol	Ergotamine tartrate
Ethanesulfonyl chloride, 2-chloro-	Ethanol, 1,2-dichloro-. Acetate	Ethion

TABLE 7.1 *(Continued)*
Partial List of Extremely Hazardous Substances

Compound	Compound	Compound
Athoprophos	Ethylbis(2-chloroethyl)amine	Ethylene oxide
Ethylemediamine	Ethyleneimine	Ethylthiocyanate
Fenamiohos	Fenitrothion	Fensulfothion
Fluenetil	Fluorine	Fluoroacetamide
Fluoroacetic acid	Floroacetyl chloride	Fluoroantimonic acid
Fluorouracil	Fonofos	Formaldehyde

Source: United States Environmental Protection Agency, USEPA. 2015. *List of Lists*. EPA 55-B-15–001. Office of Solid Waste and Emergency Response. Washington, DC. 125p.

7.5 POLLUTION PREVENTION ACT

The Pollution Prevention Act of 1990 created a national policy to prevent or reduce pollution at the source whenever possible. As discussed in the previous section, it also expanded the toxics release inventory under SARA and EPCRA. The PPA focused on reducing and preventing pollution, not just for industry but also the government itself and the general public to reduce waste and encouraged and emphasized recycling efforts (USEPA 2021f).

Up to this point in the rapid enactments of environmental laws beginning just 20 years before the PPA, the focus was on compliance. Now, with the PPA, there was a shift in focus from compliance to prevention and reducing amounts of wastes generated. However, as we discussed earlier in this book, it was in 1971 when public service announcements first were aired on television drawing attention to waste disposal and recycling. Why did it take so long? The answer is, of course, like many other aspects of human interaction with nature, complex and has much to do with how and to what extent humans impact the natural environment, which we are still learning.

The PPA is also considered the first real legislative effort in the United States to take up the subject of sustainability. Emphasis was placed on preventing pollution from being produced; in other words, preventing pollution at the source was preferred through technological innovation and advancements in manufacturing. If it could not be prevented, then the focus should be on recycling in an environmentally safe manner. The PPA also goes on to state that disposal of waste or other types of releases into the environment should be conducted only as a last resort (USEPA 2021f).

7.6 BROWNFIELD REVITALIZATION ACT

The Brownfield Revitalization Act of 2006 amended CERCLA and provided funding to assess and clean up brownfields, clarified CERCLA liability protections, and provided funds to enhance state and Native American tribal response programs. USEPA defines a **brownfield** as a property, the expansion, redevelopment, or reuse of which may be complicated by the presence or potential presence of a hazardous substance, pollutant, or contaminant (USEPA 2021g, 2021h). USEPA estimates that there 450,000 to 1 million brownfield sites in the United States.

The Brownfield Revitalization Act was enacted in large part so that these properties, many of which are impacted by hazardous substances, can be investigated, cleaned up, and redeveloped (USEPA 2021g, 2021h). Essentially, a brownfield site is typically an abandoned industrial facility in an urban area impacted by hazardous substances but not at high enough levels to qualify the site under Superfund.

However, CERCLA had to be revised, because under CERCLA, United States courts have ruled that a buyer, lessor, or lender may be held responsible for remediation of hazardous substance residues, even if a prior owner caused the contamination. Therefore, CERCLA was amended to include what is termed "All Appropriate Inquiry," which outlined procedures for the performance of a Phase I environmental site assessment (ESA). By conducting a Phase I ESA, a new property owner may create a safe harbor or protection from liability, known as the "Innocent Landowner Defense," for new purchasers or lenders for transactions involving properties. In the United States, Phase I ESAs have become the standard type of environmental investigation employed when initially investigating a property (USEPA 2005, 2021g).

Standards for conducting Phase I ESAs were originally published by the American Society for Testing Materials (ASTM) in 1993 and were revised in 1997, 2000, and 2005 (ASTM 2005). On November 1, 2006, the USEPA published federal standards for conducting Phase I ESAs termed "Standards and Practices for All Appropriate Inquiries" (AAI) by amending the Comprehensive, Environmental Response, Compensation and Liability Act of 1980, commonly known as the Brownfield Revitalization Act. As stated, the Phase I environmental site assessment is typically the first environmental investigation conducted at a specific property (Rogers 2020).

A Phase I ESA is conducted with the objective of qualitatively evaluating the environmental conditions and potential environmental risk of a property or site. A **property** is defined here as a parcel of land with a specific and unique legal description. A **site** is defined here as a parcel of land including more than one property or easement and typically refers to an area of contamination potentially affecting more than one property. As the first environmental investigation, the Phase I ESA is often regarded as the most important activity because all subsequent decisions concerning the property are, in part, based on the results of it (Rogers 1992). Therefore, great care, scrutiny, scientific inquiry, and objectivity should be exercised while conducting the Phase I ESA. According to USEPA (2005) requirements, an environmental professional, such as a geologist or environmental scientist, must conduct the Phase I ESA. General requirements for conducting a Phase I ESA include:

- Extensive review of current and historical written records, operations, and reports
- Extensive site inspection
- Interviews with knowledgeable and key onsite personnel
- Assessment of potential environmental risks from offsite properties
- Data gap or data failure analysis

Hazardous Materials Transportation Act

A Phase I ESA is typically a non-invasive assessment, performed without sampling or analysis. In some instances, limited sampling may be conducted on a case-by-case basis if, in the professional judgment of the person conducting the assessment—or due to other requests or mitigating factors—sampling is justified. In most cases, collecting and analyzing samples is usually deferred to the Phase II investigation, but if it does occur, the sampling conducted during a Phase I ESA may include the following (Rogers 2020):

- Sediment
- Surface water
- Waste material
- Lead-based paint
- Mold

- Drinking water
- Groundwater
- Soil
- Suspect asbestos-containing material
- Radon

The environmental professional conducting the Phase I ESA must perform extensive research and review all available written records. These research activities apply not only to the property in question but also to the surrounding properties within a radius of up to 1 mile of the investigated property or site. The research/review process typically involves (Rogers 2020):

- Title search and environmental liens
- Historical chemical ordering documents
- Historical aerial photographs
- Engineering diagrams
- Previous environmental incident reports
- Environmental compliance documents
- Fire insurance maps
- USGS topographic maps
- USGS geologic maps
- Native American tribal records
- Building permits

- Historical operations documents
- Historical photographs

- Engineering reports
- Historical environmental reports
- Safety data sheets

- Hazardous substance inventories
- Soil Conservation Service maps
- USGS investigations reports
- Environmental agency maps
- Local governmental records
- Construction diagrams and blueprints

The site inspection consists of a walk-through of the property or site. Items to evaluate and document during the site inspection include (Rogers 1992, 2020):

- Interviews with key personnel
- Areas absent of vegetation
- Aboveground storage tanks (ASTs)
- Chemical storage areas
- The "back 40" (the rear of the facility)

- Stressed vegetation
- Stained soil or pavement
- Underground storage tanks
- Back doors
- Signage

- Storage sheds
- Special labeling
- Recent excavations or land disturbance
- General topography
- Mold
- Insects
- Sumps
- Broken concrete
- Weather conditions
- Nearest water body
- Potential asbestos-containing materials
- Utilities
- Offsite inspection
- Potential contaminant migration pathways
- Refuse storage and containers
- Evidence of fill or mounding
- Depressions in the land surface
- Wetlands
- Animal scat
- Pits and trenches
- Floor and roof drains
- Areas not inspected or inaccessible
- Recent precipitation events
- Evidence of wells or borings
- Potential septic tanks
- Onsite and offsite dumping or fill
- Soil type(s)
- Potential ecological and human receptor pathways

Once the environmental professional has completed the data collection and site inspection portion of the Phase I ESA, an evaluation of whether there is evidence of an existing release, a past release, or a material threat of a release of any hazardous substance or petroleum is conducted. If a product has been released and made its way into structures on the property or into the ground, groundwater, or surface water of the property, then this situation is termed a recognized environmental condition (REC). A **recognized environmental condition** is defined as the presence or likely presence of any hazardous substance or petroleum product on a property or site under conditions that may materially affect or threaten the environmental condition of the property or site, human health, or the environment (USEPA 2005).

If a REC is discovered, further investigation will likely be recommended to evaluate its potential significance. Many sites have more than one REC, and many of these RECs may require further evaluation. An example of a REC with significant amounts of oil-stained soil, stressed, and dead vegetation is presented in Figure 7.5.

FIGURE 7.5 Recognized environmental condition (photograph by Daniel T. Rogers).

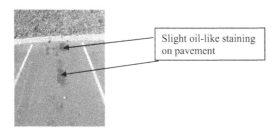

FIGURE 7.6 Example of a de minimis release (photograph by Daniel T. Rogers).

RECs are intended to exclude *de minimis* conditions generally not presenting a threat to human health or the environment and typically would not be the subject of an enforcement action if brought to the attention of appropriate governmental agencies. However, there may be impacts encountered whose perceived severity falls between a *de minimus* condition and a REC. In these situations, the term **environmental concern** is applied. Items listed as environmental concerns become RECs if left unattended or lead to a release. Figure 7.6 is a photograph of a paved parking space with a residual amount of what appears to be a petroleum product discharged from an automobile. In the opinion of the environmental professional who conducted a Phase I ESA at this property, this condition was not evaluated to be a REC but was characterized as *de minimis*.

Historical aerial photographs are effective sources of information and often help with environmental investigations. In urban areas especially, aerial photographs from several sources are readily available. These sources include (Rogers 2020):

- Private local companies specializing in aerial photography
- Private national companies specializing in aerial photography
- Local and state historical societies
- Local and state agencies
- Utility companies
- Local companies
- Federal agencies, such as USEPA, USGS, Soil Conservation Service, National Forest Service, Bureau of Land Management, National Park Service

Figures 7.7a and 7.7b demonstrate how historical aerial photographs can help identify RECs. On the left, Figure 7.7a shows a particular property as farmland with no identifiable REC. Analysis of an earlier aerial photograph taken a few years earlier of the same property (Figure 7.5b to the right) indicates the property was used as a landfill.

7.7 SUMMARY AND CONCLUSIONS

The environmental regulations discussed in this chapter each played a part in improving the overall effectiveness of environmental regulations. In addition, TSCA, PPA, and BRA also paved the way for environmental regulations to bridge directly into

FIGURE 7.7a Aerial photograph showing farmland (photograph by Daniel T. Rogers).

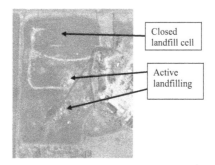

FIGURE 7.7b Earlier aerial photograph showing landfilling activities (photograph by Daniel T. Rogers).

sustainability. Specifically, the regulations discussed in this chapter accomplished the following:

- Improved the effectiveness of RCRA and CERCLA
- Provided improved safety and communications on the risks of contaminant exposure
- Provided a mechanism for banning the most harmful chemicals
- Provided a process and funding to address brownfield sites

In the next chapter, we will discuss other land-related environmental regulations that address the land either directly or indirectly.

7.8 REFERENCES

American Society for Testing Materials. 2005. *Standard Practice for Environmental Site Assessments*. E1527–05. ASTM. West Philadelphia, PA.

Eckerman, I. 2005. *The Bhopal Saga—Causes and Consequences of the World's Largest Industrial Disaster*. Universities Press. London, UK. 284p.

Markell, D. 2014. *An Overview of TSCA, Its History and Key Underlying Assumptions, and Its Place in Environmental Regulation.* New Directions in Environmental Law. Washington University Journal of Law and Policy. Washington, DC. Vol. 32. No. 448. pp. 333–376.

National Law Review. 2016. What's New about the Revised TSCA—Toxic Substance Control Act. www.natlawreview.com/article/tsca. (Accessed March 24, 2021).

Rogers, D. T. 1992. The Importance of Site Observation and Follow-Up Environmental Site Assessment—A Case Study. In: *Groundwater Management Book 12.* National Groundwater Association. Dublin, OH. pp. 563–573.

Rogers, D. T. 2020. *Urban Watershed; Geology, Contamination, Environmental Regulations, and Sustainability.* CRC Press. Boca Raton, FL. 606p.

Schierow, L. J. 2013. *The Toxic Substance Control Act (TSCA): A Summary of the Act and Its Major Requirements.* Congressional Research Service (CRS) Report to Congress. Washington, DC. 16 pages.

United States Department of Transportation. 2013. Federal Motor Carrier Safety Administration (FMCSA) Placard Requirements. www.fmcsa.dot.gov. (Accessed March 24, 2021).

United States Department of Transportation. 2021. *Summary of the Federal Hazardous Material Transportation Act.* Office of Hazardous Materials Safety. www.USDOT.gov./HMTA. (Accessed March 24, 2021).

United States Environmental Protection Agency (USEPA). 2005. *Standards and Practice for All Appropriate Inquiries.* 40 Code of Federal Regulation (CFR), Part 312. US Government Printing Office. Washington, DC.

United States Environmental Protection Agency (USEPA). 2014. The Toxic Substance Control Act: History and Implementation. USEPA, Washington, DC. https://www.epa.gov/tsca/summary. (Accessed June 14, 2021).

United States Environmental Protection Agency, USEPA. 2015. *List of Lists.* EPA 55-B-15-001. Office of Solid Waste and Emergency Response. Washington, DC. 125p.

United State Environmental Protection Agency. 2021a. Hazardous Materials Transportation Act. www.epa.gov/HMTA/overview. (Accessed March 24, 2021).

United States Environmental Protection Agency. 2021b. Summary of the Toxic Substance Control Act. https://epa.gov/laws-regulations/summary-toxic-substances-control-act. (Accessed March 24, 2021).

United States Environmental Protection Agency (USEPA). 2021c. Superfund Amendments and Reauthorization Act (SARA). www.epa.gov/SARA. (Accessed March 24, 2021).

United States Environmental Protection Agency (USEPA). 2021d. Emergency Planning and Community-Right-to-Know Act (EPCRA). www.epa.gov/EPCRA. (Accessed March 24, 2021).

United States Environmental Protection Agency (USEPA). 2021e. Regional Screening Levels. www.epa.gov/RSLs. (Accessed March 24, 2021).

United States Environmental Protection Agency (USEPA). 2021f. Pollution Prevention Act. www.epa.gov/pollution-prevention-act. (Accessed March 24, 2021).

United States Environmental Protection Agency. 2021g. Innocent Landowner Defense. www.epa.gov/enforcement/innocent-landowner. (Accessed March 24, 2021).

United States Environmental Protection Agency. 2021h. Brownfields. www.epa.gov/brownfields/all-appropriate-inquiry. (Accessed March 24, 2021).

8 Land-Related Environmental Regulations of the United States

8.1 INTRODUCTION

This chapter addresses the remaining environmental regulations related to land we have not discussed up to this point. The regulations covered in this chapter are very important and cover a wide range of interests and topics that fall into five general categories (USEPA 2021a, 2021b):

1. Workplace Safety

   - Occupational Safety and Health Act

2. Food Safety

   - Insecticide Act
   - Food, Drug, and Cosmetic Act
   - Insecticide, Fungicide, and Rodenticide Act
   - Food Quality Protection Act

3. Energy Generation

   - Federal Power Act of 1935
   - Federal Power Act of 1954
   - Nuclear Waste Policy Act
   - Energy Policy Act of 1992 and 2005

4. Wildlife, Historic Landmarks, and Parks

   - Lacey Act
   - Antiquities Act
   - Migratory Bird Treaty Act
   - National Parks Act
   - National Historic Preservation Act
   - Endangered Species Act

DOI: 10.1201/9781003150107-8

172 Environmental Compliance Handbook

5. Natural Resource Exploitation
  * Mineral Leasing Act
  * National Forest Management
  * Surface Mining Control and Reclamation Act

We shall discuss each act separately in the following sections.

8.2 INSECTICIDE ACT; FEED, DRUG, AND COSMETIC ACT; AND FEDERAL INSECTICIDE, FUNGICIDE, AND RODENTICIDE ACT

The Insecticide Act of 1910; the Feed, Drug, and Cosmetic Act of 1938; and the Federal Insecticide, Fungicide and Rodenticide Act of 1972 are discussed together in this section because they are closely related and evolved together in the development of protecting human health from chemical ingredients in pesticides, fungicides, rodenticides, and cosmetics.

8.2.1 THE INSECTICIDE ACT OF 1910

The Insecticide Act of 1910 was established to protect the sale of pesticides from misbranding and established a set of minimal standards for pesticide products sold within the United States (USEPA 2021c). Specifically, the chemicals referred to as Paris green, lead arsenate, and various fungicides were targeted. This act was passed in response to concerns from the United States Department of Agriculture and many farm groups about the sale of fraudulent or substandard pesticide products listed, which was common at the time. This act represented the first set of regulations that required that listed ingredients must be within certain percentages contained within the product. Pesticide containers were required to be marked with a serial number that was specific to the manufacturer. The labels indicated that the product was guaranteed under the act and were often embossed with a brand and identity of the product. The purpose of the act was to stop what was termed "home-brew" operations (Conn et al. 1984).

8.2.2 THE FOOD, DRUG, AND COSMETIC ACT

The Food, Drug, and Cosmetic Act of 1938 gave authority to the Food and Drug Administration (FDA) to oversee the safety of food, drugs, and cosmetics. The act was passed as a result of the deaths of over 100 patients due to prescribed medication that turned out to be poison. The act was amended in 2002 under Section 408 which authorized the FDA to set tolerance limits, or maximum residue limits, of pesticides in food (USEPA 2021d). To determine what a tolerance or residue limit should be requires USEPA to evaluate the toxicity of the pesticide itself and its breakdown products (USEPA 2021d). The Office of Pesticide Programs (OPP) regulates the use of all pesticides in the United States and has established maximum levels for pesticide amounts in food with the intended purpose of safeguarding the nation's food supply (USEPA 2021d).

Land-Related Environmental Regulations 173

The projects and programs managed by OPP include (USEPA 2021d):

- Federal Insecticide, Fungicide, and Rodenticide Act
- Pesticide Registration and Extension Act
- Key parts of the Food Quality Protection Act
- Key parts of the Food, Drug, and Cosmetic Act
- Key Parts of the Endangered Species Act
- Assessing Pesticide Risks
- Bed Bugs
- Biotechnology under FIFRA
- Endangered Species Protection Program
- Ingredients Used in Pesticides
- Insect Repellents
- Integrated Pest Management in Schools
- Mosquito Control
- Pesticide Labels
- Pesticide Registration
- Pesticide Registration Review
- Pesticide Tolerances
- Pesticide Environmental Stewardship Program
- Protecting Pets from Fleas and Ticks
- Protecting Pollinators
- Reducing Pesticide Drift
- Worker Safety Protection of Pesticides

8.2.3 The Insecticide, Fungicide, and Rodenticide Act of 1947

Little changed occurred in the regulation of pesticides until after World War II, when the use of synthetic organic pesticides in agriculture became widespread in the United States. In response to the large increase in use of pesticides, the United States passed the Federal Insecticide, Fungicide, and Rodenticide Act of 1947. This act amended and consolidated the previous regulations for the distribution, sale, and use of pesticides (USEPA 2021b).

FIFRA required manufacturers to demonstrate that when using the specific product according to its labeled specifications and applications, it will not generally cause unreasonable adverse effects on the environment beyond that which it was targeted to control. As defined under FIFRA, the term **pesticide** is generally defined as any substance or mixture of substances intended for preventing, destroying, repelling, or mitigating any pest and any substance or mixture of substances intended for use as a plant regulator, defoliant, or desiccant. The term **pest** is defined as any living organism (plant or animal) occurring where it is not wanted or causing damage to crops, humans, animals, or plants. Specifically, FIFRA requires that pesticide products have a certain label that provides information to the user that includes (Conn et al. 1984):

- Manufacturer's name
- Manufacturer's address

174 Environmental Compliance Handbook

- List of ingredients
- Directions for proper use
- Warning statements to protect users, the public
- Potential effects to non-targeted species or plants and animals

FIFRA was significantly amended in 1972 by USEPA and was retitled the Federal Environmental Pesticide Control Act (FEPCA). The most significant revision to FIFRA made by FEPCA was the requirement that manufacturers of pesticides demonstrate that any product would not cause unreasonable adverse effects on the environment (USEPA 2021c).

FEPCA defines the term "unreasonable adverse effects on the environment" to mean (USEPA 2021c):

1. Any unreasonable risk to humans or the environment, taking into account the economic, social, and environmental costs and benefits of the use of any pesticide.
2. A human dietary risk from residues that result from a use of a pesticide in or on any food inconsistent with the standard under Section 408 of the Federal Food, Drug, and Cosmetic Act of 1938.

The significant revisions to FIFRA in 1972 were the direct result of a book by Rachel Carson titled *Silent Spring* (1962), which documented the harmful effects of dichlorodiphenyl-trichloroethane (DDT). DDT is a synthetic organic compound first developed in 1874. The insecticide properties of DDT were not discovered until 1939. DDT was commonly used during World War II for prevention of malaria, typhus, body lice, and bubonic plague. Reportedly, due to its use, cases of malaria dropped from 400,000 in 1946 to virtually none in 1950 (USEPA 2021c). Once it was discovered that DDT worked effectively as an insecticide, its use and applications increased significantly. From 1950 until 1970, more than 40,000 tons were used each year worldwide. In addition, it is estimated that 1.8 million tons or more have been produced globally since 1940 (USEPA 2021d). DDT is still in use in South America, Africa, and Asia as an insecticide to prevent malaria.

8.3 OCCUPATIONAL SAFETY AND HEALTH ACT

The Occupational Safety and Health Act was passed in 1970 and was designed to protect people from injury in the workplace. Authority for enforcing workplace health and safety fell to the Occupational Safety and Health Administration, which is within the United States Department of Labor (2021). However, the first law addressing workplace safety was a state law in Massachusetts passed in 1877. This law focused on elevator safety and establishing fire exits in factories (United States Department of Labor 2021).

OSHA first focused its efforts on safety hazards in the workplace in five different industries (United States OSHA 2021):

- Marine cargo handling
- Roofing and sheet metal work

Land-Related Environmental Regulations

- Meat and meat products
- Transportation equipment (primarily mobile homes)
- Lumber and wood products

Along with establishing safety rules in the workplace, OSHA also addressed health hazards in the workplace that included (United States OSHA 2021):

- Asbestos
- Lead
- Silica
- Carbon monoxide
- Cotton dust

The Occupational Safety and Health Act grants OSHA the authority to issue workplace health and safety regulations. These regulations include (United States OSHA 2021):

- Limits on chemical exposure
- Employee access to hazard information
- Requirements to prevent falls
- Preventing hazards from operating dangerous equipment
- Preventing trench cave-ins
- Preventing exposure to infectious diseases
- Ensuring safety of workers entering confined spaces
- Machine guarding
- Providing respirators and other safety equipment
- Job training

For chemical or substance health risks, OSHA established what is termed a permissible exposure limit. A PEL is a legal limit for exposure of an employee to a chemical substance or physical agent such as noise. A PEL is usually expressed as a time-weighted average (TWA). However, there is also an additional exposure measurement termed short-term exposure limit (STEL). The TWA is usually measured over a time period of 8 hours, which is a typical work shift. Units of measure are usually expressed as parts per million (ppm), milligrams per kilogram (mg/kg), or milligrams per cubic meter (mg/m^3) (United States OSHA 2021).

OSHA also requires that employers must do the following (United States OSHA 2021):

- Find and correct safety and health problems
- Eliminate or reduce hazards by changing working conditions
- Use personal protective clothing when elimination is not feasible
- Use less harmful chemicals
- Enclose processes that emit harmful chemicals
- Use proper ventilation
- Inform workers of chemical hazards
- Provide appropriate training
- Record keeping
- Provide personal protective equipment

176　　　Environmental Compliance Handbook

- Provide hearing exams
- Post OSHA citation
- Conduct appropriate testing such as air sampling
- Post illness and injury summary data annually
- Notify OSHA within 8 hours of a workplace fatality
- Notify OSHA within 24 hours of all work-related inpatient hospitalizations, all amputations, and all losses of an eye
- Prominently display the official OSHA Job Safety and Health "It's the Law" poster that describes the rights of employees and responsibilities of employers under the Occupational Safety and Health Act
- Not retaliate or discriminate against workers for using their rights under the law, including their right to report a work-related injury or illness

Note that OSHA requires employers to evaluate whether the use of less harmful chemicals is possible in an effort to add protection to employee safety. This item is significant not only as an OSHA requirement but as an important protective action for the environment and will be described in greater detail later in this chapter, as well as in subsequent chapters.

8.4 LACEY ACT OF 1900

The Lacey Act of 1900 is a conservation law that prohibits trade in wildlife, fish, and plants that have been illegally taken, possessed, transported, or sold. Essentially, the Lacey Act made poaching illegal. The act also regulated the introduction of birds and other animals to places where they have never existed. This is now commonly referred to as invasive species. The Lacey Act is still active and has been amended several times, the latest occurring in 2008 (United States Fish and Wildlife Service 2021a).

8.5 ANTIQUITIES ACT OF 1906

The Antiquities Act of 1906 gave the president of the United States the power to create national monuments with the purpose to protect historic landmarks, historic and prehistoric structures, and other objects of historic or scientific interest. The Antiquities Act was passed as a direct result of concerns over theft from destruction of archaeological sites and was designed to provide an expeditious means to protect federal lands and resources. President Theodore Roosevelt used the authority in 1906 to establish Devil's Tower in Wyoming as the first national monument and used the act 36 times during his presidency, which is the most for any president. Sixteen of the 19 presidents since have used this authority and have created 151 monuments. Some worthy of note include (United States National Park Service 2021a):

- Grand Canyon National Park
- Grand Teton National Park
- Zion National Park
- Olympic National Park
- Statue of Liberty

Land-Related Environmental Regulations

Since it has been enacted, hundreds of millions of acres of land have been set aside and protected, including marine environments (United States National Park Service 2021a).

8.6 MIGRATORY BIRD TREATY ACT 1916

The Migratory Bird Treaty Act of 1916 is an environmental treaty between the United States and Canada. The act makes it unlawful to pursue, hunt, take, capture, kill, or sell birds that appear on the protected list. The act does not discriminate between live or dead birds and also grants full protection to any bird parts, including feathers, eggs, and nests. Currently, there are over 800 species of birds on the list (United States Fish and Wildlife Service 2021b).

8.7 NATIONAL PARKS ACT OF 1916

The National Parks Act of 1916 is also known as the Organic Act and established the National Park Service so that management of all the national parks would be conducted under one authority (United States National Park Service 2021b). These areas include a diverse variety of land of national importance and include (United States National Park Service 2021b):

- National Parks
- National Memorials
- National Battlefields
- National Parkways
- National Seashores
- National Scenic Trails
- National Heritage Areas

- National Monuments
- National Military Parks
- National Historic Sites
- National Recreation Areas
- National Scenic Riverways
- National Cemeteries
- Nationals Scenic Rivers

From the time Yellowstone National Park was created in 1872, each national park or other lands deemed of national importance was managed separately. The passage of the National Parks Act of 1916 combined management of all these protection lands and locations under the newly created National Parks Service.

8.8 MINERAL LEASING ACT OF 1920

The Mineral Leasing Act of 1920 governs leasing public lands for developing deposits of coal, petroleum, natural gas, and other hydrocarbons, in addition to phosphates, sodium, sulfur, and potassium. Previously, the General Mining Act of 1872 authorized citizens to freely prospect for minerals on public lands and allowed a discoverer to stake claims to both minerals and surrounding lands for development. This open-access policy enabled a major oil rush in the western United States and quickly became too much for the government to handle, thus prompting the passage of the Act in 1920 (United States Bureau of Land Management 2021).

The Bureau of Land Management (BLM), a division of the Department of the Interior (DOI), is the principal administrator of the Mineral Leasing Act. BLM evaluates areas for potential development and awards leases based on whoever pays the highest price during a period of competitive bidding.

8.9 FEDERAL POWER ACT OF 1935

The Federal Power Act of 1935 grew out of the Federal Water Power Act of 1920 and was established to more effectively coordinate the development of hydroelectric projects in the United States. Prior to this, the authority to develop hydroelectric power was left up to individual states. Before passage of the act, competing interests between private and public interest groups and companies inhibited any progress. One group desired to keep shipping and the other desired damming and hydroelectric development, while others pursued preservation and others recreation (United States Bureau of Reclamation 2021).

8.10 ATOMIC ENERGY ACT OF 1954

The Atomic Energy Act of 1954 amended an earlier effort in 1946 and covered both civilian and military uses of nuclear materials. It covered the development, regulation, and disposal of nuclear materials. The amended act made it possible for private companies to develop nuclear power generation plants, which was not part of the original 1946 act (United States Nuclear Regulatory Commission 2021).

8.11 NATIONAL HISTORIC PRESERVATION ACT OF 1966

The National Historic Preservation Act of 1966 was intended to preserve historical and archaeological sites in the United States. The act created the National Register of Historic Places. Although the act was not officially passed until 1966, many landmarks in the United States have been protected, such as George Washington's home, Mount Vernon, and Thomas Jefferson's home. Protection of these historic places did not seem appropriate under the Antiquities Act of 1906. Supporters and historians desired a more specifically tailored act that was specific to historical places (United States Advisory Council on Historic Preservation 2021).

8.12 ENDANGERED SPECIES ACT OF 1973

The Endangered Species Act of 1973 was designed to protect animal species from extinction as a consequence of human interference from economic growth and development untempered by adequate concern and conservation. The act was established in an attempt to protect identified endangered species from extinction at any cost (United States Fish and Wildlife Service 2021c). The act is administered by the Fish and Wildlife Service (FWS) and the National Oceanic and Atmospheric Administration.

Land-Related Environmental Regulations

Listing status is a sliding criterion that indicates the degree to which a certain species is threatened by extinction. The listing status used in the United States is as follows (United States Fish and Wildlife Service 2021c):

- E = Endangered. Endangered is the highest of the alerts, meaning the closest to extinction in the United States, and generally means any species that is in danger of extinction throughout all or a significant portion of its natural habitat range.
- T = Threatened. Threatened is any species that is likely to become endangered within the foreseeable future throughout all or a significant portion of its natural habitat range.
- C = Candidate. Candidate is a species that is under consideration due to a drop in population or loss of a significant part of its natural habitat.

The first list of endangered species in the United States included 14 mammals, 36 birds, 6 reptiles and amphibians, and 22 fish. According to the United States Fish and Wildlife Service, there are currently a total of 711 animal species on the list and 941 plant species, for a total of 1,652 just in the United States (United States Fish and Wildlife Service 2021c). Some of the animal species in the United States that are or were on the endangered species list include the following (United States Fish and Wildlife Service 2021c):

- Bald Eagle—removed in 2007
- California Condor
- Peregrine Falcon—removed in 1999
- Mexican Wolf
- Grey Whale
- California Southern Sea Otter
- Hawaiian Goose
- Black-Footed Ferret
- Whooping Crane
- Kirkland's Warbler
- Gray Wolf
- Red Wolf
- Grizzly Bear—removed in 2007
- Florida Key Deer
- Virginia Big-Eared Bat
- Florida Grasshopper Sparrow

Worldwide, there are thousands of species of animals and plants that are in danger of extinction because of direct human involvement, including development, habitat loss and destruction, poaching, pollution, and other factors. The World Wildlife Fund maintains a general list of those animals that are in danger and also includes an additional category different than that of the United States called critically endangered. Some of the animals currently listed by the World Wildlife Fund as critically endangered or endangered include (World Wildlife Fund 2021):

- Amur Leopard
- Cross River Gorilla
- Malayan Tiger
- South China Tiger
- Sumatran Rhino
- Black Rhino
- Eastern Lowland Gorilla
- Sumatran Orangutan
- Sumatran Elephant
- Sumatran Tiger
- Bornean Orangutan
- Javan Rhino
- Orangutan
- Mountain Gorilla
- Amur Tiger

180   Environmental Compliance Handbook

- Amur Leopard
- Asian Elephant
- Bluefin Tuna
- Chimpanzee
- Hectors Dolphin
- Indochinese Tiger

- Sea Lion
- Tiger

- Black Rhino
- Western Lowland Gorilla
- Blue Whale
- Borneo Pygmy Elephant
- Ganges River Dolphin
- North Atlantic Wright Whale
- Snow Leopard
- Black Spider Monkey

- Bornean Orangutan
- African Elephant
- Bonobo
- Galapagos Penguin
- Indian Elephant
- Red Panda

- Sri Lanka Elephant
- African Giraffe

This list includes only a few of the more familiar animals that are now considered threatened by extinction. Unless major efforts from the world community are made immediately, the loss of these animals and countless others is likely inevitable. The following list includes some North American species that have gone extinct just since 1500 AD (International Union on the Conservation of Nature [IUCN] 2021):

- Great Auk—1852
- California Golden Bear—1922
- Labrador Duck—1878
- Cascade Mountain Wolf—1940
- Mexican Grizzly Bear—1964
- Sea Mink—1860
- South Rocky Mountain Wolf—1935
- Carolina Parakeet—1918
- Heath Hen—1932
- Bachman's Warbler—1988

- Passenger Pigeon—1914
- Ivory-Billed Woodpecker—1942
- Eastern Cougar—2011 declared
- Eastern Elk—1887
- Newfoundland Wolf—1911
- Caribbean Monk Seal—1952
- Southern California Kit Fox—1903
- Dusky Seaside Sparrow—1987
- Eskimo Curlew—1981
- Tacoma Pocket Gopher—1970

According to the World Wildlife Fund (2021) and the International Union for Conservation of Nature (2021), hundreds of species of animals are now extinct directly due to humans in several different ways, including hunting, habitat destruction, poisoning, development, and pollution. According to the United Nations, more than 1 million species of animals and plants are now extinct.

8.13 NATIONAL FOREST MANAGEMENT ACT OF 1976

The National Forest Management Act of 1976 was enacted to establish guidelines for the management of renewable resources on national forest land. The act specifically addresses timber harvesting within national forests. The purpose of the act was to protect national forests from permanent damage from excessive logging and clear cutting practices (United States Forest Service 2021). Throughout the history of the United States, there have been numerous legal battles and conflicts between environmental groups and the timber industry.

The demand for timber for home building, farming, and other uses in the United States has always seemed to be high and has placed significant pressure on the

Land-Related Environmental Regulations   181

environment through habitat destruction. William Penn was aware of importance of healthy forests in 1681 when he required that 1 acre remain preserved for every 5 acres of forest cleared (Pennsylvania Historical Association 2021).

One of the more contentious conflicts was over the northern spotted owl in the Pacific Northwest. The northern spotted owl experienced a decrease in population levels that threatened its extinction. A major study was undertaken to evaluate the causes of the population decrease, and the results indicated that there were three major threats to the owl population decrease that involved timber harvesting: (1) variability of birth and death rates through time, (2) loss of genetic variation, and (3) random catastrophes (United States Fish and Wildlife Service 2021b).

The passage of the act required the National Forest Service to conduct an inventory of all national forests and to use a systematic and interdisciplinary approach to resource management, which also included monitoring biological effects of timber harvesting (United States Forest Service 2021). Although deforestation in the United Sates has not been as great as we will discover in other countries of the world in the next chapter, the decrease is still significant. In 1620, approximately 4.1 million square kilometers of the United States was forest. The low point was in 1926 at 2.9 million square kilometers. Since 1926, the amount of forested land in the United States is 3.0 million square kilometers (United States Department of Agriculture 2014).

Figure 8.1 shows the amount of forest in the United States in 1620, and Figure 8.2 shows the amount in 1926.

FIGURE 8.1   Amount of forest in the United States in 1620 (from United States Department of Agriculture. *U. S. Forest Resource Facts and Historical Trends.* www.fia.fs.fed.us/library/brochures/docs/2012/ForestFacts_1952-2012_English.pdf [accessed March 25, 2021], 2014).

FIGURE 8.2 Amount of forest in the United States in 1926 (from United States Department of Agriculture. *U. S. Forest Resource Facts and Historical Trends.* www.fia.fs.fed.us/library/brochures/docs/2012/ForestFacts_1952-2012_English.pdf [accessed March 25, 2021], 2014).

8.14 SURFACE MINING CONTROL AND RECLAMATION ACT OF 1977

The Surface Mining Control and Reclamation Act of 1977 was enacted for the purpose of addressing the environmental effects of coal mining. The act created two programs (United States Office of Surface Mining 2021):

- Regulating active coal mines
- Reclaiming abandoned mine lands

The act also created the Office of Surface Mining within the United States Department of Interior to ensure consistency and fair application of the requirements under the act (United States Office of Surface Mining 2021).

The act was developed because of environmental concerns with the effects of surface coal mining or strip mining. Coal has been mined in the United States since the 1740s, but surface or strip mining did not become widespread until the 1930s and by the mid-1970s accounted for more than 60% of coal mining. The act established the following (United States Office of Surface Mining 2021):

- Uniform Standards of Performance. This provision standardized what was required nationwide and eliminated confusion and differences in regulations from state to state.

Land-Related Environmental Regulations 183

- Permitting. The act required that each proposed mine obtain a permit that described:
  - Pre-mining environmental conditions and land use
  - Proposed mining activities and area
  - Post-mining reclamation
  - How the mine will achieve act requirements
  - How the land will be used following completion of mining activities

- Bonding. The act required that mining companies post a bond sufficient to cover the cost of reclaiming the mined land.
- Inspection and Enforcement. The act permitted inspections and enforcement of violations by the federal government.
- Land Restrictions. The act prohibited mining activities within national parks and wilderness areas. It also allowed for citizen challenges to proposed mining developments.

The act also created a reclamation fund to assist in funding reclamation for mines closed before the passage of the act. Some funds are used to pay for emergencies such as landslides, land subsidence, fires, and high-priority cleanups (United States Office of Surface Mining 2021).

8.15 NUCLEAR WASTE POLICY ACT OF 1982

The Nuclear Waste Policy Act of 1982 established a program with the intention of providing a safe and permanent location for the disposal of highly radioactive wastes. During the 40 years before enactment of the law, there was no regulatory framework in place for disposal of nuclear wastes generated in the United States (Nuclear Regulatory Commission [NRC] 2021).

Some of the waste that was generated had a half-life of more than a million years and was temporarily stored in various locations and types of containers. Most of the wastes generated were as a result of manufacturing nuclear weapons. Approximately 77 million gallons of nuclear waste had been generated and was being stored in South Carolina, Washington, and Idaho. In addition, there were 82 nuclear power plants that produced electricity and also generated nuclear waste in the United States in 1982. The nuclear waste at the nuclear power plants consisted of spent fuel rods, which were stored in water at the reactor sites, and many of the plants were in danger of running out of storage space (NRC 2021).

The Nuclear Waste Policy Act created a timetable and procedure for establishing a permanent repository for high-level radioactive waste by the mid-1990s and providing temporary storage for those sites that were running out of space. The USEPA was directed to establish public health and safety standards for potential releases of radioactive materials, and the NRC was required to promulgate regulations that covered construction, operation, and closure of repository locations. Generators of nuclear waste were to be charge a fee to cover the costs of disposal of nuclear wastes (NRC 2021). Another provision of the act required the secretary of energy

184 Environmental Compliance Handbook

to issue guidelines for selecting sites of two permanent underground nuclear waste repositories. The Department of Energy (DOE) studied sites in Washington near the Hanford Nuclear Reservation; in Nevada near the nuclear testing site; and in Utah, Texas, Louisiana, and Mississippi. Other sites were studied along the east coast of the United States but were quickly ruled out as possible storage locations (NRC 2021).

In 2002, Yucca Mountain was selected as the only site to be characterized as the permanent repository for all the nation's nuclear waste. However, the state of Nevada used its veto power, rejected the recommendation, and challenged the recommendation in court. The United States Court of Appeals upheld the state of Nevada's appeal, ruling that USEPA's 10,000-year compliance period for site selection was too short and was not consistent with the National Academy of Sciences' recommended compliance period of 1 million years.

USEPA subsequently revised the standard to 1 million years. At issue is selecting a disposal site that will safely contain the nation's nuclear waste for 1 million years. A license application selecting the Yucca Mountain location was submitted in 2008 and is still under review by the NRC. As of 2010, several lawsuits have been filed in federal courts to contest the legality of the selection process and proposed location of Yucca Mountain as the selected permanent repository site (NRC 2021; USEPA 2021e). In the meantime, nuclear waste is being stored at various locations in temporary containers.

8.16 ENERGY POLICY ACT OF 1992 AND 2005

The Energy Policy Act of 1992 was enacted to set goals and mandates to increase clean energy use, improve overall energy efficiency, and lessen the nation's dependence on imported energy. The act set standards for many different sectors of the economy, including establishing standards for buildings, utilities, equipment standards, renewable energy, and alternative fuels. (National Energy Institute 2005). The act was amended in 2005 to provide tax incentives for alternative energy development such as wind energy (see Figure 8.3). The act also made it easier for a new technology to expand. such as oil fracking, by exempting waste fluids generated from the Clean Air Act, Clean Water Act, Safe Drinking Water Act, and CERCLA (National Energy Institute 2005).

An unforeseen disadvantage to wind turbines is that they kill birds, especially raptors, migratory birds, and bats. Some estimates indicate 250,000 to several million birds are killed annually in the United States, and this will likely increase as more wind turbines are installed and old ones are replaced with wind turbines that are much larger and kill even more birds because of increased reach and spin speed of the turbine (Eveleth 2013; Curry 2014). However, vertical wind turbines are much more bird friendly because they are more compact and can be installed in urban areas (see Figure 8.4).

8.17 FOOD QUALITY PROTECTION ACT OF 1996

The Food Quality Protection Act (FQPA) was enacted to standardize the way USEPA would manage the use of pesticides and amended FIFRA and FFDCA. It mandated a

Land-Related Environmental Regulations 185

FIGURE 8.3 Wind farm in Washington state (photo by Daniel T. Rogers).

FIGURE 8.4 Vertical wind turbine in London, England (photograph by Daniel T. Rogers).

health-based standard for pesticides used in foods, provided for special protection for babies and infants, streamlined the approval process of safe pesticides, established incentives for the creation of safer pesticides, and required pesticide registrations remain current. The FQPA established a new safety standard that must be applied to all food commodities, "reasonable certainty of no harm." In addition, USEPA was required to consider this new risk standard as it applies to babies and infants. The FQPA required that re-testing of all existing pesticide tolerance levels be conducted within 10 years and that USEPA must account for "aggregate risk" and "cumulative exposure" to pesticides with similar mechanisms of toxicity.

As a result of the FQPA, USEPA banned methyl parathion and azinphos methyl because of the risks posed by these two pesticides to children. In 2000, USEPA banned an additional pesticide, chloropyrifis, that was common in agriculture, household cleaners, and commercial pest control products because of concerns about children's health (USEPA 2021f). FQPA further raised debate over using clinical studies of pesticides on humans. In 2004, the National Academy of Science released a report supporting the use of clinical studies under very strict regulations, citing that the benefit outweighed the risk to the individual. In 2005, USEPA adopted the Human Studies Regulation that allows human studies, with the exception of pregnant women or children, and mandates a strict ethical code (USEPA 2021f).

8.18 SUMMARY OF ENVIRONMENTAL REGULATIONS OF THE UNITED STATES

This chapter has summarized the remaining land-related environmental regulations we had not discussed up to this point. The regulations covered in this chapter are very important and cover a wide range of interests and topics that fall into five general categories:

1. Workplace Safety

   - Occupational Safety and Health Act

2. Food Safety

   - Insecticide Act
   - Food, Drug, and Cosmetic Act,
   - Insecticide, Fungicide, and Rodenticide Act
   - Food Quality Protection Act

3. Energy Generation

   - Federal Power Act of 1935
   - Federal Power Act of 1954
   - Nuclear Waste Policy Act
   - Energy Policy Act of 1992 and 2005

4. Wildlife, Historic Landmarks, and Parks

   - Lacey Act
   - Antiquities Act

Land-Related Environmental Regulations 187

- Migratory Bird Treaty Act
- National Parks Act
- National Historic Preservation Act
- Endangered Species Act

5. Natural Resource Exploitation

- Mineral Leasing Act
- National Forest Management
- Surface Mining Control and Reclamation Act

The next several chapters will focus on land environmental regulations and pollution worldwide. We will focus on industrialized nations and a few developing nations with large populations, including India and China. We will not be examining countries with what are considered extreme hardships, including Somalia, Syria, Afghanistan, Iran, and others.

We will evaluate how countries' environmental laws compare to those of the United States. Some findings might be rather surprising in that it will become apparent that a few countries are more advanced at protecting human health and the environment than the United States. More surprising still might be that impression that some countries either lack environmental laws or the will to enforce the environmental laws they have due to other overriding social or economic challenges. As we shall see in the next several chapters, it is not enough just to have environmental regulations. Many other factors are required in order to have an effective environmental regulatory framework that actually accomplishes the ultimate goal of protecting human health and the environment.

8.19 REFERENCES

Conn, R. L., Leng, M. L. and Solga, J. R. 1984. *Pesticide Regulatory Handbook*. Executive Enterprises Publishing Company. New York, NY. 450p.

Curry, A. 2017. *Will Newer Wind Turbines Mean Fewer Bird Deaths*. The National Geographic. www.nationalgeographic.com/news/energy/2014/04/140427altamont-pass. (Accessed March 25, 2021).

Eveleth, R. 2013. *How Many Birds Do Wind Turbines Really Kill?* Smithsonian Magazine. Washington, DC. www.smithsoniation.com/smart-new/how-many-birds-do-wind-turbines-kill. (Accessed March 25, 2021).

International Union on the Conservation of Nature (IUCN). 2021. Extinct Species in North America since 1500 AD. www.iucn.org. (Accessed March 25, 2021).

National Energy Institute. 2005. Summary of the Energy Policy Act of 1992 and 2005. www.nei.org/summary-of-energy-policy-act. (Accessed March 25, 2021).

Nuclear Regulatory Commission. 2021. Nuclear Waste Policy Act of 1982. www.nrc.gov/NWPA. (Accessed March 25, 2021).

Pennsylvania Historical Association. 2021. The Vision of William Penn. http://explorepahistory.com/story.php?storyId=1-9-3&chapter=1. (Accessed March 25, 2021).

United States Advisory Council on Historic Preservation. 2021. National History Preservation Act. www.achp.gov. (Accessed March 25, 2021).

United States Bureau of Land Management (BLM). 2021. Summary of the Mineral Leasing Act. https://.www.blm.gov/mineral-leasing-act.html. (Accessed March 25, 2021).

United States Bureau of Reclamation. 2021. Summary of the Federal Power Act. www.usbr.gov/federal-power-act/html. (Accessed March 25, 2021).

United States Department of Agriculture. 2014. United States Forest Resource Facts and Historical Trends. www.fia.fs.fed.us/library/brochures/docs/2012/ForestFacts_1952-2012_English.pdf. (Accessed March 25, 2021).

United States Department of Labor. 2021. History of the Occupational Safety and Health Administration. www.osha.gov./history. (Accessed March 25, 2021).

United States Environmental Protection Agency (USEPA). 2021a. Laws and Regulations. www.epa.gov/laws-regulations/regulations. (Accessed March 25, 2021).

United States Environmental Protection Agency. 2021b. Summary of Environmental Regulations of the United States. https://epa.gov/summary-environmental-laws. (Accessed March 25, 2021).

United States Environmental Protection Agency. 2021c. Pesticide Laws and Regulations. www.epa.gov./pesticide. (Accessed March 25, 2021).

United States Environmental Protection Agency (USEPA). 2021d. Summary of the Food Quality Protection Act. www.epa.gov/fqpa-summary. (Accessed March 25, 2021).

United States Environmental Protection Agency (USEPA). 2021e. Summary of the Nuclear Waste Policy Act. www.epa.gov/laws-regulations/summary-nuclear-waste-policy-act. (Accessed March 25, 2021).

United States Environmental Protection Agency (USEPA). 2021f. Addressing Major Public Policy Concerns: Accomplishments under the Food Quality and Protection Act. www.epa.gov/fqpa/policy-concerns. (Accessed March 25, 2021).

United States Fish and Wildlife Service. 2021a. Summary of the Lacey Act. https://.fws.gov/us-conservation-laws/lacey-act-html. (Accessed March 25, 2021).

United States Fish and Wildlife Service. 2021b. Summary of the Migratory Bird Treaty Act. www.nps.gov/migratory-bird-act. (Accessed March 25, 2021).

United States Fish and Wildlife Service. 2021c. Endangered Species Act. www.fws.gov/endangered/laws-policies/. (Accessed March 23, 2021).

United States Forest Service. 2021. Forest Conservation and Management Act. www.nfs.gov/FCMA. (Accessed March 25, 2021).

United States National Park Service. 2021a. Summary of the Antiquities Act. www.nps.gov/history/antiquities-act-html. (Accessed March 25, 2021).

United States National Park Service (NPS). 2021b. History of the National Park Service. www.nps.gov/history. (Accessed March 25, 2021).

United States Nuclear Regulatory Commission. 2021. Summary of the Atomic Energy Act. www.nrc.gov/atomic-energy-act/html. (Accessed March 25, 2021).

United States Occupational Safety and Health Administration. 2021. www.osha.gov. (Accessed March 25, 2021).

United States Office of Surface Mining and Reclamation. 2021. Surface Mining Control and Reclamation Act. www.osmra.gov. (Accessed March 25, 2021).

World Wildlife Fund (WWF). 2021. Species List of Endangered, Vulnerable, and Threatened Animals. www.wwf.org/species/endangered-list. (Accessed March 25, 2021).

9 Review of Global Assessments and Standards for Land Contamination

9.1 INTRODUCTION

Pollution of the planet encompasses numerous aspects of the land, air, and water. However, there are what are called sinks or final resting places of pollution, and the largest sink on Earth is the oceans. In fact, in the geological sciences, there is a common saying that states, "whatever is created eventually ends up in the ocean." Other sinks include the land and in some instances the atmosphere.

International organizations such as the WHO and the UN have begun to assess, on a global scale, the quality of fresh water and air on Earth. A global assessment of the land has not yet been completed. The UN and WHO have both have published guidelines for assessing, improving, or maintaining acceptable air and water quality standards as far back as 1983 (WHO 2013; United Nations 2016a, 2016b). According to WHO (2013, 2015a, 2015b) and the United Nations (2016a, 2016b), approximately 2.8 billion humans lack access to basic sanitation and improved drinking water. While this number is improving, it still represents nearly 30% of the planet's human population.

In addition, according to WHO (2021), 9 out of 10 people on Earth breathe air that is considered unacceptable because it contains high levels of pollution. The United Nations estimates that most land pollution is caused by agricultural activities, such as grazing, use of pesticides, fertilizers, irrigation, plowing, and confined animal facilities (United Nations 2016c).

Significant effort was made to include data tables for land pollution, remediation goals, and others for every chemical compound for every country evaluated. However, this was just not possible to keep this book manageable. Therefore, only select tables listing specific chemicals are provided, and website addresses are provided for those not included. Table 9.1 lists the countries that will be evaluated by continent in the next six chapters.

9.2 CLIMATE AND GEOGRAPHIC INFLUENCES

Global driving forces for issues related to the land include climate change, increasing water scarcity, population growth, demographic changes, and urbanization. These driving forces are expected to place additional stress on almost all activities

DOI: 10.1201/9781003150107-9   189

TABLE 9.1
Countries and Areas in Which Environmental
Regulations Are Examined

North America	Europe
Canada	Austria
Mexico	Belgium
South America	Bulgaria
Argentina	Croatia
Brazil	Cyprus
Peru	Czech Republic
Chile	Denmark
Asia	Estonia
China	Finland
India	France
Japan	Germany
Saudi Arabia	Greece
South Korea	Hungary
Indonesia	Ireland
Malaysia	Italy
Oceania	Latvia
Australia	Lithuania
New Zealand	Luxemburg
Africa	Malta
Egypt	Netherlands
Kenya	Norway
Tanzania	Poland
South Africa	Portugal
Antarctica	Romania
The Oceans	Russia
	Slovakia
	Slovenia
	Spain
	Sweden
	Switzerland
	Turkey
	United Kingdom

conducted on land. As climate change scenarios become increasingly reliable, existing infrastructure will need to be adapted, and planning of new systems and services will need to be updated. Extreme weather conditions are also reflected in the increased frequency and intensity of natural disasters and will become the new normal (WHO 2013). Figure 9.1 shows the climate regions of the world.

The surface of the Earth is approximately 510,100,000 square kilometers (Coble et al. 1987). Approximately 70%, or 361,800,000 square kilometers, is covered by water, and 30%, or 148,300,000 square kilometers, is land. The amount of land on Earth that is farmed is estimated at 40%, or 59,320,000 square kilometers (United

Review of Global Assessments 191

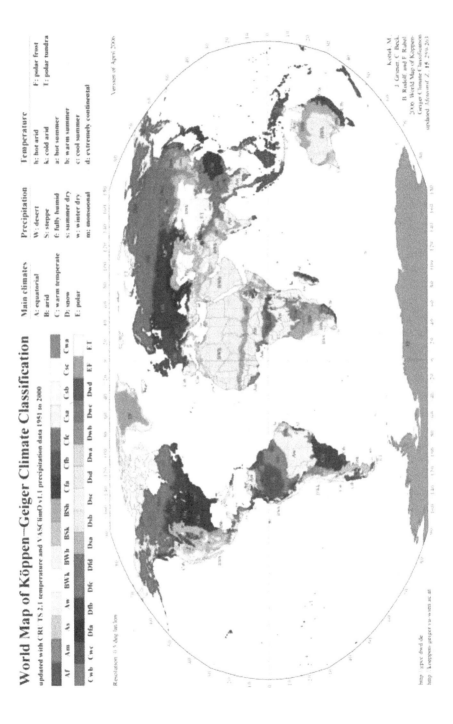

FIGURE 9.1 Climate regions of the world (from National Oceanic and Atmospheric Administration. *Koppen-Geiger Climate Changes*. https://sos.noaa.gov/datasets/koppen-geiger-climate-changes [accessed June 30, 2021], 2021).

Nations 2016c). This is a huge number, especially since the amount of urbanized land on Earth is estimated to be only 2.7% and is home to roughly half the current human population (United Nations 2016c). The amount of land currently used for farming becomes an even larger value since 10.4% of Earth's land, or 15,600,000 square kilometers, is covered by ice and is not farmable; 20%, or 29,660,000 square kilometers, is mountainous; and another 20%, or 29,660,000 square kilometers, is desert and has little or no topsoil (Coble et al. 1987). When factoring out those land areas not suitable for farming, only 7% of land remains available on Earth.

9.3 GLOBAL ASSESSMENTS AND ENVIRONMENTAL STANDARDS FOR LAND POLLUTION

Land pollution is defined as a degradation or even destruction of the Earth's surface, soil, rock, or subsurface as a result of anthropogenic activities. The expansion of housing developments, businesses, industry, infrastructure, and agriculture, all in response to an increasing population over the last several decades, accounts for humans modifying over 50% of the Earth's topsoil (United Nations 2016c). Land also becomes contaminated from the migration of contaminants through the soil column or when contaminants change media, such those carried by the atmosphere perhaps hundreds or even thousands of miles and deposited on the land or water, perhaps in a different country.

The World Health Organization and United Nations have not established guidelines for soil or groundwater. However, most countries that have not developed standards for the remediation of land that includes, soil, sediment, or groundwater commonly refer to a set of standards established by the Dutch in 1987 and revised in 1994, 2001, and 2015, commonly referred to as the Dutch Intervention Values (DIVs) (Lijzen et al. 2001; Netherlands Ministry of Infrastructure and the Environment 2015).

The DIVs are based on potential risks to human health and ecosystems and are technically evaluated. The values presented usually have two integers for each chemical listed. One is a target level, which is considered optimal and likely will result in no or limited health or ecological effects, and the other is a level or concentration above which some intervention is necessary to lower risk posed by site-specific exposure pathways.

The DIVs have been used worldwide as a benchmark for cleanup of contaminated sites in many areas of the world where cleanup values have not been established. In addition, even if there are established values, the Dutch Intervention Values are used for comparison purposes. Therefore, in many instances, the Dutch Intervention Values represent the benchmark for cleanup standards worldwide.

Table 9.2 presents a partial list of Dutch Intervention Values. A complete list can be viewed at www.sanaterre.com/guidelines/dutch.html (Lijzen et al. 2001; Netherlands Ministry of Infrastructure and the Environment 2015).

9.4 SUMMARY OF GLOBAL ASSESSMENTS AND STANDARDS

It may not be clearly noticeable at this point, but there is a general lack of enforcement of international standards for dealing with pollution. As we shall see during our journey through the environmental regulations of select countries of the world, many

TABLE 9.2
Dutch Intervention Values for Select Compounds for Soil and Groundwater

Compound	Soil		Groundwater	
	Target Value mg/kg	Intervention Value mg/kg	Target Value ug/l	Intervention Value ug/l
Arsenic (As)	29.0	55.0	10	60
Barium (Ba)	160	625	50	625
Cadmium (Cd)	0.8	12	0.4	6
Chromium (Cr)	100.0	380	1	30
Copper (Cu)	36.0	190	15	75
Nickel (Ni)	35.0	210	15	75
Lead (Pb)	85.0	530	15	75
Mercury (Hg)	0.3	10.0	0.05	0.3
Silver (Ag)	None	15	None	40
Selenium (Se)	0.7	100	None	40
Zinc (Zn)	140	720	65	800
Chloride	None	None	100,000	None
Cyanide	None	20	5	1,500
Cyanide (complex)	None	50	10	1,500
Thiocyanate	None	20	None	1,500
Benzene	None	1.1	0.2	30
Ethyl benzene	None	110	4	150
Toluene	None	320	7	1,000
Xylenes (sum)	None	17	0.2	70
Styrene (vinylbenzene)	None	86	6	300
Phenol	None	14	0.2	2.000
Cresols (sum)	None	13	0.2	200
Naphthalene	None	None	0.01	70
Phenanthrene	None	None	0.003	5
Anthracene	None	None	0.0007	5
Vinyl chloride	None	0.1	0.01	5
Dichloromethane	None	3.9	0.01	1,000
Trichloroethene	None	2.5	24	500
Tetrachloroethene	None	8.8	0.01	40
Pentachlorophenol	None	12	0.04	3
Polychlorinated biphenyl (sum)	None	1	0.01	0.01
Monochloroanilines (sum)	None	50	None	30
Dioxin (sum)	None	0.00018	None	None
Chloronaphthalene (sum)	None	23	None	6
Chlordane	None	4	0.00002	0.2
DDT (sum)	None	1.7	None	None
DDE (sum)	None	2.3	None	None
DDD (sum)	None	34	None	None
DDT/DDE/DDD sum	None	None	0.000004	None
Aldrin	None	0.32	0.000009	None
Dieldrin	None	None	0.0001	None
Lindane	None	1.2	0.009	None

(Continued)

194 Environmental Compliance Handbook

TABLE 9.2 *(Continued)*

	Soil		Groundwater	
Compound	Target Value mg/kg	Intervention Value mg/kg	Target Value ug/l	Intervention Value ug/l
Atrazine	None	7.1	0.00029	150
Carbofuran	None	0.017	0.0009	100
Asbestos	None	100	None	None

mg/kg = milligram per kilogram

ug/l = microgram per liter

Source: Netherlands Ministry of Infrastructure and the Environment. 2015. Dutch Pollutant Standards. www.government.nl/ministry-of-infrastructure-and-the-environment or at www.sanaterre.com/guidelines/dutch.html (accessed June 30, 2021), 2015.

countries struggle with meeting existing standards, in part because enforcement is either absent or substandard. This is significant because, as we already know, pollution does not respect political boundaries.

We will now turn our attention from global standards to those of select countries. We shall begin with North America. Since the United States was covered in its own chapter, we will start with a short introduction of North America and then describe the environmental regulations and challenges dealing with pollution in Canada and Mexico. We will then move to Africa, Asia, Oceania, South America, and conclude with Antarctica and the oceans.

9.5 REFERENCES

Coble, C. R., Murray, E. G. and Rice, D. R. 1987. *Earth Science.* Prentice-Hall Publishers. Englewood Cliffs, NJ. 502p.

Lijzen, J. P., Baars, A. J., Otte, P. F., Rikken, M. G., Swartjes, F. A., Verbruggen, E. M. and van Wezel, A. P. 2001. *Technical Evaluation of the Intervention Values for Soil/Sediment and Groundwater.* Institute of Public Health and the Environment. The Netherlands. 147p.

National Oceanic and Atmospheric Administration (NOAA). 2021. Koppen-Geiger Climate Changes. https://sos.noaa.gov/datasets/koppen-geiger-climate-changes. (Accessed June 30, 2021).

Netherlands Ministry of Infrastructure and the Environment. 2015. Dutch Pollutant Standards. www.government.nl/ministry-of-infrastructure-and-the-environment. or at www.sanaterre.com/guidelines/dutch.html. (Accessed June 30, 2021).

United Nations. 2016a. *Global Drinking Water Quality Index and Development and Sensitivity Analysis Report.* UNEP Water Programme Office. Burlington, ON, Canada. 58 pages.

United Nations. 2016b. *World Air Pollution Status.* United Nations News Center. New York, NY. www.un.org/sustainabledevelopment/2016/09. (Accessed June 30, 2021).

United Nations. 2016c. Human Development Report. United Nations Development Programme. www.hdr.undr.org/sites/2017. (Accessed June 30, 2021).

World Health Organization. 2013. Water Quality and Health Strategy 2013–2020. www.who.int/water_sanitation_health/dwq/en/. (Accessed June 30, 2021).

Review of Global Assessments

World Health Organization (WHO). 2015a. *Guidelines for Drinking-water Quality.* Geneva, Switzerland. 516 pages.

World Health Organization. 2015b. *Progress on Drinking Water and Sanitation.* World Health Organization. New York, NY. 90p.

World Health Organization. 2021. *Nine out of Ten People Breathe Unhealthy Air.* Geneva, Switzerland. www.who.int/news-room/detail/02-02-2018. (Accessed June 30, 2021).

10 Land Pollution Regulations of North America

10.1 INTRODUCTION TO NORTH AMERICA

North America is a continent entirely within the northern hemisphere and almost entirely within the western hemisphere. North America covers an area of approximately 24,709,000 square kilometers (9,540,000 square miles) and covers 4.8% of the total surface area of Earth and 16.5% of Earth's land surface (Marsh and Kaufman 2015).

North America is the third-largest continent on Earth, after Asia and Africa, and is fourth in population after Asia, Africa, and Europe. As of 2021, the estimated population of North America was 565 million people in 23 independent states, or about 7.5% of the total human population on Earth (United Nations 2021). The United States at 327 million, Mexico at 130 million, and Canada at 33.4 million make up the majority (approximately 85%) of the human population of North America. Estimates of human habitation of North America range from approximately 40,000 to 17,000 years ago. Together with South America, this age range of human habitation is rather recent compared with other continents, excluding Antarctica (United Nations 2021).

The North American Free Trade Agreement (NAFTA) addresses the issues of environmental protection of the three participating countries, the United States, Canada, and Mexico. However, it leaves establishing environmental rules and standards to the three participating countries, with two exceptions (United States Department of State 2021):

1. They are to comply with existing treaties between themselves.
2. They are not to reduce their environmental standards as a means of prompting investment in business.

As we shall see in the following sections, the three countries that primarily make up North America that we will examine may have similar environmental regulations, but they differ in practice significantly due to social concerns, corruption, lack of enforcement, and physical geography.

10.2 CANADA

Canada is the second-largest country in land area on Earth (9,984,670 square kilometers), with significant natural resources including timber, petroleum, and

DOI: 10.1201/9781003150107-10

198   Environmental Compliance Handbook

minerals. The population of Canada estimated in 2009 was 33,487,208, which is about 10% of the population of the United States. It is a significant trade partner of the United States; in fact, approximately 75% of exported goods from Canada are to the United States, and the majority of the population of Canada lives within 200 kilometers of the US–Canadian border (Canadian Government 2021). Canada closely resembles the United States in its free market-oriented economic system and pattern of industrial production. Canada and the United States also share several industries (Ganadian government 2021). Since the early 1970s, provincial governments have adopted several environmental laws and regulations pertaining to the protection or enhancement of the environment. These laws and regulations require duties and obligations on individuals or companies that cannot be ignored and, in the event of non-compliance or of an environmental occurrence, release, or incident, can potentially lead to individual or company liability not only under the environmental laws but also in a civil or common law context. In addition, the same can be said for the laws that have been adopted at the federal level (Canadian Government 2021).

Canada was given the right to self-govern from England in 1849 and officially became its own country in 1867. The Constitution Act of 1867 described which level of government (federal or provincial) can do within its jurisdiction. However, environmentally related issues had not yet been contemplated. As a result, legislative and enforcement authority over the environment is split between the federal and provincial governments, each of which has adopted its own set of rules (Canadian Government 2021). This has led to an extra level of complexity, with each province being different, similar to that of the United States.

The federal government has exclusive jurisdiction over criminal law; coastal and inland fisheries; navigation; federal works; and laws for the peace, order, and good government. The provinces have jurisdiction over property and civil rights as well as matters of a local and private nature. Provinces can also delegate many environmental powers to municipalities. As a result of this delegation of power, municipalities regulate matters such as noise, nuisances, pesticides, sewers, and land use planning. As an example, the provincial government of Quebec has delegated responsibility over air emissions and waste water discharges to the sewer system and waterways within its boundary to the Montreal Metropolitan Community (Canadian Government 2021).

10.2.1 Environmental Regulatory Overview of Canada

The main environmental laws at the federal level in Canada include (Environment Canada 2021):

- Canadian Environmental Protection Act (CEPA) of 1999
- Fisheries Act of 2012
- Transportation of Dangerous Goods Act
- Species at Risk Act
- Canadian Environmental Assessment Act

Land Pollution Regulations, North America

The provinces and territories of Canada have adopted some form of the environmental legislation at the federal level. The following is a general list of each province and dominant environmental law (Environment Canada 2016; Canadian Environmental Protection Act 1999):

- Province of Alberta: the main environmental regulations are included in the Environmental Protection and Enhancement Act (Province of Alberta 2021).
- Province of British Columbia: the Environmental Management Act (Province of British Columbia 2021).
- Province of Manitoba: the Environment Act (Province of Manitoba 2021).
- Province of New Brunswick: the Clean Environment Act (Province of New Brunswick 2021).
- Provinces of Newfoundland and Labrador: the Environment Act (Province of Newfoundland and Labrador 2021).
- Province of Nova Scotia: the Environment Act (Province of Nova Scotia 2021).
- Province of Ontario: the Environmental Protection Act, Ontario Water Resources Act, Safe Drinking Water Act, Endangered Species Act, and the Environmental Assessment Act, which constitutes the majority of legislation in Ontario (Province of Ontario 2021).
- Province of Quebec: the main environmental legislation is called the Environmental Quality Act, under which a series of subsections address hazardous materials, biomedical wastes, residual materials, air quality, and contaminated sites (Province of Quebec 2021).
- Province of Saskatchewan: the Environmental Management and Protection Act (Province of Saskatchewan 2021).
- Northwest Territories: the Environmental Protection Act (Province Northwest Territories 2021).
- Yukon Territory: the Environment Act (Province Yukon Territory 2021).

In Canada, the purpose of environmental law is to promote protection of the environment, particularly in the context of sustainable development. In Canada, the "environment" or "natural environment" is defined as the air, land, water (including groundwater), and all other external conditions or influences under which humans, animals, and plants live, are developed, or have dynamic relations. A "contaminant" is defined as any solid, liquid, gas, odor, heat, sound, vibration, radiation, or any combination of these resulting directly or indirectly from human activities (Canadian Environmental Protection Act 1999). Environmental laws in Canada create a prohibition against contaminating the environment and establish duties to protect the quality of the environment and obligations associated with permits and approvals that allow impacts (discharges) to the environment but in a controlled and regulated manner in order to prevent harm to the environment (Environment Canada 2016). Failure to comply may result in sanctions, fines, and administrative orders, much the same as in the United States.

10.2.2 Solid and Hazardous Waste

In Canada, hazardous waste and hazardous recyclable materials are defined as those that are flammable, corrosive, or inherently toxic and are regulated under the Canadian Environmental Protection Act of 1999. Canada acknowledges that many household items may be hazardous, including batteries, computers and other electrical equipment, cleaners, paints, pesticides, herbicides, and others items. Providing a separate definition for hazardous waste and hazardous recyclable material provides regulators with more flexibility in managing waste (Canadian Environmental Protection Act 1999). The Canadian Environmental Protection Act (1999) includes authority to:

- Set criteria to assess the environmentally sound management of wastes and hazardous recyclable materials and refuse to permit import or export if criteria are not met
- Require exporters of hazardous waste to submit export-reduction plans
- Regulate the export and import of prescribed non-hazardous wastes for final disposal
- Control inter-provincial movements of hazardous wastes and hazardous recyclable materials

The Canadian Environmental Protection Act (1999) lists nine different hazardous waste classes:

1. Explosives
2. Gases
3. Flammable liquids
4. Flammable solids
5. Oxidizers
6. Toxics and infectious
7. Radioactive
8. Corrosives
9. Miscellaneous

The basic approach relies on the nine different hazardous wastes classes and analytical tests such as the toxic characteristic leaching procedure (TCLP), which is commonly applied in the United States to evaluate whether a waste should be classified as hazardous (Canadian Environmental Protection Act 1999).

10.2.3 Remediation Standards

Similar to the United States, Canada has developed remediation standards for many chemical compounds. As in the United States, the federal government of Canada has established guidelines so that each province or territory can develop remediation criteria for specific chemical compounds based on toxicity, exposure assumptions, and land use. Remediation standards for the province of Ontario are listed at www.ontario.ca/page/soil-ground-water-and-sediment-standards-use-under-part-xvl-environmental-protection-act (Province of Ontario Canada 2011).

Perhaps the most interesting and significant difference between the Canadian remediation goals and those in the United States is that Canada has established cleanup criteria for agricultural soil, and the United States has not. In most circumstances,

Land Pollution Regulations, North America 201

agricultural land is exempt from environmental regulations under USEPA in the United States. In the United States, circumstances where agricultural land is included in environmental assessments are when there is a proposed land use change and the agricultural land is then developed as residential, commercial, or industrial. It is under this circumstance that the land is evaluated and often investigated for the presence of contamination. The detection of contamination may fall within a gray zone as to whether any detected contamination is the result of agricultural activities, which may be exempt or from other activities such as open dumping, chemical spills, leaking tanks, or other activities not directly associated with agricultural activities.

10.2.4 SUMMARY OF ENVIRONMENTAL REGULATIONS OF CANADA

Canada's population is 10% of the United States, but it has larger land area and is located north of the continental United States. These geographic and demographic factors have influenced environmental laws within Canada and also how they are applied. This is, in part, due to the northern portions of Canada's sensitivity to pollution and human habitation and to much of Canada's natural resources, including oil and natural gas, forest products, and minerals, being located in the northern portions of Canada. Canada has to deal with many pollution issues, namely air and water, caused by the United States since Canada shares its southern border with the United States. This is because pollution does not respect political borders, which is a subject that will repeatedly appear in this chapter.

In general, the regulations of Canada are robust and have been in place long enough to be effective. Canada also has vast areas that are relatively pristine from industrial and urban development, leaving much of the country with limited human-caused environmental impacts. However, having large areas of uninhabited land also places stress on providing infrastructure and pollution prevention activities by increasing the cost caused by distance, isolation, and remoteness.

10.3 MEXICO

The official name of what we commonly refer to as Mexico is the United Mexican States (Organization for the Economic Co-operation and Development [OECD] 2015). It's the world's eighth-largest country by land size at 1,972 million square kilometers, with a population estimated at 120 million as of 2015, which is approximately 35% of the population of the United States and almost four times that of Canada. Historically, Mexico has received criticism for not instituting effective environmental laws and regulations, and then once Mexico developed laws and regulations addressing growing environmental concerns, it was again criticized for not enforcing its own laws (OECD 2015). Over the past 20 years, that has changed. As we shall see, Mexico has made significant strides in improving protection of human health and the environment. In fact, Mexico has streamlined many of the requirements for industry into a single operating permit, which is more than we can say the United States has accomplished. However, Mexico still has a long way to go in two major areas of the environment, water and air. As we shall see in the next few sections, availability of safe drinking water and improving air quality, especially in and around Mexico City, remain key challenges for Mexico.

10.3.1 ENVIRONMENTAL REGULATORY OVERVIEW OF MEXICO

Environmental regulations are present in Mexico and are patterned after the United States, specifically California. The federal government of Mexico, through the Secretariat of the Environment and Natural Resources (SEMARNAT), has sole jurisdiction over acts that affect two or more states, acts that include hazardous waste, and procedures for the protection and control of acts that can cause environmental harm or serious emergencies to the environment. SEMARNAT's main roles include (OECD 2015):

- Develop environmental policy
- Enforce environmental policy
- Assist in urban planning
- Develop rules and technical standards for the environment
- Grant or deny licenses
- Authorize permits
- Decide on environmental impact studies
- Offer opinions and assist states with environmental programs

SEMARNAT enforces the law, regulations, standards, rulings, programs, and limitations issued through the National Environment Institute and the Federal Attorney Generalship of Environmental Protection (PROFEPA). Environmental laws in Mexico address air, water, hazardous waste, pollutants, pesticides, and toxic substances, along with environmental protection, natural resource protection, environmental impact statements, risk evaluation and determination, ecological zoning, and sanctions (OECD 2015).

10.3.2 SOLID AND HAZARDOUS WASTE

In Mexico, the definition of hazardous waste is very similar to that of a characteristic waste in the United States; it is defined as a substance that is explosive, ignitable, corrosive, chemically reactive, or toxic (directly or indirectly) to plants, animals, or humans. Air, water, solid waste, and hazardous waste are all handled similarly to the United States (i.e., California in particular), with one exception: stormwater (Basurto and Soza 2007; Rodriguez et al. 2015; Mexico Environmental and Natural Resource Ministry 2021a). Within many Mexican states, stormwater is of little or no concern due to dry conditions, especially in the northern and western regions of the country. However, that will likely change soon.

Permits, authorizations, and documentation likely required at manufacturing facilities in Mexico include the following (Basurto and Soza 2007; Rodriguez et al. 2015; Mexico Environmental and Natural Resource Ministry 2021a):

- Operating license
- Waste water discharge registration

Land Pollution Regulations, North America

- Hazardous waste generator's manifest
- Monthly log of hazardous waste generation
- Ecological waybills for the incineration and/or exportation of hazardous materials and wastes
- Semi-annual report on hazardous wastes sent to recycling, treatment, or final disposition
- Accidental hazardous waste spill manifest
- Delivery, transport, and receipt of hazardous waste manifests
- Environmental impact studies

In Mexico, the operating license is a key document/permit and must be obtained before operations begin. However, in some instances, this may be difficult to do far in advance since many items may change that will affect the details in the license and render it invalid even before operations begin (Mexico Environmental and Natural Resource Ministry 2021a). In addition, if any changes are made within the facility, the operating license may no longer be valid and must be amended, much as it is in the United States and, as we shall see, in most of the world. In the United States, for instance, with an air permit, if a modification in operations is made that increases or has the potential to increase emissions, then a permit modification is necessary, and in some circumstances a new permit may be required, a construction permit. This same principle applies in most other countries of the world.

In Mexico, except in relation to the petroleum and petrochemical industries or treatment facilities for hazardous waste, an operating license may be obtained through a single application. The single operating license integrates the various permits, licenses, and authorizations for matters of environmental impact, water service, emissions of atmospheric contaminants, and generation and treatment of hazardous waste (Mexico Environmental and Natural Resource Ministry 2021a). Obtaining a single license that covers all environmental matters is very different from how environmental permits are handled in the United States. A single operating license or permit that covers all aspects of potential environmental impact a facility may have upon the environment is far more manageable in nearly all aspects than having separate permits for every source. With the exception of Title V permits for air emissions, the United States issues permits separately.

10.3.3 REMEDIATION STANDARDS

In 2004, Mexico passed what is termed Mexico's General Law for Prevention and Integral Management of Wastes, commonly referred to as Mexico's Waste Law. The passage of this law significantly advanced Mexico's efforts to develop a detailed regulatory framework for dealing with legacy sites and the discovery of new or unknown sites of environmental contamination (Mexico Environmental and Natural Resource Ministry 2021b). It mirrors the United States Brownfields Revitalization Act and All Appropriate Inquiry. Figures 10.1, 10.2, and 10.3 show some of the challenges of contaminated sites in Mexico and mirror many aspects of similar issues that were faced in the United States in the early 1970s, with uncontrolled and unregulated

FIGURE 10.1 Example of uncontrolled surface waste disposal (photograph by Daniel T. Rogers).

FIGURE 10.2 Example of improper waste disposal (photograph by Daniel T. Rogers).

waste disposal that resulted in high contaminant levels in soil and groundwater and also affected surface water quality.

The Mexican Waste Law created strict liability against owners and possessors (e.g., operators) of a contaminated site. Previously, parties who caused the contamination were held responsible for cleanup. The Mexican Waste Law holds that one does not have to have caused the contamination to be held liable for its cleanup. To protect new landowners, the law mandates disclosure of known contamination by hazardous materials or wastes from owners to potential third-party buyers or tenants (Mexico Environmental and Natural Resource Ministry 2021b).

FIGURE 10.3 Example of open burning of solid waste (photograph by Daniel T. Rogers).

In addition, the Mexican Waste Law does not allow for the transfer of a site contaminated with hazardous materials or wastes without express authorization from Mexico's environmental ministry. Last, the environmental ministry or SEMARNAT will not grant approval until the contamination is remediated or until an plan has been approved by all parties. The impact of the Mexican Waste Law and its regulation has been realized in situations where properties contain known or suspected contamination (Mexico Environmental and Natural Resource Ministry 2021b). As one can imagine, potential purchasers are reluctant to take the title of contaminated property without recourse to prior studies that characterize the potential contaminant issue(s) and define the responsibility and costs associated with addressing the contamination. Sellers, on the other hand, are reluctant to conduct environmental investigations because of the potential to discover contamination that may require notice to SEMARNAT and cost significant time and resources to address.

The most common method to achieve cleanup objectives in Mexico is to simply excavate the contaminated media, usually soil, and dispose of the material in a landfill, if possible. The Mexico Environmental and Natural Resource Ministry (2004, 2005, 2015) has published remedial objectives at www.Mexicana.nom-147-SEMARNAT/ss-2003, www.Mexicana.nom-138-SEMARNAT/ss-2003, and www.Mexicana.nom-133-SEMARNAT/ss.2003 Mexico Environmental and Natural Resources Ministry 2004, 2005, 2015).

Cleanup levels for soil have been divided into groups and are based on exposure scenarios and land use similar to the United States and other countries.

The categories include residential, agricultural, commercial, and industrial. Occasionally, some of the groups are combined under one heading. Note that in Mexico, agricultural land is specifically noted and has applicable cleanup objectives, unlike in the United States (Mexico Environmental and Natural Resource Ministry 2021b).

The objective of remediation in Mexico is much the same as that throughout the world: to either eliminate or reduce contaminant concentrations or to control levels of contamination through institutional controls or other acceptable methods such that the risk posed by the presence of the contamination no longer puts human health in danger or adversely affects the environment. Remedial efforts in Mexico should accomplish the following (Mexico Environmental and Natural Resource Ministry 2021b):

- Permanently reduce contaminant concentrations
- Reduce contaminant bioavailability, solubility, or both
- Avoid contaminant dispersion in the environment
- Establish institutional controls, if necessary

There are two regulatory agencies that are responsible for overseeing compliance with the investigation and remediation of contaminated sites in Mexico: the Procuraduria Federal de Proteccion al Ambiente, which is a regulatory division of SEMARNAT, and the Health Secretariat, which is involved in cases of risk to human health (Mexico Environmental and Natural Resource Ministry 2021b).

10.3.4 SUMMARY OF ENVIRONMENTAL REGULATIONS IN MEXICO

Mexico has made significant progress in environmental regulations in the last two decades but lacks significant infrastructure, such as wastewater treatment, in Mexico City and other locations. Other significant impediments include corruption and lack of political priorities on issues pertaining to the environment. Mexico also struggles with consistent enforcement of environmental regulations in parts of the country.

10.4 SUMMARY AND CONCLUSION

Environmental regulations of the United States, Canada, and Mexico are very similar. The development of environmental regulations in Canada and Mexico was based on the USEPA basic platform. One significant difference is that Canada and Mexico do not have robust regulations for stormwater. In a way, it makes sense because Mexico is relatively dry and also does not have effective sanitary treatment. Canada has a much lower population compared to the United States and Mexico, and most of its residents live within 160 kilometers of the US border.

Environmental regulations for land in Canada and the United States may not be perfect but are effective compared to Mexico. Mexico struggles with applying consistency in enforcement and is plagued with corruption and poverty, especially in Mexico City and in many rural areas.

The next continent that will be evaluated is Europe.

10.5 REFERENCES

Basurto, D. and Soza, R. 2007. Mexico's Federal Waste Regulations. *Journal of the Air and Waste Management Association*. pp. 7–10.

Canadian Environmental Protection Act. 1999. Canadian Environmental Protection Act. www.ec.gc.ca/Lcpe-cepa. (Accessed June 30, 2021).

Canadian Government. 2021. Canada: A Brief Overview. www.cic.gc.ca/english/newcomers. (Accessed June 30, 2021).

Environment Canada. 2016. Overview of Environmental Act, Regulations and Agreements. www.ec.gc.ca/environmental-acts. (Accessed June 30, 2021).

Marsh, W. B. and Kaufman, M. M. 2015. *Physical Geography*. University of Cambridge Press, Cambridge, UK. 647 pages.

Mexico Environmental and Natural Resource Ministry (SAMARNAT). 2004. Official Mexican Remediation Objective Concentrations for Heavy Metals. www.Mexicana.nom-147-SEMARNAT/ss-2003. (Accessed June 30, 2021).

Mexico Environmental and Natural Resource Ministry (SAMARNAT). 2005. Official Mexican Remediation Objective Concentrations for Hydrocarbons. www.Mexicana.nom-138-SEMARNAT/ss-2003. (Accessed June 30, 2021).

Mexico Environmental and Natural Resource Ministry (SAMARNAT). 2015. Official Mexican Remediation Objective Concentrations for Hydrocarbons. www.Mexicana.nom-133-SEMARNAT/ss-2003. (Accessed June 30, 2021).

Mexico Environmental and Natural Resource Ministry. 2021a. Mexico's Waste Law. www.gob.mx/semarnat. (Accessed June 30, 2021).

Mexico Environmental and Natural Resource Ministry. 2021b. Remediation of Contaminated Sites in Mexico. www.gob.mx/semarnat. (Accessed June 30, 2021).

Organization for Economic Co-operation and Development (OECD). 2015. *Summary Report of Mexico's Environmental Laws and Status*. Paris, France. 27pages. www.OECD.org/Mexico/environmental. (Accessed June 30, 2021).

Province of Alberta Canada. 2021. Environmental Laws and Regulations. www.environment.gov.ab.ca/legislation. (Accessed February 4, 2021).

Province of British Columbia Canada. 2021. Environmental Laws and Regulations. www.environment.gov.bc.ca/legislation. (Accessed February 4, 2021).

Province of Manitoba Canada. 2021. Environmental Laws and Regulations. www.environment.gov.mb.ca/legislation. (Accessed February 4, 2021).

Province of New Brunswick. 2021. Environmental Laws and Regulations. https://www2.gnb.ca. (Accessed February 4, 2021).

Province of Newfoundland and Labrador Canada. 2021. Environmental Laws and Regulations. www.environment.gov.nf.ca/legislation. (Accessed February 6, 2021).

Province Northwest Territories Canada. 2021. Environmental Laws and Regulations. www.oag-bvg.ca. (Accessed February 4, 2021).

Province of Nova Scotia Canada. 2021. Environmental Laws and Regulations. www.environment.gov.ns.ca/legislation. (Accessed April 6, 2017).

Province of Ontario Canada. 2021. Environmental Laws and Regulations. www.ontario.ca/environmental-laws. (Accessed February 6, 2021).

Province of Ontario Canada. Ministry of Environment, Conservation and Parks. 2011. www.ontario.ca/page/soil-ground-water-and-sediment-standards-use-under-part-xv1-environmental-protection-act. (Accessed June 30, 2021).

Province of Quebec Canada. 2021. Environmental Laws and Regulations. https://legisquebec.gouv.qc.ca. (Accessed February 6, 2021).

Province of Saskatchewan Canada. 2021. Environmental Laws and Regulations. https://www.saskatchewan.ca/legislation/environmental-protection. (Accessed February 6, 2021).

Province Yukon Territory. 2021. Environmental Laws and Regulations. www.env.gov.yk.ca/environment. (Accessed February 6, 2021).

Rodriguez, A., Castrejon-Godinez, M. L., Ortiz-Hernandez, M. L. and Sanchez-Salinas, E. 2015. Management of Solid Waste in Mexico. *Fifteenth International Waste Management and Landfill Symposium.* CISA Publisher. Cagliari, Italy. 7p.

United Nations. 2021. World Population Prospects. United Nations Department of Economic and Social Affairs. Population Division. https://esa.un.org/unpd/wpp/data. (Accessed June 30, 2021).

United States Department of State. 2021. Office of the United States Trade Representative. https://ustr.gov/nafta. (Accessed June 30, 2021).

11 Land Pollution Regulations of Europe

11.1 INTRODUCTION

We will discuss Europe's environmental laws and regulations for land first through the 28-member European Union (EU) and then move on to other European countries that are not members of the EU but need to be covered. These other countries are Russia, Norway, Switzerland, and Turkey. We will discuss the United Kingdom (UK) as part of the EU, since the UK is still operating environmentally as if it were still an EU member.

11.2 EUROPEAN UNION

The EU is a collection of 27 countries since the exit of the United Kingdom. The EU represents 4,324,773 square kilometers in area, the seventh largest in the world, with a population of 507,416,607 residents as of a 2014 census (European Union 2021). The EU is considered by many to have the most extensive and strictest environmental laws of any international organization. The EU's environmental legislation addresses issues such as (European Environment Agency [EEA] 2021a):

- Acid rain
- Thinning of the ozone layer
- Air quality
- Noise pollution
- Solid and hazardous waste
- Water pollution
- Sustainable energy

The Institute for European Environmental Policy (2021) estimated that the body of EU environmental laws totals over 500 directives, regulations, and decisions.

11.2.1 EUROPEAN UNION ENVIRONMENTAL REGULATORY OVERVIEW

The 1972 Paris Summit meeting of heads of state and government of the European Economic Community (EEC) is often used to pinpoint the beginning of the EU's environmental policy making and declarations (EEA 2021a).

The countries of the EU are (European Environment Agency 2021a):

- Austria
- Belgium
- Bulgaria
- Croatia
- Cyprus
- Czech Republic
- Denmark
- Estonia
- Finland
- France
- Germany
- France

DOI: 10.1201/9781003150107-11

210   Environmental Compliance Handbook

- Hungary   - Ireland   - Italy   - Latvia
- Lithuania   - Luxembourg   - Malta   - Netherlands
- Poland   - Portugal   - Romania   - Slovakia
- Slovenia   - Spain   - Sweden

The EU is composed of two major institutional groups. The first group is composed of the following (EEA 2021a):

- European Commission, which initiates and implements EU law, represents the driving force, and is the executive body
- Council for the EU, which provides legislative approval to laws from the Commission and represents the governments of each of the member states
- Court of Justice, which ensures compliance with the law and serves as an appellate judicial body when questions of EU law arise
- Court of Auditors, which controls the EU budget

The second group consists of the following (EEA 2021a):

- European Economic and Social Committee, which expresses the opinions of organized civil society on economic and social issues
- Committee of the Regions, which represents the local and regional authorities in Europe
- European Central Bank, which is responsible for monetary policy and managing the currency of the EU, the Euro
- European Ombudsman, which investigates EU institutions and bodies
- European Investment Bank, which assists in EU objectives by financing government projects

The EU has three forms of binding legislation as part of its institutional framework (EEA 2021a):

- Directives. Directives are binding on the member states to which they are addressed regarding the results to be achieved. The member states, however, may choose how to bring about those results. A directive does not take effect until a member state passes national legislation to implement the provisions of European law, which it must do within 2 years of the date of passage of any specific directive. A directive normally enters into force on the date specified in the directive or on the 20th day after publication in the EU's official journal. Directives are the most frequently used EU laws.
- Regulations. Regulations are binding in their entirety and apply directly to all member states, similar to national laws. They are stronger and much less common than directives. In addition, regulations go into effect in the member states immediately, without national implementing legislation. Approximately 10% of EU laws are regulations. They supersede any conflicting

national laws and are not allowed to be transposed into national law, even if the law is identical to the regulation. Regulations are binding the day they come into force, which, like directives, is either specified in the document itself or is the 20th day after publication in the EU official journal.
- Decisions. Decisions are binding in their entirety on those to whom they are addressed. Decisions are individual legislative acts that differ from directives or regulations in that they are specific in nature and are used to specify detailed administrative requirements or update technical aspects of regulations or directives. They may be addressed to a certain government, an enterprise, or to even an individual.

Other forms of legislation include recommendations and opinions, which are not binding but are expected to be taken into account when decisions are made. In addition, resolutions are nonbinding statements by which the Council of Ministers expresses a political commitment to a specific objective. Often what happens is that after some time passes, usually a few years, a resolution gives way to either a directive or a regulation (EEA 2021a).

The European Commission, which is the executive of the EU, consists of one commissioner from each member state and is headed by a president who serves a 5-year term. Commission members are fully independent and are not permitted to take instructions from the government with which they originate (EEA 2021a).

The responsibilities of the European Commission include (EEA 2021a):

- Guardianship of treaties and the initiation of infringement proceedings against member states and others who disobey the treaties and other community law
- Negotiation of international agreements
- Adoption of technical measures to implement legislation adopted by the Council
- Matters relating to the environment, education, health, consumer affairs, the development of trans-European networks (TENS), research and development of policy, culture, and economic and monetary union
- Preparing the groundwork for incorporating former Soviet bloc countries into the EU, identifying areas where these groups can align their policies with those of the EU, and a procedure of implementation
- To initiate legislation only in areas where the EU is better placed than individual member states to take effective action, thus taking the principle of "subsidiarity" into account
- Taking action when necessary against those who do not respect their treaty obligations by perhaps bringing them before the European Court of Justice when compliance is not voluntary
- Managing policies and negotiating international trade and cooperation agreements
- Acting as a mediator between conflicting interests of member states
- Sustaining, managing, and developing agricultural and regional development policies

212 Environmental Compliance Handbook

- Developing cooperation with other European nations that are not members of the EU and countries in Africa, Caribbean, and the Pacific Rim
- Promoting research and technological developmental programs

The EEA plays an important role in EU environmental policy by providing the Commission with information used in setting EU environmental policy. The EEA aims to support sustainable development, which helps to improve Europe's environment through the provision of timely, targeted, relevant, and reliable information to policy making agents and the public. The EEA provides research, clarifies Europe's environmental issues and challenges, and employs a whole range of tools to assist policymakers and the public to address environmental quality and sustainable development issues.

In effect, the EEA is separate from other institutions and is charged with providing objective information and analysis. Other duties of the EEA are to provide objective information and analysis including a catalog of data sources and projects and prototype called the European Information and Observation Network (EINETICS), which is a computerized network for environmental data (EEA 2021a).

Current EEA members include the 28 member nations plus Bulgaria, Romania, Turkey, Iceland, Norway, and Liechtenstein. A membership agreement has been entered into with Switzerland, Albania, Bosnia and Herzegovina, Croatia, the former Yugoslav Republic of Macedonia, and Serbia and Montenegro. EEA's priorities include (EEA 2021a):

- Air quality
- Flora and fauna
- Land use
- Waste management
- Coastal protections
- The states of the soil
- Biotopes
- Natural resources
- Noise emissions
- Hazardous chemicals

The need for EU environmental protection has grown through time as industrial and population growth has occurred. The Single European Act of 1987 provided the first express legal basis for EU environmental policy and is outlined in four basic principles (EEA 2021a):

- Applying the principle of "polluters pay," which is similar to the policy adopted in the United States under the passage of CERCLA in 1980
- Pollution prevention, which is similar to the Pollution Prevention Act of the United States enacted 3 years later
- Rectifying environmental degradation at the source, when possible
- Integration of environmental protection into other EU policies

11.2.2 SOLID AND HAZARDOUS WASTE

Europe is densely populated, and nowhere is this more evident than in the EEA's approach to solid and hazardous waste. The EEA's waste management approach is

Land Pollution Regulations of Europe

heavily reliant on reuse and recycling (EEA 2021b). Over the last 20 years, European countries have shifted focus from disposal methods to prevention and recycling and also examining methods to reduce packaging waste. The EEA has set a target of recycling at 65% of all waste by 2030 (EEA 2021b).

The EEA (2021c) regulates generators of hazardous waste through a "cradle to grave" method, much like the United States. The classification of wastes into hazardous or non-hazardous is based on methods of generation and characteristics, much like the United States, that include whether the waste material is explosive, flammable, corrosive, toxic, infectious, a gas, oxidizer, or radioactive (EEA 2021c). Hazardous waste limits are listed at http://ec.eurpoa.eu/environment/waste/hazardous_index.html (EEA 2021c).

11.2.3 REMEDIATION

In the EU, land pollution has affected an estimated 2.5 to perhaps as high as 3.5 million sites (EEA 2021d). Cleaning up these sites will take decades and cost on average 0.5% to 2% of GDP per year, as estimated by the EEA (2021b). Therefore, Europe seems to have similar challenges with contaminated land as the United States. Figure 11.1 shows contamination to soil from landfilling that occurred during World War I in France. In the EU, contaminated land is treated as two groupings, land and groundwater. The EEA lists cleanup criteria for many chemical compounds as target levels

FIGURE 11.1 Impacted soil from landfilling during World War I in France (photograph by Daniel T. Rogers).

and intervention values. These values are commonly referred to as Dutch Intervention Values (Netherlands Ministry of Infrastructure and the Environment 2015).

The soil intervention values indicate when the functional properties of the soil for humans, plants, and animals are considered seriously impaired or threatened. A concentration of a single chemical compound at or exceeding its respective intervention value is considered a serious case of soil or groundwater contamination if the affected volume exceeds 25 cubic meters of soil or 100 cubic meters of groundwater (Netherlands Ministry of Infrastructure and the Environment 2015). A comparison of the DIVs to cleanup levels in the United States indicates that in some instances, the cleanup criterion for individual chemical compounds is lower in the EU than in the United States (i.e., vinyl chloride) (Netherlands Ministry of Infrastructure and the Environment 2015).

11.2.4 Summary of Environmental Regulations of the European Union

Environmental regulations in the European Union are considered by many the most comprehensive set of environmental regulations currently known to the world. From what started at the Paris Summit in 1972 establishing the EU's environmental policy and basing its regulatory framework upon that of the United States Environmental Protection Agency, it has now become the system that sets the example. No more is this evident than in the field of sustainability, where we shall see examples of the EU's commitment to environmental stewardship.

11.3 RUSSIA

Russia is the largest country in the world, encompassing much of eastern Europe and the whole northern portion of Asia. The total land area of Russia is 17,098,246 square kilometers (6,601,670 square miles). Due to its size and geographic distribution, Russia has many different climate zones, including tundra, coniferous forest, mixed and broadleaf forest, grassland, and semi-desert (Blinnikiv 2011). Russia occupies 11% of the world's land surface, stretches through 11 time zones, and has 40 UNESCO biosphere reserves. The population of Russia as of 2016 was estimated at 143.4 million. Interestingly, the population of Russia in 1991 was estimated to be 148.5 million (United Nations 2021b).

According to the United Nations (2021b), the population of Russia has stabilized and has started to increase since 2015. Russia's largest city by population is its capital, Moscow, with a population 11.5 million, followed by Saint Petersburg, with a population of 4.8 million (United Nations 2021a). For centuries, Russia has built its economy on its vast and rich natural resources that include large regions of wilderness and timber, mineral and energy resources, and fresh water (Newell and Henry 2017).

The discussion of the environmental regulations of Russia will be more in depth than most other countries. This is because Russia is a very large country with enormous natural resources, particularly in oil and gas, forestry products, minerals (much of which have not been fully developed), and significant biodiversity. How Russia decides to implement and enforce its environmental regulations will likely have either future positive or negative implications for the rest of the planet.

Land Pollution Regulations of Europe

11.3.1 Environmental Regulatory Overview and History of Russia

Russia's relationship with the environment has historically not been considered good, with many troubling concerns that have been a carry-over from its Soviet days. From the aftermath of the Chernobyl nuclear disaster (in what is Ukraine today); the drying up of the Aral Sea; unchecked industrial emissions and discharges; and its dependence on mining, oil, and gas, Russia has generated plenty of attention from environmental organizations worldwide (Brain 2016). According to the OECD (2021), environmental policies and regulations in Russia continue to suffer from an implementation gap that continues to this day. In addition, many regulatory agencies have undergone multiple reorganizations, continue to be fragmented, and have sometimes brought many agencies to the brink of institutional paralysis.

Environmental policy in Russia strives to achieve a balance between protecting the natural world and economic development. Although this objective may sound pleasing, it has been difficult to achieve in practice or actually begin, especially since these ideologies are often times in direct conflict. There are numerous environmental issues in Russia. Many of these issues have been attributed to policies of the former Soviet Union, which was a period of time when officials within the government felt that pollution control was an unnecessary hindrance to economic development and industrialization (OECD 2021). As a result, 40% of Russia's territory began experiencing symptoms of significant ecological stress by the 1990s that have been traced in large part back to a number of environmental issues (National Intelligence Council 2012):

- Deforestation
- Energy irresponsibility
- Unchecked disposal of hazardous wastes
- Surface mining
- Improper nuclear waste disposal

For most of the Soviet period, the task of environmental protection was fragmented and shared by more than 15 ministries, each of which was responsible for a particular economic sector, but many environmental issues had overlap with other sectors, which resulted in confusion over who was responsible and who had the actual power to effect change and ultimately begin to protect human health and the environment (Henry and Douhovnikoff 2008).

The Chernobyl nuclear accident in 1986 and Mikhail Gorbachev's glasnost, or policy of openness, were catalysts for change in Russia and opened the doors for environmentalism in Russia for the first time. In addition, it also gave environmental watch groups and the international community some insight and data on just how large the disregard for the environment had been during the days of the Soviet Union (Henry and Douhovnikoff 2008).

In 1988, the Russian State Committee on Environmental Protection was established and had the authority to conduct environmental reviews for all new government projects (Russian Federal Law on Environmental Protection 2002). Environmental protection gained further popularity and importance with the collapse of the Soviet Union. One of the first laws passed by the newly formed Russian Federation was the 1991 Federal Act on the Protection of the Natural Environment. Russia also announced that it had committed itself to the principle of sustainable development in the early 1990s (Henry and Douhovnikoff 2008).

Along the lines of openness, it was announced in 1993 through a commission on the environment chaired by Aleksei Yablokov, who at the time was President Yeltsin's advisor on the environment, that the Soviet Union had disposed of 2.5 million curies of radioactive waste in the Sea of Japan starting in 1965 (Henry and Douhovnikoff 2008).

From 1991 to 2000, environmental protection officials were largely ineffective because of difficult conditions due to lack of resources, bureaucratic infighting, and lack of real authority to effect change. Intense lobbying by industrial groups eroded environmental protection over time. Other significant issues including widespread corruption, lack of funding, constant reorganization, lack of clear legal direction and laws, and pressure for economic development. Confronted with these obstacles, it was no surprise that the environmental movement in the newly formed Russian Federation failed. Since 1995, the little power that the environment authorities had gradually disappeared. In 1996, President Yeltsin reduced that status of the Ministry of Environment to committee status, and in 2000, the new president, Vladimir Putin, dissolved the Committee of Environment and the Federal Forest Service and passed on the responsibility to the Ministry of Natural Resources (MNR). The motivation behind this action was to encourage exploitation of Russia's natural resources in order to ignite economic development due to the Russian financial crisis of 1998 (OECD 2021; Henry and Douhovnikoff 2008).

In Russia, environmental management relies on a set of command and control, economic, and information instruments. Russia began setting hygiene standards in 1922, at the very start of the Soviet era. Most of the environmental quality standards in Russia are a carryover from the Soviet Union. Since environmental protection in Russia was passed to the MNR in 2000 by presidential decree, the MNR operates as the executive authority and is responsible for the following (Newell and Henry 2017):

- Public policy and statutory regulation for the study, use, renewal, and conservation of natural resources, including hunting, hydrometeorology, environmental monitoring, and pollution control.
- Public environmental policy and statutory regulation as it pertains to production and waste management, conservation, and state environmental assessments.
- Ensuring compliance with international treaties on environmental matters.
- Working with other federal agencies, states, and local governmental authorities on environmental matters.

Russia is a signatory to the following select international conventions (OECD 2021):

- 1972 Convention on the International Regulations for Preventing Collisions at Sea
- 1972 Convention on the Prevention of Marine Pollution by Dumping of Wastes and Other Matter

Land Pollution Regulations of Europe 217

- 1973 Convention to Regulate International Trade in Endangered Species of Wild Flora and Fauna
- 1974 International Convention for the Safety of Life at Sea
- 1978 Convention for the Prevention of Pollution at Sea
- 1979 United Nations Convention on Long-range Trans-boundary Air Pollution
- 1982 United Nations Convention on the Law of the Sea
- 1985 Vienna Convention for the Protection of the Ozone Layer
- 1987 Montreal Protocol on Substances that Deplete the Ozone Layer
- 1989 Basel Convention on the Trans-boundary Movements of Hazardous Wastes and disposal
- 1992 Convention on the Protection of the Black Sea from Pollution
- 1992 Convention on Biological Diversity
- 1992 United Nations Framework Convention on Climate Change
- 1992 Convention on the Protection and Use of Trans-boundary Watercourses and International Lakes
- 1997 Kyoto Protocol on Climate Change
- 2001 Stockholm Convention on Persistent Organic Pollutants

Facilities subject to federal environmental supervision are those facilities that exceed one or more of the following (OECD 2021):

- A waste disposal site that accepts 10,000 tons of hazardous waste or more
- A facility that discharges 15 million cubic meters of wastewater per year into a water body
- A facility that emits 500 tons a year or more of polluting substances into the ambient air

Environmental laws in Russia cover eight main topics (OECD 2021):

- Air emission management
- Chemical management
- Water management
- Waste management
- Safety management
- Environmental impact assessment
- Energy management
- Dangerous goods—hazardous materials management

Each of these eight main topics is similar to the United States and other countries of the world.

11.3.2 Solid and Hazardous Waste

In Russia, the law that addresses waste management was enacted in 1998 and is termed the Industrial and Consumption Waste is Russian Federal Law No. 89-FZ, 1998 (Russian Ministry of Natural Resources 2021). Solid waste generation increased significantly as the Russian Federation became an independent nation and as many Russian residents adopted Western-style consumption. The vast majority of solid waste in Russia is disposed of in landfills, with a small percentage being incinerated.

Russia has begun to emphasize pollution prevention initiatives through reducing and minimizing the amounts of waste, particularly industrial waste during the production process. Many facilities are required to employ best management practices to reduce the amount of industrial wastes generated. As in the United States, Russia requires each facility to keep records on the generation and disposal of industrial wastes. In addition, Russia requires industrial facilities to obtain a permit to generate hazardous wastes, sets limits on the volumes of wastes that can be generated, and sets analytical limits on wastes similar to those of the United States and European Union (Russian Federal Law on Environmental Protection 2002; Russian Ministry of Natural Resources 2021). The definition of hazardous waste in Russia is similar to that in the United States and the rest of the developed world; it is defined as waste containing harmful substances with hazardous properties making it flammable, toxic, explosive, or reactive and also includes substances containing potentially infectious agents or those representing a direct or indirect danger to human health or the environment either alone or in combination with other substances.

Solid waste in Russia is also classified into five environmental hazard classes according to Russian Federal Law 7-FZ, 2002 (Russian Ministry of Natural Resources 2021):

- Class I—Very hazardous
- Class II—Highly hazardous
- Class III—Moderately hazardous
- Class IV—Low hazard
- Class V—Virtually harmless

Handling and transportation of hazardous waste is regulated under Russian Federal Law 89-FZ, 1998. All hazardous wastes must be properly classified on the basis of their composition and characteristics. The owner and generator of the waste is responsible for assuming the cost for proper handling of the waste (Russian Ministry of Natural Resources 2021).

11.3.3 Environmental Impact Assessments

Environmental impact assessments are generally required under Article 32 of the Russian Federal Law No. 7-FZ for projects if there is a potential that operation of the project could have an adverse effect on the environment directly or indirectly. The procedure and nature of the work to be done in order to conduct the EIA, as well as the necessary documentation, depend on a multitude of factors similar to that of many other countries and the United States. The EIA process in Russia involves several stages that are as follows (OECD 2021; Arctic Center 2021):

- The facility must first prepare all documentation containing a general description of the proposed project, including its purpose and alternatives, and submit them to the applicable regulatory authority.

Land Pollution Regulations of Europe

- Inform the general public of the project, including the location and nature of the project, by using publications or announcements in the central or regional press.
- Evaluate all environmental risks.
- Evaluate and analyze the state of the territory on which the project is intended to be located.
- Develop any measures that may be required to restore the environment.
- Calculate the consumption of natural resources that may be necessary to the project.

The legal or physical person responsible for the project must certify and approve the final version of the EIA before submitting it together with all necessary documents to the state environmental expert committee for consideration. Regulations are in effect since 2010 that require environmental audits of the progress and impact of the project to determine whether the project meets the requirements set forth in EIA and Russian environmental protection laws (OECD 2021).

11.3.4 Summary of Environmental Regulations and Protection in Russia

Russia is a country that has vast resources but has not performed well when it has come to environmental protection. Russia appears to have a robust set of environmental regulations but has difficulty in following its own rules. Not unlike other parts of the world, including the United States, Russia has had its share of environmental incidents that have caused huge negative impacts on human health and the environment. However, what may be different in Russia more than any other country examined in the world is that the occurrence of environmental incidents causing great harm has not resulted in change for the positive.

The Russian environmental movement, which began in the late 1980s and into the early 1990s under Gorbachev's reforms, has for the most part crumbled due to economic hardship and political instability largely due to the current administration. Currently, the Putin administration has labeled many internal environmental groups as "anti-Russian" and has used aggressive tactics to quash openness when subject matter is focused on environmental protection (Newell and Henry 2017). This reality, together with corruption, poor environmental enforcement, and general lack of environmental awareness and care, indicates that environmental protection will not improve in the short term. According to Newell and Henry (2017), Russia's most significant environmental challenge is the illegal and unregulated use of its natural resources.

11.4 NORWAY

Norway is located on the Scandinavian peninsula and also includes the island of Jan Mayen and the archipelago of Svalbard. Norway occupies a total area of 385,252 square kilometers (148,747 square miles) and as of January 2017 has a population of 5,258,317 (United Nations 2021a). Norway is bordered by Finland and Russia to the northeast, Sweden to the east, the Atlantic Ocean to the west, and the Skagerrak to the south, with Denmark beyond the strait.

220   Environmental Compliance Handbook

Norway maintains a combination of market economy and a Nordic welfare model that includes universal health care and a comprehensive social security system. Norway has extensive natural resource reserves that include petroleum, natural gas, minerals, timber, seafood, fresh water, and hydropower. Norway's petroleum industry accounts for approximately 25% of its GDP and, on a per capita basis, is the world's largest producer of oil outside the Middle East (United Nations 2021a). In July 2013, the Norwegian Climate and Pollution Agency and the Norwegian Directorate for Nature Management merged into what is now called the Norwegian Environment Agency (NEA), under the direction of the Ministry of Climate and Environment (Norwegian Environment Agency 2021a).

11.4.1 Environmental Regulatory Overview of Norway

The Norwegian Environment Agency has been assigned key tasks to achieve national objectives that include (Norwegian Environment Agency 2021a):

- A stable climate and strengthened adaptability
- Biodiverse forests
- Unspoiled mountain landscapes
- Rich and varied wetlands
- An unpolluted environment
- An active outdoor lifestyle
- Well-managed cultural landscapes
- Living seas and coasts
- Healthy rivers and lakes
- Effective waste management and recycling
- Clean air and less noise pollution

Norway has set goals to be achieved by 2020 that include (Norwegian Environment Agency 2021a):

- Pollution of any type will not cause injury to health or environmental damage
- Releases of substances that are hazardous to health or the environment will be eliminated
- The growth of wastes will be less than economic growth
- Wastes will be fully used for recycling and energy recovery
- To ensure safe air quality
- Noise will be reduced by 10% of 1999 values

11.4.2 Solid and Hazardous Waste

Not unlike much of the EU, Norway struggles with solid waste and limited land-fill space. Norway also incinerates a portion of its solid waste for energy recovery

Land Pollution Regulations of Europe 221

(Norwegian Environment Agency 2021b). Norway recycles up to 97% of its plastic, which is higher than any other nation measured. The recycling effort is successful because there is an environmental tax on plastic producers and an incentive for citizens to recycle, which is equivalent of up to a $0.15 to $0.30 rebate per container depending on size (Norwegian Environment Agency 2021b). In Norway, hazardous waste is defined as a waste that has the potential to cause serious pollution or involve risk of injury to people or animals. Solid and hazardous waste in Norway is handled using the EU standards (Norwegian Environment Agency 2021b).

11.4.3 REMEDIATION

Norway has not ignored impacts to the environment that have resulted from oil and gas, forestry, mining, agriculture, shipping, and other activities. Given that most of Norway's inhabitants reside along its coastal areas and it is located in a far northern latitude with significant cold weather climate patterns, there are additional difficulties in addressing its environmental remediation efforts.

Norway follows risk assessment guidelines for evaluating and remediating sites of environmental contamination and has established remediation standards for different types of land use. Remediation standards in Norway are in some instances much lower than the United States (Norwegian Environment Agency 2021c). For example, the soil value for lead in Norway is set at 60 mg/kg, whereas an acceptable level for lead in the United States may be as high as 400 mg/kg depending on the exposure route (Norwegian Environment Agency 2021c).

Remediation standards are listed at the Norwegian Environmental Agency (2021c) at www.miljodirektoratet.no/old/klif/publikasjoner/andre/1691/ya1691.pdf.

11.4.4 SUMMARY OF ENVIRONMENTAL REGULATIONS OF NORWAY

In Norway, sustainability is a fundamental principle for all development. The government's strategy is based on the principles of equitable distribution, international solidarity, the precautionary principle, polluters-pay principle, and the principle of common commitment. In part, this is achieved through an awareness and value of ecosystems and their connection with sustainability. Adequate knowledge and study of the condition of ecosystems is a necessary precondition for good nature management (Norwegian Environment Agency 2021a).

Sustainability also includes efforts to address climate change where Norway has been divided into sectors with each sector having its own climate change action plan with established targets for reducing greenhouse gases and integrating the action plan into specific sustainability targets (Norwegian Environment Agency 2021a).

11.5 SWITZERLAND

Switzerland is officially known as the Swiss Federation. Switzerland is located in west-central Europe. Switzerland has a land area of 41,285 square kilometers (15,940 square miles) and is a landlocked country that is located in the Alps and

Swiss Plateau. As of 2016, the population of Switzerland was 8,401,201. The human density of Switzerland is 195 people per square kilometer, which is very dense considering that much of the country is uninhabitable because of the mountainous terrain. For comparison purposes, the United States has a density of 33 humans per square kilometer (United Nations 2021a). Human occupation in Switzerland dates back nearly 150,000 years. Evidence of farming dates to 5,300 BC (United Nations 2016).

Switzerland is one of the most developed countries on Earth. Switzerland has the highest nominal wealth per adult and the eighth-highest GDP per capita. Switzerland ranks near the top of many metrics of national performance, including government transparency, civil liberties, quality of life, economic competitiveness, and human development. Zurich and Geneva consistently each rank among the top cities in the world with respect to quality of life (Bowers 2011).

11.5.1 ENVIRONMENTAL REGULATORY OVERVIEW OF SWITZERLAND

Switzerland is associated with lakes, mountains, and clean air, along with a good quality of life. The natural environment is an integral part of Switzerland's identity. Along these lines, Switzerland has been proactive at (OECD 2016):

- Protecting natural resources, especially forests
- Higher-density urban planning
- Reducing carbon dioxide and other greenhouse gas emissions
- Preserving water quality
- Maintaining biodiversity
- Improving air quality
- Preserving soil
- Cleaning up contaminated sites

Switzerland is under intense environmental pressure due to its population density and amount of land surface that is uninhabitable because of its mountainous nature (see Figure 11.2). During the 1970s and 1980s, ambitious environmental policies were implemented that included substantial government funding in promoting environmental awareness due to recognizing the environmental decline of land areas within the country, notably its forests (European Environment Agency 2015a).

Switzerland has designed and implemented pollution abatement policies with ambitious objectives. Most of these objectives have been achieved with remarkable success and include air pollution emission rates among the lowest in Europe and very high levels of waste water infrastructure and waste management facilities. This success was realized by means of a very active regulatory approach combined with rigorous enforcement, strong support from the public, and a considerable financial effort. The environmental policy applied in Switzerland rests on an extremely comprehensive body of federal environmental regulations. Their enactment is not without careful forethought and public input, active research, and referendums that are voted upon by the citizens on major environmental policy issues (OECD 2016).

Land Pollution Regulations of Europe

FIGURE 11.2 Switzerland's mountainous landscape (photograph by Daniel T. Rogers).

Although Switzerland spends approximately 1.7% of its GDP on environmental issues, which is high compared to other European countries, much work remains to be accomplished, which includes (OECD 2016):

- Meeting air pollution target for NO_x, VOCs, and ozone
- Maintaining and upgrading waste water treatment infrastructure
- Upgrading solid waste management infrastructure
- Cleaning up contaminated sites
- Source identification and cleanup of non-point source water pollution
- Continuing to restore forests and natural areas

11.5.2 Solid and Hazardous Waste

Solid and hazardous waste is regulated under the Switzerland Environmental Protection Act (2021). Since Switzerland is a small mountainous country, landfill space is at a premium. Therefore, Switzerland is strict on the amounts and types of wastes that are permitted and is dedicated to efforts of waste avoidance. For instance, products intended for once-only or short-term use may be prohibited because of limited landfill space. In addition, Switzerland requires any manufacturers to avoid generation of any waste, especially hazardous waste, that does not readily degrade under natural conditions and to treat any waste so that it contains as little bound carbon as possible and is insoluble (Switzerland Office for the Environment 2021).

To discourage the generation of waste, Switzerland imposes high costs on generators of solid and hazardous waste and under certain circumstances requires prepaid disposal fees (Switzerland Office for the Environment 2021).

11.5.3 Remediation

Remediation of contaminated sites in Switzerland is regulated under the Switzerland Environmental Protection Act (2021). Switzerland has adopted the "polluters pay"

224 Environmental Compliance Handbook

principle, similar to the United States. The methodology for evaluating potential contaminated sites uses a risk assessment approach similar to those of the EU and United States. Switzerland also requires that an environmental impact study be conducted for any new proposed development and also requires that all such reports be made public (Switzerland Office for the Environment 2021).

11.5.4 SUMMARY OF ENVIRONMENTAL REGULATIONS OF SWITZERLAND

Switzerland's environmental regulations are very comprehensive and also show a commitment from the government and residents. Switzerland has realized that it must be strict on environmental pollution matters since the country is small, mountainous, and heavily populated. Luckily, Switzerland does not have an industrial base as large as many other nearby countries but does have to deal with the challenges of pollution from agricultural sources and urbanization.

11.6 TURKEY

Turkey is a transcontinental country in Eurasia. Portions of Turkey are located within what is considered Asia and southeastern Europe. For the purposes of this book, we will treat Turkey as a part of Europe. Turkey is bordered by eight countries: Greece and Bulgaria to the northwest; Georgia to the northeast; Armenia, Iran, and Azerbaijani to the east; and Iraq and Syria to the south. Turkey is also bordered by seas on three sides: the Aegean Sea to the west, the Black Sea to the north, and the Mediterranean Sea to the south. Ankara is the capital, but Istanbul is the largest city in the country. Turkey covers 783,356 square kilometers (302,455 square miles) and has a population of 79,814,871 as of 2016, which equals a human density of 102 per square kilometer (United Nations 2021a; Turkish Statistical Institute 2021).

Turkey is a charter member of the United Nations, an early member of the North Atlantic Treaty Organization (NATO), and the OECD. Turkey's growing economy and diplomatic initiatives have led to its recognition as a regional power, while its location has given it geopolitical and strategic importance throughout history (Porter 2009). Turkey faces some significant environmental issues that include (Anderson 2011; European Environment Agency 2013):

- Conservation of its biodiversity
- Air pollution
- Water pollution
- Greenhouse gases
- Land degradation

11.6.1 ENVIRONMENTAL REGULATORY OVERVIEW OF TURKEY

Current environmental laws in Turkey are similar to those in the EU and the United States and cover air pollution, water pollution, and land pollution and target improving human health and protecting its biodiversity. First enacted on August 11, 1983,

Land Pollution Regulations of Europe 225

the objective was to (Turkey Ministry of Environment and Urbanization 2021a, 2021b):

> improve the environment which is the common asset of all Turkish citizens; make better use of, and preserve the land and natural resources in rural and urban locations; prevent water, land and air pollution; by preserving vegetative and livestock assets and natural and historical richness, organize all arrangements and precautions for improving and securing health, civilization and life conditions of present and future generations in conformity with economical and social development objectives, and based on certain legal and technical principles.

However, Turkey is facing significant economic hardships that greatly influence its ability to address its environmental issues. Therefore, Turkey faces significant challenges in improving environmental quality in the short and long term if it does not use control measures to improve and protect human health and the environment. This is especially acute since the biodiversity of plant life in Turkey is so significant (Anderson 2011; European Environment Agency 2013; Environmental Health 2021). There are more than 3,000 endemic plant species in Turkey, making it one of the richest biodiversity areas on Earth. This is due to Turkey's wide variety of habitats and unique position between three continents and three seas (Anderson 2011). In addition, Turkey has not made significant progress at remediating sites of environmental contamination, and remediation standards are either lacking or not enforced (European Environment Agency 2013).

11.6.2 Solid and Hazardous Waste

Rapid growth in Turkey has resulted in the generation of significant amounts of waste materials. Turkey has not kept up the pace at establishing the required infrastructure for solid waste management. For example, in 2015, it was estimated that over 80% of the solid waste generated in Turkey was not landfilled properly and was discarded by other means (Akkoyuniu et al. 2017). In addition, there are over 2,000 open dump sites scattered throughout the country (Berkun 2011).

11.6.3 Remediation

As stated, Turkey has environmental laws in place but currently lacks enforcement abilities due to other overriding social and economic factors. Land degradation is also severely stressed from uncontrolled development, leading to unchecked erosion, which has also affected water quality. Turkey is also facing significant environmental deterioration of land due to overuse, lack of urban green space, poor urban planning, overdevelopment, deforestation, population growth, and drought (European Environment Agency 2015b).

11.6.4 Summary of Environmental Regulations in Turkey

Turkey is facing some of the most significant environmental deterioration compared to other countries we have examined thus far. The main causes appear to be centered

226 Environmental Compliance Handbook

on its economic and financial hardships, which have also caused a significant degree of political unrest. This indicates that if the human population at large is experiencing hardship, the natural environment is likely experiencing hardship and stress at an even larger scale and severity. As mentioned, this threatens its biodiversity and ecosystems with collapse if improvements are not enacted.

11.7 SUMMARY AND CONCLUSIONS

The European Union countries have some of the more robust and complete environmental regulations in the world. The effectiveness of those environmental regulations varies between each country. In general, those countries in eastern Europe with a lower-than-average gross national product have more difficulty in enforcing and coping with the complexity of environmental regulations. Previous Soviet bloc countries have more difficulty that the countries of western Europe, largely due to poor economies, lack of quality infrastructure, and technology gaps. Another good example is Turkey, which struggles greatly with protecting human health and the environment due to many internal struggles that include a high population, water scarcity, corruption, poverty, and other obstacles.

In the next chapter, we turn our attention to the environmental regulations of several countries in Africa.

11.8 REFERENCES

Akkoyuniu, A. Avsar, Y. and G. O. Erguven. 2017. Hazardous Waste Management in Turkey. *Journal of Hazardous, Toxic, and Radioactive Waste.* Vol. 21. No. 4. 7p.

Anderson, Sean. 2011. Turkey's Globally Important Biodiversity Crisis. *Journal of Biological Diversity.* Vol. 144. Elsevier Publishers. New York, NY. pp. 2752–2769.

Arctic Center. 2021. Environmental Impact Assessment Processes in Northwest Russia. www.articcenter.org/RussianEIA/process. (Accessed June 30, 2021).

Berkun, M., Aras, E. and Amlan, T. 2011. Solid Waste Management in Turkey. *Journal of Materials and Waste Management.* Vol. 13. No. 1. pp. 305–313.

Blinnikiv, M. 2011. *A Geography of Russia and Its Neighbors.* The Guilford Press. London, UK. 425 pages.

Bowers, S. 2011. *Swiss Top of the Rich List.* The Guardian. London, UK. www.Theguardian.com/features/2011qualityoflife. (Accessed September 16, 2017).

Brain, S. 2016. *Environmental History of Russia. Oxford Research Encyclopedias.* Oxford University Press. Oxford, UK. 237 pages.

Environmental Health. 2021. *Air Pollution and Health in Turkey.* Health and Environment Alliance (HEAL). Brussels, Belgium. www.env-health.org/imf/pdf/150220. (Accessed June 30, 2021).

European Environment Agency. 2013. Turkey Air Pollution Fact Sheet. www.eea.europe.eu/themes/air/air-pollution-fact-sheets. (Accessed June 30, 2021).

European Environment Agency. 2015a. Environmental Status of Switzerland. www.eea.europa.eu/switzerland. (Accessed June 30, 2021).

European Environment Agency. 2015b. Turkey Country Briefing—The European Environment—State and Outlook 2015. https://eea.europe.eu/soer-2015/countries/turkey. (Accessed June 30, 2021).

European Environment Agency (EEA). 2021a. Overview of Environmental Regulations within the European Union. www.eea.europa.eu/overview. (Accessed June 30, 2021).

European Environment Agency (EEA). 2021b. Resource Efficiency and Waste. www.eea.europa.eu/themes/waste. (Accessed June 30, 2021).

European Environment Agency. 2021c. Hazardous Waste. http://ec.europa.eu/environment/waste/hazardous_index.html. (Accessed June 30, 2021).

European Environment Agency (EEA). 2021d. Soil Contamination in Europe. www.eu.europa.soil-contamination. (Accessed June 30, 2021).

European Union. 2021. European Countries in Brief. www.eu/european-union/about-eu/countries/member-countries_eu. (Accessed June 30, 2021).

Henry, L. A. and Douhovnikoff, V. 2008. Environmental Issues in Russia. *Journal of Annual Review of Environmental Resources*. Vol. 33. No. 1. pp. 437–460.

Institute for European Environmental Policy. 2021. Institute for European Environmental Policy. https://ieep.eu/. (Accessed June 30, 2021).

National Intelligence Council. 2012. The Environmental Outlook in Russia. www.dni.org/nic/special_russianoutlook.html. (Accessed June 30, 2021).

Netherlands Ministry of Infrastructure and the Environment. 2015. Dutch Pollutant Standards. www.government.nl/ministry-of-infrastructure-and-the-environment. or at www.sana-terre.com/guidelines/dutch.html. (Accessed June 30, 2021).

Newell, J. P. and Henry, L. A. The State of Environmental Protection in the Russian Federation: A Review of the Post-Soviet Era. *Journal of Eurasian Geography and Economics*. doi:10.1080/15387216.2017.1289851. http://dx.doi.org/10.1080/15387216.2017.1289851. (Accessed June 30, 2021).

Norwegian Environment Agency. 2021a. Ministry of Climate and Environment. https://miljo-direktoratet.no/no/Om-Miljodirektoratet?Norwegian-Environment-Agency. (Accessed June 30, 2021).

Norwegian Environment Agency. 2021b. Solid and Hazardous Waste. www.miljodirektoratet.no/en/legislation1/regulations/waste-regulations/chapter11/. (Accessed June 30, 2021).

Norwegian Environment Agency. 2021c. Guidelines for Risk Assessment of Contaminated Sites. www.miljodirektoratet.no/old/klif/publikasjoner/andre/1691/ya1691.pdf. (Accessed June 30, 2021).

Organization for Economic Co-Operation and Development (OECD). 2016. *Summary Report of Switzerland's Environmental Laws*. Paris, France. 21pages. www.OECD.org/Switzerland/environmental. (Accessed June 30, 2021).

Organization for Economic Co-Operation and Development (OECD). 2021. Summary Reports of Russia. www.oecd.org/russia/publicationsdocuments. (Accessed June 30, 2021).

Porter, M. E. 2009. *Turkey's Competitiveness: National Economic Strategy and the Role of Business. Institute for Strategy and Competitiveness*. Harvard Business School. Cambridge, MA. www.ics.hbs.edu. (Accessed June 30, 2021).

Russian Federal Law on Environmental Protection. 2002. www.rospotrebnadzor.ru. (Accessed June 30, 2021).

Russian Ministry of Natural Resources (MNR). 2021. Ministry of Natural Resources and Environment of the Russian Federation. www.mnr.ru/english. (Accessed June 30, 2021).

Switzerland Environmental Protection Act. 2021. Federal Act on the Protection of the Environment. www.admin.ch/opc/en/classified-compilation/19830267/index.html. (Accessed June 30, 2021).

Switzerland Office for the Environment. 2021. Reporting for Switzerland on the Protocol for Water and Health. www.unece.org/fileadmin/DAM/env/water/Protocol/reports/pdf. (Accessed June 30, 2021).

Turkey Ministry of Environment and Urbanization. 2021a. Turkey Environmental Law. https://csb.gov.tr/. (Accessed June 30, 2021).

Turkey Ministry of Environment and Urbanization. 2021b. Directorate of Environmental Management. Department of Marine and Coastal Management. http://mavikart.cevre.gov.tr/en/Haberler.aspx. (Accessed June 30, 2021).

Turkish Statistical Institute. 2021. Census of the Population of Turkey 2017. https:///www.turkstat.gov.tr. (Accessed June 30, 2021).

United Nations. 2016. Human Development Report. United Nations Development Programme. www.hdr.undr.org/sites/2017. (Accessed June 30, 2021).

United Nations. 2021a. World Population Prospects. United Nations Department of Economic and Social Affairs. Population Division. https://esa.un.org/unpd/wpp/data. (Accessed June 30, 2021).

United Nations. 2021b. Population Trends of Russia. Department of Economic and Social Affairs. Division of Population. www.UN.org.unpd. (Accessed June 30, 2021).

12 Land Pollution Regulations of Africa

12.1 INTRODUCTION

Africa is known as the location where humans are believed to have evolved nearly 2 million years ago (Smithsonian 2021). Africa is the second-largest continent and second most populous. Africa occupies approximately 30.3 million square kilometers (11.7 million square miles) and covers 6% of Earth's total surface area and 20.6% of Earth's land area (United Nations 2021a). Africa's population in total is approximately 1.22 billion and has an average age of 19.7, which is the youngest average age of any continent (United Nations 2021a).

Africa has a large diversity of ethnicities, cultures, and languages and has 54 separate identified countries, nine territories, and two independent states with limited or no recognition. Some locations within Africa are developed, while many of the rural areas are not (see Figures 12.1, 12.2, and 12.3). Africa has very diverse environmental climates, economics, historical ties, and governmental systems. This diversity has hindered its development and contributes to its environmental degradation and lack of environmental controls of pollution (Mwambazambi, K. 2010; United States Environmental Protection Agency 2013).

Africa faces significant environmental degradation as a continent caused by many factors, including (Chikanda 2009; University of Michigan 2021):

- Overpopulation
- Lack of pollution controls
- Urban sprawl and unchecked development
- Deforestation
- Invasive species
- Soil degradation and erosion
- Water degradation
- Lack of basic sanitation
- Political unrest and armed conflicts
- Poaching of wildlife, which is significant in most Africa counties

USEPA has been involved in a collaborative effort to stabilize environmental issues in many parts of Africa, specifically sub-Saharan Africa, relating to a growing population and industrial pollution issues that have impacted native population health and wellness, particularly vulnerable populations such as children and the elderly and economically disadvantaged. Areas of focus have centered on air quality, water quality, and reducing exposure to toxic chemicals (USEPA 2021). USEPA has not been involved significantly or at all in collaborating with

DOI: 10.1201/9781003150107-12

FIGURE 12.1 Typical east central Africa living and working conditions (photograph by Daniel T. Rogers).

FIGURE 12.2 Photo demonstrating nomadic livestock herding practices in east central Africa (photograph by Daniel T. Rogers).

northern countries of Africa above or within the Sahara for numerous reasons, some of which include territorial disputes and political unrest. Specific programs where USEPA is assisting sub-Saharan countries in Africa include (USEPA 2021):

- Air quality management
- Safe drinking water practices
- Lead paint abatement
- Reducing and managing methane emissions

Land Pollution Regulations of Africa

FIGURE 12.3 Urban area in Africa lacking adequate building safety and basic sanitation (photograph by Daniel T. Rogers).

- Improving environmental governance
- Improving sanitation
- Improving hazardous waste management

12.2 SOUTH AFRICA

South Africa is located at the southern tip of the African continent and occupies 1,220,813 square kilometers, making it the 25th-largest country by size. It is bordered to the north by Namibia, Botswana, Zimbabwe, and Mozambique. Swaziland and Lesotho are two countries that are located within South Africa. The Atlantic Ocean is located to the west and southwest, and the Indian Ocean is located to the east and southeast. The population of South Africa as estimated in 2018 exceeds 57 million (Statistics South Africa 2021).

South Africa is largely located in a dry climatic region of Africa, with most of its western region located in a semi-desert. Rainfall increases toward the east and falls primarily in summer. Recently, the Cape Town region of South Africa has experienced drought conditions that have now threatened its potable water supply. The central cause of the drought has been blamed on significant increases in population, which now exceeds 4 million, and on climate change (Welch 2021). Figure 12.4 shows Cape Town.

12.2.1 Environmental Regulatory Overview of South Africa

South African environmental law attempts to protect and conserve the environment of South Africa. South African environmental law encompasses natural resource conservation and utilization, as well as land-use planning and development and enforcement. In South Africa, the National Environmental Management Act (NEMA) of 1998 forms the framework for the protection of human health and the environment.

FIGURE 12.4 Cape Town, South Africa (photograph by Daniel T. Rogers).

The South Africa National Environmental Management Act covers environmental subjects that include (South Africa Government Gazette 2021):

- Sustainable development
- Environmental justice
- Public trust
- Preventative principle
- Local-level governance
- Intergenerational equity
- Environmental rights
- Precautionary principle
- Polluters-pay principle
- Common differentiated responsibility

Although some names may be different, NEMA is very similar to what we are accustomed to in the United States and most other countries, especially in the following ways (South Africa Department of Environmental Affairs 2021a):

- The preventive principle, which relates to treatment or capture of contaminants before they are emitted or released into the environment (i.e., installation and operation of air pollution control equipment). This is the fundamental notion of NEMA regulating generation, treatment, storage, and disposal of hazardous waste and the use of pesticides.
- The precautionary principle is designed such that lack of full scientific certainty shall not be used as a reason for postponing measures to prevent environmental degradation.
- A polluters-pay principle, which is analogous to Superfund in the United States.
- Environmental justice is intended to safeguard unfair environmental discrimination toward any person or persons.
- Environmental rights does not mean to imply that the environment has rights but is intended to address the rights of the individual to an environment that is safeguarded.

Land Pollution Regulations of Africa

- Local-level governance requires that decisions affecting local municipalities should be made by local municipalities.

NEMA defines the "environment" as the surroundings in which humans exist, including (South African Department of Environmental Affairs 2021b):

- The land
- The water
- The atmosphere of the Earth
- Micro-organisms,
- Plant life
- Animal life
- Any part or combination of the items listed previously and the interrelationships among and between them
- The physical, chemical, aesthetic, and cultural properties and conditions that influence human health and well-being

Furthermore, NEMA also goes on to define the environment as:

The aggregate of surrounding objects, conditions and influences that impact the life and habits of man or any other organism or collection of organisms.

NEMA addresses the following distinct but interrelated areas of general concern (South African Department of Environmental Affairs 2021a):

- Land-use planning and development
- Resource conservation and utilization
- Waste management and pollution control

One of the main purposes of NEMA was to set up a national environmental management system that would outline procedures for cooperative governance within governmental agencies such that the act does not impose overly burdensome requirements on the private sector (South Africa National Environmental Management Act 1998). Other prominent environmental legislative laws in South Africa include (South Africa Department of Environmental Affairs 2021a):

- Air Quality Act No. 39 of 2004
- Protected Areas Act No. 31 of 2004
- Protected Areas Amendment Act No. 31 of 2004
- Biodiversity Act No. 72 of 2005
- Clean Development Regulations Act of 2004
- Protected Areas Act No. 57 of 2004
- Environmental Conservation Act No. 50 of 2004
- Marine Living Resources Act of 2004
- Biodiversity Act: Threatened or Protected Species Amendment Act of 2011
- Environmental Impact Assessment Regulations of 2010
- Environmental Management Regulations of 2010

12.2.2 WASTE MANAGEMENT

In South Africa, solid and hazardous waste are differentiated and are regulated by the National Environmental Waste Act No 59 of 2009 (South African Department of Environmental Affairs 2009). Implementation guidelines were published in 2012 (South Africa Department of Environment 2012). Hazardous wastes are also termed priority wastes in South Africa. Solid and hazardous waste management in South Africa relies heavily on landfill disposal of wastes.

According to the most recent information available, South Africa produced approximately 108 million metric tons of solid waste in 2011, of which 98 million metric tons were disposed in landfills. The remaining 10 million metric tons (or just under 10%) was recorded as being recycled. The largest volumes of waste are produced by the industrial and mining sectors (South Africa Department of Environmental Affairs 2012).

In South Africa, hazardous waste is defined as (South Africa Department of Environmental Affairs 2012):

> Any waste that contains organic or inorganic elements or compounds that may, owing to the inherent physical, chemical or toxicological characteristics of that waste, have a detrimental impact on health and the environment.

Hazardous wastes are classified according to the following criteria (South Africa Department of Environmental Affairs 2012):

- Reactive
- Corrosive
- Flammable
- Characteristic
- Explosive
- Gases
- Radioactive
- Organic (for halogenated organic wastes)
- Infectious
- Miscellaneous dangerous substance (examples of wastes in this category may include asbestos, dry ice, and other environmentally hazardous substances that do not fall into any of the previously listed categories)

Classifying solid wastes by characteristic in South Africa is conducted by comparing concentration of contaminants within the waste in two ways: leachable fraction and total concentration. If the concentration of the specific substance exceeds either or both criteria, then it is considered hazardous. Leachable concentration is to be determined by collecting a representative sample of the waste material and analyzing the waste using the toxic characteristic leaching procedure similar to that outlined by the USEPA. Compounds and concentration limits, which are similar to those in the United States, EU, and Australia, are provided at https://environment.gov.sa/legislative/acts/regulations (South Africa Department of Environmental Affairs 2012).

Generators of waste must select the potential chemical contaminants that are either known to exist or may be present within the waste material. This may require

Land Pollution Regulations of Africa 235

knowledge of site activities, site history, or the processes which created the waste. Generators must be able to justify the chemical contaminants selected for analysis and keep records of that decision for 3 years. If chemical contents of a waste are unknown, then it is generally recommend to conduct a comprehensive analysis, which may include (South Africa Department of Environment 2012):

- Volatile organic compounds
- Polycyclic aromatic hydrocarbons
- Semi-volatile organic compounds
- Phenols
- Polychlorinated biphenyls
- Heavy metals

12.2.3 REMEDIATION STANDARDS

Remediation standards for South Africa were established in 2013 (South Africa Department of Environmental Affairs 2013). South African remediation standards are divided into categories using the following definitions:

- Contaminant is defined as any substance present in an environmental medium at concentrations that exceed natural background concentrations.
- Informal residential means an unplanned settlement on land that has not been proclaimed as a residential and consists mainly of makeshift structure(s) not erected according to approved architectural plans.
- Remediation is the management of a contaminated site to prevent, minimize, or mitigate damage to human health or the environment.
- Soil Screening Value 1 are soil quality values that are protective of both human health and eco-toxicological risk for multi-exposure pathways and inclusive of migration to a water source of either surface water or groundwater.
- Soil Screening Value 2 are soil quality values that are protective of risk to human health in the absence of a water resource or ecological exposure.
- Standard residential means a settlement that is formally proclaimed and serviced and generally developed with formal permanent structures, including land parcels.

Remediation values are similar to those of the United States and EU and are listed at https://environment.gov.sa/legislative/acts/regulations (South Africa Department of Environmental Affairs 2013).

Groundwater remediation values in South Africa follow either surface water criteria or the Dutch Intervention Values (Netherlands Ministry of Infrastructure and the Environment. 2015).

The variations in the climate of South Africa allow for a wide variety of crops that range from tropical fruit to corn and tree plantations. This in turn has led to extensive use of pesticides, herbicides, and fungicides, to the point where South Africa is one of the largest importers of these types of chemicals in all of Africa (Quinn et al. 2011). Estimates on the annual total pesticide use in South Africa exceed 2,800 metric tons. However, remediation standards for most of these chemicals are lacking.

236 Environmental Compliance Handbook

South Africa depends heavily on groundwater resources, but due to drought in many areas, the resource is experiencing stress from overexploitation, and many wells are now dry, especially in the Cape Town region. Currently dam levels are at 60% capacity and there is a water use limit placed on the population of 50 liters of water per day (South Africa Department of Environmental Affairs 2021a).

12.2.4 Summary of South Africa Land Environmental Regulations

Compared to the United States, environmental regulations in South Africa have only recently been enacted in that the National Environmental Management Act is only 20 years old, and many of the detailed regulations for air, water, and solid and hazardous waste are generally 10 years old. The framework of the regulations is predominantly based on USEPA. This may be true from a framework point of view, but many of the details are similar to the European Union and Australia. In fact, as we shall discover when we are evaluating Australia, the solid waste classification in South Africa is nearly identical to Australia.

The environmental regulations of South Africa appear to be very robust and comprehensive, even when comparing them to the United States or the European Union. However, they are still rather recent, and given the political and financial pressures and drought in South Africa, implementation and delays in enforcement have hindered and slowed progress. Evidence of this is that recycling efforts have not seen significant progress and remain at only 10% (South Africa Department of Environmental Affairs 2012). Another very real disadvantage within South Africa is lack of sufficient infrastructure due to its geography in that it is rather isolated from other developed countries and therefore must import technologies and equipment over a longer distance.

Finally, climate change appears to have impacted South Africa, as most significantly realized with the ongoing drought and the city of Cape Town experiencing a water shortage that is perhaps more severe than that of any other developed city on Earth at the moment. The impact is clear, but the response and long-term plan to address climate change are not yet certain.

12.3 KENYA

Kenya is located in east central Africa on the Indian Ocean between Somalia and Tanzania. Other countries that share a border with Kenya are Ethiopia, South Sudan, and Uganda. Kenya is 582,650 square kilometers in size and has an estimated population of 48 million (Kenya National Bureau of Statistics 2021). Nairobi is the largest city in Kenya and is considered the 10th-largest city in Africa, with an estimated population of 6.5 million residents (Kenya National Bureau of Statistics 2021). The name originates from the Maasai phrase Enkare Nyrobi, which translates to "cool water" and is a reference to the Nairobi River, which flows through the city. Central and western Kenya are located in the Kenyan Rift Valley characterized by mountains and volcanoes, some of which are considered active. Kenya's highest peak, Mount Kenya, which exceeds 5,700 meters in elevation, is located in this region (Kenya Geological Society 2021).

Land Pollution Regulations of Africa 237

The population of Kenya has increased significantly over the last 50 years, especially in Nairobi, where the population has increased from 0.5 million to 6.5 million. This has placed pressure on providing adequate infrastructure to support a population increase of that magnitude that quickly (Kenya National Bureau of Statistics 2021). The increase in population has also placed stress on the natural environment in Kenya, including water and air pollution from urban and industrial areas and agriculture, deforestation, pesticide and herbicide use, erosion, desertification, and poaching of wildlife (Kenya National Environment Management Authority 2021).

Approximately 8% of the landmass of Kenya is contained with 22 national parks and 28 national reserves (Kenya Wildlife Service 2021). Kenya's economy is dominated by agriculture, followed by manufacturing, much of which is agriculture related. Kenya is the banking capital of central Africa, and tourism is also very significant to the economy (Kenya National Bureau of Statics 2021).

12.3.1 OVERVIEW OF KENYA ENVIRONMENTAL REGULATIONS

Environmental regulations were enacted starting in 2006 and now include several acts that cover the following environmental areas (Kenya National Environmental Management Authority 2021):

- Air quality
- Industrial water discharge
- Noise
- Hazardous waste
- Chemical regulations
- Controlled substances
- Biodiversity
- Domestic water
- Wetlands
- Solid waste
- Environmental impact assessments
- Waste transport
- Coastal protection
- Land development

12.3.2 SOLID AND HAZARDOUS WASTE

Management of solid and hazardous waste in Kenya is regulated by the Environmental Management and Co-Ordination Waste Management Act of 2006 (Kenya Waste Management Act 2006). The act covers:

- Solid waste
- Hazardous waste
- Industrial waste
- Pesticides and toxic substances
- Biomedical wastes
- Radioactive wastes

The act outlines the responsibilities of the generator, transporter and disposal facility, permitting, licensing, transportation, and environmental audit procedures. It also outlines requirements for evaluation of an environmental impact assessment, training, labeling, packaging, segregation, monitoring, and classification (Kenya Waste Management Act 2006).

238 Environmental Compliance Handbook

Hazardous waste determination in Kenya is by content and percentage of certain chemicals considered hazardous by their very nature. For example, waste that contains the following is considered hazardous in Kenya (Kenya Waste Management Act 2006):

- Radio-nuclides
- Medical waste
- Pharmaceutical drugs or medicines
- Biocides, germicides, herbicides, insecticides, fungicides
- Wood-preserving chemicals
- Organic solvent waste
- Heat treatment and tempering wastes containing cyanide
- Mineral oil waste
- PCBs
- Wastes from inks, dyes, pigments, and paints
- Waste chemicals from research, development, or teaching
- Explosives
- Wastes containing metal carbonyls, beryllium, hexavalent chrome, copper, zinc, arsenic, selenium, cadmium, antimony, tellurium, mercury, thallium, lead, fluorine, phosphorus, phenol, ethers, and cyanide at 1% or more by weight
- Waste at a pH of less than 2
- Waste at a pH greater than 11.5
- Wastes containing asbestos
- Halogenated organic solvents at 0.1% or more by weight
- Any congener of polychlorinated dibenzo-furan
- Any congener of polychlorinated dibenzo-p-dioxin

Other wastes that are considered hazardous not listed previously include wastes that are (Kenya Waste Management Act 2006):

- Flammable
- Oxidizers
- Infectious
- Toxic gas
- Radioactive

- Explosive
- Organic peroxides
- Corrosive
- Persistent wastes
- Carcinogens

- Combustible
- Toxic or poisonous
- Eco-toxic
- Leachate
- Medical waste

12.3.3 Remediation Standards

Remediation stands in Kenya were first enacted in 2003 and was amended in 2009 and are termed the Environmental Impact and Assessment and Audit Act (Kenya Impact Assessment and Audit Act 2003). The act requires that an environmental impact assessment be conducted before any land is developed or re-developed and must consider the following at a minimum:

- Environmental, social, cultural, economic, and legal considerations
- Identify environmental impacts and scale of impacts

Land Pollution Regulations of Africa

- Identify and analyze alternative
- Develop an environmental management plan
- Consult with the regulatory agency on a regular basis
- Seek public comment
- Hold at least three public meetings
- Prepare a detailed report

After regulatory review and acceptance, a license will be issued. Following development, environmental audits and monitoring shall be conducted at intervals outlined in the license.

The environmental impact assessment is a detailed environmental study similar to that in the United States and usually includes extensive investigation and testing for the presence of contamination, and, if discovered, the nature and extent of impacts must be defined. Following the completion of testing, the assessment is required to evaluate whether there are any risks posed by the presence of contamination and lower those exposure risk to an acceptable level, if necessary (Kenya Impact Assessment and Audit Act 2003).

12.3.4 SUMMARY OF LAND ENVIRONMENTAL REGULATIONS OF KENYA

Kenya does, in fact, have a very robust set of environmental regulations in place. The regulations have not been enacted for very long; most are less than 15 years old. Therefore, it should come as no surprise that the current major obstacle is implementation and enforcement, along with cooperation between ministries responsible for the environmental interpretation and enforcement. Kenya's air and water environmental regulations are similar to those of the United States and the European Union. However, the solid and hazardous waste regulation are unique to Kenya and are strict. This is likely due to a combination of factors that include Kenya's reliance on its national parks and biodiversity for tourism, lack of a significant industrial base, and perhaps disincentives for additional industrial development. An example of the strict solid and hazardous waste regulations in Kenya is that Kenya considers any plastic waste hazardous.

12.4 TANZANIA

Tanzania is located in east central Africa on the Indian Ocean, located along the eastern border of the country. Kenya is located immediately to the north and Mozambique, Malawi, and Zambia are located along the southern border. The Democratic Republic of Congo, Burundi, and Rwanda are located immediately to the east.

Tanzania has an estimated population of slightly more than 57 million and is 947,300 square kilometers in size (Tanzania Bureau of Statistics 2021). Tanzania's highest elevation is Mount Kilimanjaro at an elevation of 5,895 meters. There are numerous national parks, conservation areas, and game reserves within Tanzania, including the Serengeti National Park, Ngorongoro Crater Conservation Area, and Selous Game Reserve (Tanzania Bureau of Statistics 2021). Lake Victoria, Africa's largest lake, is located in the northwest portion of the country. Northern and central portions of Tanzania are mountainous and are part of the African Rift Valley. Most of Tanzania's population is located in the northern portion of the country and the eastern border with the Indian Ocean (Tanzania Bureau of Statistics 2021).

The largest city in Tanzania is Dar es Salaam, with an estimated population of over 5.5 million, the former capital of the country. The capital of Tanzania is now Dodoma. Dodoma has an estimated population of greater than 2.2 million and includes outlying areas (Tanzania Bureau of Statistics 2021). The population of Tanzania has increased significantly since 1963, when the population was estimated at 11 million. This has created pressure to provide adequate infrastructure to support a population increase of that magnitude (Tanzania Bureau of Statistics 2021). Tanzania has the second-largest economy in central Africa, second to Kenya. The economy is dominated by agriculture, which employs approximately 50% of the workforce. Approximately one third of the population of Tanzania lives at or below the poverty level (Tanzania Bureau of Statistics 2021).

12.4.1 Environmental Regulatory Overview of Tanzania

Not unlike Kenya, Tanzania's increase in population has placed stress on the natural environment, including water and air pollution, deforestation, pesticide and herbicide use, erosion, desertification, and poaching of wildlife. Water-borne illnesses, such as malaria and cholera, account for over half of the diseases affecting the population (Tanzania Bureau of Statistics 2021).

12.4.2 Solid and Hazardous Waste

Solid and hazardous waste is regulated through the Tanzania Environmental Management Act of 2008, more commonly referred to as the Hazardous Waste Control and Management Regulations (2008). The Hazardous Waste Control and Management Regulations cover the following topics:

- Classification procedures
- Packaging
- Labeling
- Handling
- Transporting
- Storage
- Permitting

In Tanzania, wastes are considered hazardous if they fit the following categories (Tanzania Environmental Management Act 2008)

- Characteristics such as being explosive, flammable, corrosive, or reactive or having the potential to produce a leachate considered toxic
- Are generated from certain types of operations, such as:
  - Medical waste
  - Pharmaceuticals, drugs, or medicines
  - Biocide production

Land Pollution Regulations of Africa 241

- Wood-preserving chemicals
- Organic solvent production
- Heat treatment and tempering operations that use cyanide
- Used mineral oils
- Oil and water mixtures
- Waste containing PCBs
- Ink, dye, pigment, paint, lacquer, and varnish operations
- Waste from resins, latex, plasticizers, glues, and adhesives
- Waste chemical substances from research, development, or teaching
- Residues from waste treatment operations

- Waste with the following constituents:

  - Metals such as beryllium, hexavalent chromium, copper, zinc, arsenic, selenium, cadmium, antimony, tellurium, mercury, thallium, lead
  - Inorganic cyanides
  - Asbestos
  - Phenols
  - Halogenated organic solvents
  - Ethers
  - Any congener of polychlorinated dibenzo-furan
  - Any congener of polychlorinated dibenzo-p-dioxin

12.4.3 REMEDIATION STANDARDS

Soil quality standards were established in Tanzania in 2007 with the Environmental Management Act that established maximum allowable concentrations of many compounds in soil. Groundwater standards have not been established. The intention of the Environmental Management Act of 2007 was to make clear that any intentional disposal of solid refuse or putrid solid matter onto the ground is prohibited (Tanzania Environmental Management Act 2007).

12.4.4 SUMMARY OF LAND ENVIRONMENTAL REGULATIONS OF TANZANIA

The environmental regulations in Tanzania are similar to those of its neighbor Kenya. Differences between Kenya and Tanzania focus on the fact that Kenya is more developed and is further along in implementation of its regulations. Tanzania appears to struggle with providing basic sanitation and supplying water to its population. This is largely due to the lack of basic infrastructure within the country. In addition, approximately one-third of the population of Tanzania lives below the poverty level. Therefore, it should come as no surprise that Tanzania is also struggling with maintaining its natural environment and the health of its residents.

12.5 EGYPT

Egypt is located in northern Africa and has an estimated population of 99 million and a land area of 1,001,449 square kilometers (United Nations 2021a). The Mediterranean Sea forms Egypt's northern border; Libya is located to the west,

Sudan to the south, and Jordan and Saudi Arabia to the east. Egypt is the third most populous country in Africa behind Nigeria and Ethiopia. Approximately 95% of the population of Egypt lives along the Nile River, in the Nile Delta, or along the Suez Canal. These regions are among the most densely populated areas of the world. An estimated 75% of the population of Egypt is under the age of 25, making it one of the most youthful populations in the world (United Nations 2021b). Cairo, located along the Nile River, is the largest city in Egypt, with an estimated metropolitan population of over 20 million residents (United Nations Populations Programme 2021b).

12.5.1 Environmental Regulatory Overview of Egypt

In Egypt, the environment is regulated by the Ministry of Environment. Environmental standards in Egypt were first enacted in 1994 and were amended in 2009 and are simply known and referred to as the Environmental Law. They cover protection of the land, air, water, and marine environment (Egypt Environmental Affairs Agency 2021).

12.5.2 Solid and Hazardous Waste

Solid and hazardous waste is regulated through the Environmental Law of 2009 and administrated through the Egypt Environmental Affairs Agency (2021). In Egypt, all solid and hazardous wastes are regulated, and requirements include (Egypt Environmental Affairs Agency 2021):

- No wastes shall be disposed of in unlicensed locations
- All treatment of wastes must be conducted under an applicable permit and license
- No wastes shall be imported
- Appropriate health and safety measures for all those who handle or ship wastes
- All waste generators shall be registered

The Egyptian solid and hazardous waste regulations include (Egypt Environmental Affairs Agency 2021):

- Transportation
- Storage
- Licensing
- Treatment

- Labeling
- Disposal
- Operations
- Disposal

- Manifesting
- Characterization
- Packaging

In Egypt, a hazardous waste is defined as (Egypt Environmental Affairs Agency 2021):

Wastes of activities and processes or their ashes that maintain their harmful properties and have no subsequent original or substitutive uses.

Land Pollution Regulations of Africa

Unlike most countries of the world, no characteristics of hazardous wastes, processes, waste streams, or constituents of wastes have been identified or defined (Ramadam and Nadim 2014).

12.5.3 REMEDIATION STANDARDS

Egypt requires any new proposed development to conduct an environmental impact assessment, which must be conducted in accordance to the elements, designs, specifications, bases, and pollutant loads determined by the Egyptian Environmental Affairs Agency (Egypt Environmental Affairs Agency 2021). The focus of the EIA is not on current soil or groundwater quality but only addresses future potential impacts.

12.5.4 SUMMARY OF EGYPTIAN LAND ENVIRONMENTAL REGULATIONS

Egypt is a very urbanized country with very dense population centers. Together with an increasing population, this has placed enormous stress on the health of its population (World Health Organization 2021). Air pollution, sanitation, remediation of contaminated sites, and deterioration of ancient historical relics are of immediate and pressing concern. Near-future progress on solving environmental issues does not look promising due to recent political strife and violence within the country.

12.6 SUMMARY AND CONCLUSION

Africa is a developing continent, and because of this, it struggles with infrastructure, with a population increase that generally exceeds the rest of the world. Africa is also struggling with war, poverty, drought, lack of education, lack of sanitation, poor water quality, human rights issues, poaching, and disease. Each of these social and cultural issues takes priority over the environment. Most countries in Africa have established environmental regulations that all too often do not get enforced.

12.7 REFERENCES

Chikanda, A. 2009. Environmental Degradation in Sub-Saharan Africa. In: *Environment and Health in Sub-Saharan Africa: Managing an Emerging Crisis*. Springer Publishing. New York, NY. pp. 79–94.

Egypt Environmental Affairs Agency (EEAA). 2021. Environmental Protection Law. www. eeaa.gov.eg/en-us/laws/envlaw.aspx. (Accessed June 30, 2021).

Kenya Geological Society. 2021. Summary of the Geology and Minerals Resources of Kenya. www.apsea.or.ke/index/geological-society-of-kenya. (Accessed February 6, 2021).

Kenya Impact Assessment and Audit Act. 2003. *National Environmental Management Authority*. Nairobi, Kenya. www.nema.go.ke/NEMA/Impactassessmentandauditact. (Accessed June 30, 2021).

Kenya National Bureau of Statistics. 2021. Estimated Population of Kenya in 2021. www. knbs.or.ke/. (Accessed June 30, 2021).

Kenya National Environment Management Authority. 2021. Kenya State of the Environment and Outlook 2010. www.nema.go.ke/downloads. (Accessed June 30, 2021).

Kenya Waste Management Act. 2006. *National Environmental Management Authority.* Nairobi, Kenya. www.nema.go.ke/NEMA/Wastemanagementact. (Accessed June 30, 2021).

Kenya Wildlife Service. 2021. National Parks and National Wildlife Reserves in Kenya. www.go.ke/national-parks. (Accessed June 30, 2021).

Mwambazambi, Kalemba. 2010. Environmental Problems in Africa: A Theological Response. *Ethiopian Journal of Environmental Studies and Management.* Vol. 3. No. 2. University of South Africa. Pretoria, South Africa. pp. 54–66.

Netherlands Ministry of Infrastructure and the Environment. 2015. Dutch Pollutant Standards. www.government.nl/ministry-of-infrastructure-and-the-environment. or at www.sanaterre.com/guidelines/dutch.html. (Accessed June 30, 2021).

Quinn, L. P., de Vos, B. J., Fernandes-Whaley, M., Roos, C., Bouwman, H., Pieters, K. R. and Berg, J. 2011. Pesticide Use in Africa. www.environment.gov.sa/docs/pesticides.pdf. (Accessed June 30, 2021).

Ramadam, A. R. and Nadim, A. H. 2014. Hazardous Waste in Egypt: Status and Challenges. In: Popov, V., Itoh, H., Brebbia, C. A. and Kungolos, S. editors. *Waste Management and the Environment.* WIT Press. Southhampton, UK. pp. 125–129.

Smithsonian Museum of Natural History. 2021. *Introduction to Human Evolution. Smithsonian Institution's Human Origins Program.* Smithsonian Institution. Washington, DC. http://humanorigins.si.edu/education/introduction-human-evolution. (Accessed June 30, 2021).

South Africa Department of Environmental Affairs. 2009. National Environmental Waste Act No. 59 of 2009. Pretoria, South Africa. 309p.

South Africa Department of Environmental Affairs. 2012. *National Waste Management Strategy. South Africa Government Gazette.* Pretoria, South Africa. 77p.

South Africa Department of Environmental Affairs. 2013. *National Norms and Standards for the Remediation of Contaminated Land and Soil Quality in the Republic of South Africa. Government Gazette.* South Africa Department of Environmental Affairs. Pretoria, South Africa. 15p.

South Africa Department of Environmental Affairs. 2021a. Outline of Environmental Regulations. https://environment.gov.sa/legislative/actsregulations. (Accessed June 30, 2021).

South Africa Department of Environmental Affairs. 2021b. Cape Town Drought; Current Status. www.capetowndrought.com. (Accessed June 30, 2021).

South Africa Government Gazette. 2021. South Africa National Environmental Management Act (NEMA). 1998. National Environmental Management Act. Act No. 107. Pretoria, South Africa. November 27, 1998.37p.

Statistics South Africa. 2021. Mid-Year 2018 Population Estimate of South Africa. www.statsa.gov.sa/?=15. (Accessed June 30, 2021).

Tanzania Bureau of Statistics. 2021. Current Population of Demographics of Tanzania. www.nbs.go.tz. (Accessed June 30, 2021).

Tanzania Environmental Management Act. 2007. *Environmental Management Act.* Soil Quality Regulations. Dar es Salaam, Tanzania. 25p.

Tanzania Environmental Management Act. 2008. *Hazardous Waste Control and Management Regulations.* United Republic of Tanzania. Dar es Salaam, Tanzania. 49p.

United Nations. 2021a. World Population Prospects. United Nations Department of Economic and Social Affairs. Population Division. https://esa.un.org/unpd/wpp/data. (Accessed June 30, 2021).

United Nations. 2021b. United Nations Population Programme For Egypt. www.egypt.unfpa.org/em/united-nations-population-egypt-2. (Accessed June 30, 2021).

United States Environmental Protection Agency (USEPA). 2013. Air Quality in Africa. http://epa.gov/international/air/africa.htm. (Accessed June 30, 2021).

United States Environmental Protection Agency (USEPA). 2021. EPA Collaboration with Sub-Saharan Africa. www.epa.gov.international-cooporation/epa-collaboration-sub-saharan. (Accessed June 30, 2021).

University of Michigan. 2021. Environmental Issues of Africa. www.globalchange.umich.edu.globalchange1. (Accessed June 30, 2021).

Welch, C. 2021. Why Cape Town is Running Out of Water and Who's Next. www.news.national geographic.com/2018/02/cape-town-running-out-of-water. (Accessed June 30, 2021).

World Health Organization. 2021. World Health Organization Country Profile—Egypt. www.who.int/countries/egy/en/. (Accessed June 30, 2021).

13 Land Pollution Regulations of Asia

13.1 INTRODUCTION

Asia is the largest and most populous continent. The land area of Asia is estimated at 43,608,000 square kilometers and represents approximately 29.4% of Earth's land surface. Asia's estimated population is greater than 4 billion and represents approximately 56% of the world's human population. China and India account for most of the population of Asia, with a combined population of 2.7 billion. Countries within the continent of Asia that we will evaluate are China, India, Japan, South Korea, Saudi Arabia, Malaysia, and Indonesia. We have already evaluated Russia, most of which lies within Asia but was evaluated as part of Europe. This is because the western part of Russia, including its capital of Moscow, is located within the continent of Europe.

13.2 CHINA

China is located in southeast Asia and is the third-largest country by area in the world, behind Russia and Canada, with a land area of 9,596,960 square kilometers. The geography of China is perhaps the most wide ranging of any country on Earth. As an example, China's lowest point is Turpan Pendi at 154 meters below mean sea level, which is one of the lowest places on Earth. In contrast, China's highest point is Mount Everest, the highest point on Earth at 8,848 meters above mean sea level. Due to the extremes in elevation differences, China's climatic range is from tropical along the southeastern portion of the country to subarctic along the western boundary. China's estimated population now exceeds 1.42 billion, making it the most populous country on Earth (United Nations 2021a). Most of China's population is located along its eastern coastline near the Yellow Sea and the East China Sea. The two largest cities in China are Shanghai and China's capital, Beijing. Shanghai has an estimated population of 24.5 million, and Beijing has an estimated population of 21 million (United Nations 2021a). Figure 13.1 shows Shanghai. Figure 13.2 is an example of rapid urbanization in China through high-rise building construction.

China's economy is now considered the largest in the world at $23.12 trillion, followed by the European Union at $19.9 trillion and then the United States at $19.3 trillion. However, China is still a relatively poor country in terms of its standard of living, at only $15,600 per person compared to the United States at $59,500 per person based on total GDP. It is staggering to imagine that China's economy is larger than that of the United States but will likely continue to grow significantly in

DOI: 10.1201/9781003150107-13

247

FIGURE 13.1 Downtown Shanghai (photograph by Daniel T. Rogers).

FIGURE 13.2 Example of rapid urbanization in China (photograph by Daniel T. Rogers).

the future. China has built its economy on relatively cheap labor for manufacturing. China's largest trading partner is the United States (United States Department of State 2021a; United States Department of Commerce 2021; United States Census Bureau 2021).

Land Pollution Regulations of Asia

13.2.1 Environmental Regulatory Overview of China

China faces significant environmental issues due to its huge population, rapid growth, and relatively low standard of living. Those environmental issues include air and water pollution, water shortages in the northern part of the country, deforestation, sanitation, soil and groundwater impacts, soil erosion, and desertification. Much of the air pollution is due to its heavy dependence on coal; the rapid increase in the number of motor vehicles, especially in its urban regions; and industry (USEPA 2021a).

China is a party to many international environmental treaties, protocols, and agreements, including signing the Kyoto Protocol to reduce carbon emissions. Others include the following (United Nations 2021b):

- United Nationals Convention to Combat Desertification
- Marine Dumping
- Ozone Layer Protection
- Endangered Species
- International Tropical Timber Agreement
- International Convention for the Regulation of Whaling
- Ship pollution
- Wetlands
- Convention on Biological Diversity
- Climate change
- Antarctic Environmental Protocol
- Nuclear Test Ban Treaty

The USEPA and China's Ministry of Ecology and Environment (MEE) have been collaborating for several years to strengthen and expand China's capabilities at mitigating pollution that now present significant challenges to human health and the environment in China (USEPA 2021a). Objectives of the collaboration have been to improve many of China's environmental protection programs, including the following (USEPA 2021a):

- Improving air quality
- Reducing water pollution
- Preventing exposure to chemicals and toxics
- Remediating soil
- Treating hazardous waste
- Disposal of hazardous and solid waste
- Improving environmental enforcement
- Promoting environmental compliance
- Improving existing environmental laws
- Enhancing environmental education
- Chemical management

USEPA has also been collaborating with China's Ministry of Science and Technology on joint research to better access air emission sources and impacts on China to better develop mitigation practices and to enhance sustainability. Areas of research have included (USEPA 2021a):

- Soil remediation
- Groundwater remediation
- Water sustainability
- Toxicology
- Motor vehicle emissions
- Clean cooking methods
- Air pollution monitoring and assessment

13.2.2 Solid and Hazardous Waste

Efforts to regulate solid and hazardous waste in China have been plagued by numerous significant social, economic, and political challenges. In 2007, the first national census of hazardous waste in China estimated that China produced 45 million metric tons of hazardous waste, but only 10 million metric tons were produced from industry (Wencong 2012). However, much information was lacking, including (Wencong 2012):

- Types of waste
- How the wastes were generated
- How the wastes were transported
- Where the wastes were disposed of

The first law addressing solid waste in China was passed in 1995 and was titled Law of the People's Republic of China on the Prevention and Control of Environmental Pollution by Solid Waste (China Ministry of Ecology and Environment [CMEE] 2021). As part of China's solid waste law, a national catalogue of hazardous waste was developed to identify and classify solid wastes as either hazardous or not hazardous. In China, as in most other countries, a waste is hazardous if it is any of the following (CMEE 2021):

- Corrosive
- Medical waste
- Explosive

- Ignitable
- Toxic
- Pesticide

- Flammable
- Reactive
- Herbicide

Other types of wastes considered hazardous in China include (CMEE 2021):

- Electronic wastes
- Waste Hg-containing thermometers
- Waste solvents
- Waste Ni-Cd battery cells
- Waste paints

- Fluorescent light bulbs
- Waste mineral oils
- Waste film and photographs
- Disinfectants

Exempted from the law in China are wastes generated by households in daily life, which is similar to that of the United States (CMEE 2021). Until recently, China did not have a reliable infrastructure for disposal of hazardous wastes. Therefore, many industrial sites stored wastes until a satisfactory transport method and disposal

Land Pollution Regulations of Asia

FIGURE 13.3 Historical burning of solid waste in China (photograph by Daniel T. Rogers).

site were established nearby that were in compliance with China's solid waste law. However, historically, solid wastes were disposed of anywhere convenient, including farm fields or next to ponds or lakes, or wastes were simply burned. Figure 13.3 is an example of burning waste. Figure 13.4 is an example of disposal of wastes near a pond. China is currently examining and will likely revise its solid and hazardous waste regulations sometime in 2019 (CMEE 2021).

13.2.3 REMEDIATION STANDARDS

In 2014, the China Ministry of Environmental Protection announced five standards of environmental protection (CMEP Notice No. 14 of 2014) that provided a framework for future regulatory development of soil and groundwater pollution prevention and remediation (CMEP 2014). The five standards were (CMEP 2014):

- Technical Guidelines for Environmental Site Investigation
- Technical Guidelines for Environmental Site Monitoring
- Technical Guidelines for Risk Assess of Contaminated Sites
- Technical Guidelines for Site Soil Remediation
- Definition of Terms of Contaminated Sites

On August 31, 2018, China passed a new law on soil pollution and control. The new law went into effect on January 1, 2019. The law stipulates that the China Ministry of Environmental Protection establish soil remediation standards and conduct a census

FIGURE 13.4 Historical disposal of solid waste in a pond (photograph by Daniel T. Rogers).

on the condition of soils nationwide every 10 years (China Ministry of Environmental Protection 2021). The law strengthens the responsibilities of governments and the responsibilities and liabilities of polluters. In addition, the law also requires the polluters of farmland to develop rehabilitation plans, provide them to the appropriate government authority, and carry out the plans.

Groundwater pollution in China is also a huge challenge, especially in the northern portion of the country since it's in this portion that groundwater is rather scarce compared to the those area in the southern portion of the country (China Research Academy of Environmental Sciences (2013).

Until the China Ministry of Environmental Protection adopts cleanup values for pollutants of its own, the Dutch Intervention Values are commonly used for comparison purposes to evaluate whether soil, sediment, and groundwater require cleanup and to what extent.

13.2.4 Summary of Environmental Regulations in China

Environmental regulations in China are similar to those of the United States and European Union, with two major exceptions:

- Requirement that each facility obtain one operating permit that covers all operating conditions, including environmental aspects
- Many regulations are either rather new (i.e., air and water) or are currently under development (i.e., land)

Land Pollution Regulations of Asia

China is a rapidly developing nation with significant air, water, sanitation, and land pollution issues. Together with its large population, China faces significant future challenges. However, China now has the largest economy in the world and will likely continue to grow and prosper from this development. In addition, China has made major strides at developing an environmental framework that can hopefully deal with improving the quality of its air, water, and land from pollution in the future. The combination of a strong and growing economy with a robust set of environmental regulations and political will means that it is likely that China will continue to improve environmentally at a fast pace.

13.3 JAPAN

Japan is an island nation composed of four main islands and thousands of smaller islands just to the east of the coast of Asia in the Pacific Ocean. The population of Japan is estimated at just over 127 million (United Nations 2021a). The capital of Japan is Tokyo, which has a population of 36 million residents and is considered the largest urban center in the world (United Nations 2021a). According the United Nations population projects, Japan's population is not expected to grow over the next ten years. In fact, Japan's population is expected to decrease by more than 2 million in the next 10 years (United Nations 2021a).

Japan is a highly developed country with the fourth-largest economy in the world behind China, the United States, and the European Union. Japan has an estimated GDP of over $5 trillion annually (United Nations 2021a). Japan is ranked the 61st-largest country, with a land area of 377,973 square kilometers, and it has 29,751 kilometers of coastline (United Nations 2021a). The population density of Japan is approximately 348 individuals per square kilometer, which is much higher that the worldwide average of 51 and ten times that of the United States, which is 33 individuals per square kilometer (United Nations 2021a).

Japan experiences natural disasters in the form of volcanic eruptions, landslides, tsunamis, and typhoons that can lead to significant environmental concerns and loss of life. For example, an earthquake in March 2011 damaged more than 270,000 buildings and was also responsible for the generation of an estimated 24 million metric tons of waste debris. In addition, the earthquake and subsequent tsunami damaged the Fukushima Nuclear Power Plant and released significant quantities of radiation that had large aerial impacts and will have lasting effects for decades (Japan Ministry of the Environment 2021a).

Japan's economy is primarily production of motor vehicles, electronics, industrial tools, steel, and other metals. Japan also has some food production that is composed of rice, sugar beets, fruits and vegetables, fish, and beef (United States Department of State 2021b). Japan has undergone rapid industrialization since the end of World War II in 1945. The United States and Japan have a very close relationship on matters including (United States Department of State 2021b):

- Development assistance
- Environmental issues
- Women's empowerment

- Health
- Resource protection
- Science and technology

- Infectious disease
- Space exploration
- International diplomatic initiatives
- Medicine
- Education

13.3.1 ENVIRONMENTAL REGULATORY OVERVIEW OF JAPAN

Environmental regulations in Japan are some of the strictest in the world and are the responsibility of the Japan Ministry of the Environment. Environmental regulations in Japan had their beginning in 1967 and were significantly revised in 1994 as Japan was experiencing significant industrialization (Japan Ministry of the Environment 2021a). Japan has significant environmental issues related to air, water, and land. Disposal of solid waste is of particular concern since Japan has a relatively little land compared to other countries and has a large population with a high density and generates large amounts of solid waste (Japan Ministry of the Environment 2021a).

Japan itself points to an incident that occurred in 1896 at the Ashio Copper Mine as the beginning of the environmental movement in Japan. A massive flood from torrential rains occurred in the region of the mine in September 1896 and caused not only the mine to flood but also the Watarase, Tone, and Edo Rivers to overflow their banks. Damage from the flood nearly destroyed 136 towns in the path of the flood and contaminated the entire affected area with heavy metals, including many agricultural areas that effectively poisoned the soil and thousands of residents in the path of flood or who ate food from farms within the affected area (Kichira and Sugai 1986; Japan Ministry of the Environment 2021a). This incident is considered a natural disaster triggered by human actions that impacted an entire region of Japan because of a general lack of appreciation for the environment that included deforestation, poor mining practices, little or no drainage control, and lack of knowledge of potential environmental impact from mining wastes. This incident highlighted the need to place more emphasis on safety and environmental integrity and started the first mass citizen movement in Japanese history (Japan Ministry of the Environment 2021a).

13.3.2 SOLID AND HAZARDOUS WASTE

Solid and hazardous waste in Japan, along with air and water, is perhaps the most important environmental issue in Japan. Being a highly urban and industrialized country with a high relative population and density, exacerbated by the fact that Japan is a relatively small country, has placed enormous pressure on Japan to control its solid and hazardous waste (Japan Ministry of the Environment 2021b). Currently, Japan estimates that it produces approximately 50 million metric tons of solid and hazardous waste per year. As a measure to limit the amount of wastes disposed of in landfills, Japan incinerates up to 80% of its solid waste and has an aggressive recycling effort for paper, plastic, and metals. As part of recycling efforts, Japan also has aggressive programs in place to reuse material and reduce the amount of waste generated (Japan Ministry of the Environment 2021b).

Land Pollution Regulations of Asia   255

Of the 50 million metric tons of solid and hazardous waste generated in Japan, approximately 42 million is from industry, and the remainder is municipal waste (Japan Ministry of the Environment 2021b). Japan requires specialized air and water pollution control equipment on all incinerators used to burn solid waste. Japan measures dioxin levels in ambient air and water because of its significant use of incinerators used to burn solid wastes, and the levels of dioxin have been significantly reduced by well over 95% since 1998 (Japan Ministry of the Environment 2021c). Japan has developed contaminant limits on the leachate from residual waste following the incineration process that is composed of what is termed cinder dust, ash, sludge, and slag; these are located at www.env.go.jp/en/soil/leachate.html (Japan Ministry of the Environment 2021d).

13.3.3 REMEDIATION STANDARDS

During the post-war period, Japan faced the reality of dealing with millions and millions of metric tons of waste and urban waste land. At the time, these wastes were dumped into the ocean or rivers and piled up in the open, causing plagues of flies and mosquitoes and the spread of infectious diseases. The Japanese government made several attempts at passing laws such as the Public Cleansing Act of 1954 and the Act on Emergency Measures of 1963 concerning the development and living environment and the Basic Act for Environmental Pollution Control enacted in 1967 (Japan Ministry of the Environment 2021b). None of these laws adequately addressed the issue of soil pollution.

It wasn't until the enactment of the Japan Soil Contamination Countermeasures Act of 2002, which had the objective to formulate measures to evaluate the degree to which land in Japan was impacted by contaminants and to provide methods and criteria to prevent further deterioration and restore impacted locations to protect human health, that Japan made significant progress at restoring and improving its land environment (Japan Ministry of the Environment 2021b). Cleanup levels for soil in Japan are listed at www.env.go.jp/en/soil/leachate.html (Japan Ministry of the Environment 2021e).

The general conditions under the Soil Contamination Countermeasures Act of 2002 are similar to those of the United States in that Japan requires those who were responsible for polluting the environment to be responsible for environmental restoration under a Japanese provision in the regulations termed the polluters pay principle.

Japan is heavily dependent on surface water as its primary source of drinking water. In fact, 89% of Japan's drinking water is from surface water sources, and 11% is from groundwater. Since Japan is densely populated, a relatively small country by land area, and an island nation, providing a reliable source of drinking water and preventing fresh water sources from contamination is a very high priority (Japan Ministry of the Environment 2021a). Therefore, Japan has an array of over 4,000 monitoring wells located throughout the country to monitor groundwater and thousands of surface water monitoring points to monitor surface water quality. The purpose of the well network is to act as an early warning system so that potential threats to water quality are addressed at the earliest stage possible (Japan Ministry of the Environment 2021f).

In addition, Japan has strict soil cleanup values and monitoring points near potential sources of contamination such as landfills and heavily industrialized areas (Japan Ministry of the Environment 2021f). Last, Japan has undertaken several water conservation and recycling efforts (Japan Ministry of the Environment 2021f). Groundwater quality values are at www.env.go.jp/en/spcl/html (Japan Ministry of the Environment 2021f). In Japan, soil quality values are the same as groundwater quality values. This is done to protect the groundwater sources (Japan Ministry of the Environment 2021f).

13.3.4 SUMMARY OF ENVIRONMENTAL REGULATIONS OF JAPAN

Japan is one of the most developed nations on Earth and, as a result, has experienced significant environmental degradation of its air, water, and land. The contributing factors to environmental degradation in Japan can be traced to wars, lack of knowledge, deforestation, erosion and drainage control, rapid industrialization, and lack of effective pollution control measures. In addition, Japan's geography and geology have greatly contributed to its environmental degradation and loss of life from volcanic eruptions, landslides, tsunamis, earthquakes, and typhoons.

However, Japan now has one of the most comprehensive and effective environmental regulatory structures and has improved its air, water, and soil quality. To maintain its quality of life and to sustain its population, Japan must continue to develop methods to protect its air, water, and land. Japan has a regulatory structure similar to that of the United States, and air, water, and soil quality criteria are strict and consistent from one medium to the next. Japan's inspection and enforcement activities are significant and robust compared to other countries.

13.4 INDIA

India is the second most-populated country on Earth, with an estimated population of over 1.2 billion, second only to China. India is the seventh-largest country on Earth, with a total area of 3,287,263 square kilometers. The largest river in India is the Ganges, which is 2,525 kilometers in length. India has varied geography, temperature, and precipitation ranges. India experiences monsoons in the summer months that often flood many urban areas. India's lowest elevation is –2.2 meters, and its highest elevation is 8,586 meters (Kanchenjunga), the third-highest peak on Earth in the Himalayan mountains along its northern border. India shares borders with Bangladesh, China, Pakistan, Nepal, Myanmar, Bhutan, and Afghanistan. Southern portions of India are located along the coastline of the Indian Ocean, which is more than 7,500 kilometers in length (Marsh and Kaufman 2015).

India faces significant pollution issues of its air, water, and land. This becomes difficult to manage because of its large human population, limited financial resources, lack of infrastructure, and current low technological capabilities. However, between 1984 and 2010, India has made significant progress in addressing environmental issues and improving environmental quality (World Bank 2011). This is reflected in India's increase in life expectancy, which was 41 years in 1940 compared to 68 in 2016 (World Bank 2016).

Land Pollution Regulations of Asia 257

India is one of the top agricultural countries in the world, and because of this, it uses large quantities of water for irrigation. Approximately 20% of its population does not have access to usable water, and 21% of disease in India can be traced back to unsafe water (World Health Organization 2021a, 2021b). The Ganges River is considered especially polluted, with untreated sewerage discharge to the river at several locations. There is a wide disparity in the standard of living in India, perhaps more than almost any other country of the world. The average per person GDP in India is $3,600 (World Bank 2016).

13.4.1 Environmental Regulatory Overview of India

One of the most significant environmental incidents in India occurred on December 3, 1984, when more than 36,000 kilograms of methyl isocyanate leaked from a Union Carbide pesticide manufacturing plant in Bhopal, India. At least 3,800 people were immediately killed, and thousands more were injured (Broughton 2005). The disaster highlighted the need for environmental and safety standards, preventative and engineering strategies, and disaster preparedness.

This incident, perhaps more than any other environmental incident in India, changed the course of environmental regulations in India and in many other parts of the world, including the United States. The disaster also demonstrated that a seemingly local incident of release of contaminants into the environment was intimately tied to global economic markets (Broughton 2005). In addition, this incident and many others that have occurred historically throughout the world through history, many of which we have mentioned and discussed in some detail, highlight the need for consistent environmental regulation and enforcement globally.

13.4.2 Solid and Hazardous Waste

In India, as in most other countries, a solid waste is considered hazardous if it has any of the following characteristics (India Ministry of Environment, Forestry, and Climate Change 2021a):

- Toxic
- Ignitable
- Corrosive
- Radioactive
- Reactive
- Explosive

Similar to that of the United States, India requires as part of its solid and hazardous waste regulations a detailed evaluation, termed an environmental impact assessment, of any facility that treats, stores, or is the final disposal location for hazardous wastes. The EIA evaluates all aspects of potential impact that the facility may pose, including (India Ministry of Environment, Forestry, and Climate Change 2021a):

- Surface and groundwater
- Wetlands
- Forested areas

- Nearby urban areas
- Noise
- Wildlife
- Marine life
- Invertebrates
- Other potential sensitive habitats

For compliance purposes in India, each facility that generates solid waste must follow the recommended steps (India Ministry of Environment, Forestry, and Climate Change 2021a):

- Identify the point of generation of the waste
- Collect data on the process that generated the waste
- Characterize the waste pertaining to the properties of ignitability, corrosivity, flammability, reactivity, and toxicity
- Evaluate whether the waste was generated under any of the 18 listed waste categories
- Evaluate whether the waste can be recycled or reused or is considered a solid waste requiring disposal or a hazardous waste requiring disposal at a licensed TSDF
- Identify the appropriate site for disposal
- Arrange for proper transportation of the waste from a licensed transporter

13.4.3 REMEDIATION STANDARDS

In India, remediation standards for cleanup of contaminated sites are generally administered through the local or provincial governmental authority and follow the procedures outlined in conducting an environmental impact assessment (India Ministry of Environment, Forestry, and Climate Change 2021b). The process of conducting an EIA in India is nearly identical to that in the United States, the EU, and many other nations. The process includes following the fundamental steps (India Ministry of Environment, Forestry, and Climate Change 2021b):

- Screening: Considered the initial stage and addresses whether there is a need to conduct an EIA
- Scoping: Once it's established that an EIA is necessary, the scoping stage is conducted and determines all relevant environmental concerns
- Public Consultation: Public consultation provides the opportunity for input from any interested stakeholders to provide input through the EIA process
- Investigation: This stage may involve several steps or iterations before the data set is complete and concerns are well understood
- Appraisal: This stage is generally considered a data gap analysis and can either accept data, reject data, or require that more data be collected
- Risk Evaluation: This stage then takes all the data and then conducts a project-specific health risk assessment. The standards used to conduct a

Land Pollution Regulations of Asia

health risk assessment are those guidelines established by USEPA or the European Union

- Remediation goals: After each of the previous stages is completed, remediation goals with specific targets are established

13.4.4 SUMMARY OF ENVIRONMENTAL REGULATIONS OF INDIA

India has a large disparity of distribution of wealth and standard of living. The average standard of living for the residents of India is less one twentieth (1/20) that of the United States. Together with its large population of well over 1 billion, lack of infrastructure, and other social and political struggles, the environmental pressures are, at times, overwhelming. However, despite these huge challenges, India has made progress at improving its environment and life expectancy since the Bhopal disaster in late 1984.

India has established a robust set of environmental regulations that compare adequately to the United States, EU, and other rather developed nations. However, lack of infrastructure and enforcement, along with the social and economic factors discussed previously, highlight the need for continued improvement.

13.5 SOUTH KOREA

South Korea is located on the Korean Peninsula in northeast Asia. Since 1950, Korea has been divided in half at the 38th parallel as a result of the Korean War. The Amnok River forms the border of North Korea, China, and Russia. The Yellow Sea is located to the east, the East China Sea and Korea Strait is to the south, and the Sea of Japan is located to the east. Korea is approximately 20% of the size of California, or about 100,000 square kilometers (United Nations 2012). The devastation caused by the Korean War in the early 1950s is still apparent today. The two Koreas have evolved from a common cultural and historical base into two very different societies with radically dissimilar political and economic systems (Asia Society 2018). North Korea is heavily influenced by Chinese and Soviet/Russian culture, and South Korea is heavily influenced by western culture. Today, the population of North Korea is approximately 25 million, while the population of South Korea is just over 51 million (United Nations 2021a).

South Korea is mountainous along its western boundary and is much less mountainous along its eastern and southern parts. In fact, South Korea is considered 70% mountainous. Rivers tend to flow from east to west, with a general descending topography toward the west and south. South Korea has a total land area of 97,230 square kilometers, with a population density of approximately 526 people per square kilometer (United Nations 2016). Since the cease fire with North Korea in 1953, South Korea has essentially been rebuilt. The capital of South Korea is Seoul, which is located in the northwestern portion of the country. Seoul has a population of greater than 10 million residents and more than 25 million in the metropolitan proper.

South Korea, which was primarily an agrarian society before the Korean War with 75% of its population living in rural areas and farming small plots of land, has

FIGURE 13.5 Intermixed land use in Korea, recreational (bottom), agriculture and industry (middle), and residential (upper) (photograph by Daniel R. Rogers).

been completely transformed to an industrial and urban nation. South Korea now has 83% of its population living in urban areas. Since much of the population in Korea is urban, land use is an intermixing of industry, residential, and agriculture, which is very different than the United States. This presents a challenge in applying environmental regulations evenly. However, the cost and time spent on transportation are much lower because many people live near where they work, and food is grown nearby (United Nations 2012). Figure 13.5 shows an example of the intermixed land use between industrial, residential, agricultural, and recreational.

The standard of living in Korea per person is just over $35,000, which is more than twice that of China but well below that of the United States. South Korea experienced rapid growth in the 1970s and remains under rapid development. The GDP of South Korea is $1.69 trillion, which is approximately 12 times less than the United States. As a result of this development, the natural environment suffered deterioration of the air, water, and land and resulted in the passage of numerous environmental laws (United Nations 2012).

13.5.1 Environmental Regulatory Overview of South Korea

The USEPA is heavily engaged in collaborating with Korea as a result of the Korea-United States Free Trade Agreement (USEPA 2021b).

Through this mechanism, the USEPA and the Korean Ministry of the Environment and partner agencies in both countries have been cooperating to strengthen (USEPA 2021b):

- Environmental governance
- Air quality

Land Pollution Regulations of Asia 261

- Water quality
- Reduce waste
- Recycling
- Reusable energy
- Protections against toxic pollutants

13.5.2 SOLID AND HAZARDOUS WASTE

The Waste Control Act of 2016 governs the disposal of solid and hazardous waste in Korea and amends previous versions of 2007, 2011, and 2015. The purpose of the Waste Control Act of 2016 is to contribute to environmental conservation and the enhancement of the people's standard of living by minimizing the production of wastes and disposing of generated wastes in an environmentally friendly manner (Korea Ministry of the Environment 2021a). Innate in the stated purpose is the desire to minimize the amount of wastes generated and to improve the quality of their disposal so as not to significantly impact the environment.

The Waste Control Act of 2016 states the following requirements (Korea Ministry of the Environment 2021a):

- Every business entity shall reduce the generation of wastes to the maximum extent possible by improving the manufacturing process and so on of products and minimize the discharge of wastes by recycling his/her own wastes.
- Every person shall take prior appropriate measures with respect to the discharge of wastes to prevent any harm to the environment or the health of residents.
- Waste treatment shall be properly managed in a manner that reduces their quantities and degree of hazard or otherwise is consistent with environmental conservation and the protection of the people's health.
- Any person who causes environmental pollution by discharging wastes shall be responsible for restoring the affected environment and bear the expenses incurred in restoring the damage caused by such pollution.
- To the extent possible, wastes originating in the Republic of Korea shall be disposed of within the Republic of Korea, and the importation of wastes shall be restrained.
- Wastes shall be recycled rather than incinerated or buried in order to contribute to the improvement of resource productivity.

Korea instituted a tracking system as part of its waste management system in 1999 to track the transportation of hazardous wastes (Korea Ministry of the Environment 2021b).

In Korea, a waste is hazardous if it a listed waste and is any of the following (Korea Ministry of the Environment 2021b):

- Explosive
- Radioactive
- Medical waste
- Corrosive
- Flammable
- Infectious
- Reactive
- Characteristic waste

A waste is characteristic hazardous waste if a leachate test indicates a concentration of the listed compounds exceeding concentrations listed at www.law.go.kr/DRF/law/wastecontrolact (Korea Ministry of the Environment 2021b).

Listed hazardous wastes in Korean are similar to other nations and include the following (Korea Ministry of the Environment 2021b):

- Waste pesticides or herbicides
- Electronic wastes
- Waste thermometers that contain mercury
- Waste oils
- Waste solvents
- Medical wastes
- Wastewater treatment sludges
- Paint waste
- Waste from pigments, glue, varnish, and printing ink

In efforts to reduce solid wastes, Korea has a vigorous recycling program that also includes food wastes. As stated by the Korean government, increased living standards have resulted in excessive convenience, the use of disposable items, and over-packaged products that cause a waste of resources, generate unnecessary waste, and bring about negative impacts to the environment. To combat this, the Korean Ministry of the Environment began a recycling program in 1994 and continues to improve the program through the following (Korea Ministry of the Environment 2021c):

- Reduction of plastics
- Reduction of packaging waste
- Reuse of paper cups and bags
- Introduction of farm produce green packaging guidelines and safety regulations
- Developing technologies for flexible packaging and paper containers
- Recycling of metals
- Recycling of wood products

13.5.3 REMEDIATION STANDARDS

Korea has established a nearly identical law to that in the United States for dealing with legacy contamination of land, groundwater, and surface water. The Korean Liability, Compensation and Relief System for Damages from Environmental Pollution Act of 2016 (Korea Ministry of the Environment 2021c). Soil and groundwater cleanup values are at www.eng/me/go.kr/eng/web/law/Actno.166922 (Korea Ministry of the Environment 2021c).

In an effort to be proactive and avoid soil and groundwater contamination, the Korea Ministry of the Environment has identified 22,868 specific facilities that are subject to soil and groundwater monitoring that include (Korea Ministry of the Environment 2021d):

- Gasoline service stations
- Certain industrial plants

Land Pollution Regulations of Asia

- Petroleum storage sites
- Abandoned metal, coal, and asbestos mines
- Hazardous waste storage and disposal facilities
- Chemical plants
- Treatment plants
- Power generating plants
- Other ecologically sensitive areas

Each site must conduct tests to ensure that no soil or groundwater is present every 1 to 5 years (Korea Ministry of the Environment 2021e). This type of proactive approach to environmental issues is more advanced that almost any country in the world, including the United States.

Through this program, Korea has identified, investigated, and remediated several sites and continues to be proactive to restore its environment (Korea Ministry of the Environment 2021e).

13.5.4 Summary of Environmental Regulations of Korea

Korea has essentially rebuilt its country in the last 70 years due to war. This has provided an opportunity to reconstruct a country with fewer political and social impediments and more opportunity for ease of modernization.

Korea perhaps has the most complete set of environmental regulations that we have examined thus far, even more comprehensive than the United States and the European Union. This is reflected in the fact that Korea regulates its farmland as strictly as other locations such as residential and industrial properties. In addition, Korea has robust recycling laws that involve all citizens, not just industry or commercial businesses. In addition, from a remediation point of view, Korea monitors its high-risk locations for the presence of contamination as an early warning system and also closely monitors its surface water and groundwater resources as well.

Korea has a standard of living of just over $35,000 per person, which is well below the average in the United States of $59,500. The population of Korea slightly exceeds 50 million residents, which is approximately seven times less than the United States. However, Korea is only around 100,000 square kilometers in size, which is approximately 100 times smaller in land area than the United States. This means that Korea has a rather large population for its overall land size.

The economic, geographical, and demographical statistical facts listed previously, along with its environmental regulations, hold insight as to why Korea has such a complete environmental system with close monitoring. With its low relative land area and high population, Korea is forced to protect its environment and must be efficient in its use and care of natural resources. This puts pressure on its economy, but Korea has found an effective balance, especially since its standard of living is twice that of China.

13.6 SAUDI ARABIA

Saudi Arabia is located in westernmost Asia and occupies the majority of the Arabian peninsula. The total land area of Saudi Arabia is 2,149,690 square kilometers, which makes it the 12th-largest country in the world. The population of Saudi Arabia is approximately 30 million (United States Department of State 2021c). Saudi Arabia is bordered by Jordan and Iraq to the north; Kuwait to the northeast; Qatar, Bahrain, and the United Arab Emirates to the east; Oman to the southeast; and Yemen to the south. It is separated from Israel and Egypt by the Gulf of Aqaba.

Oil was discovered in Saudi Arabia in 1938, and it has since developed into the largest oil producer and exporter of oil in the world, controlling the second-largest oil reserve and eighth-largest gas reserves (United States Department of State 2021c). Oil production has provided Saudi Arabia with economic prosperity and substantial political leverage in the region. The political system in Saudi Arabia is considered a monarchy and has no free elections (United State Department of State 2021c).

Saudi Arabia's geography is dominated by the Arabian Desert, associated semi-desert, and several mountain ranges. There are very few lakes and no permanent rivers. Average summer temperatures can exceed 45°C (113°F) and have reached 54°C (129°F). In summer, the temperature rarely drops below freezing. The largest city in Saudi Arabia is Riyadh, which has an estimated population of 6.5 million residents. The gross domestic product of Saudi Arabia is estimated at $646 billion, and per capita income is estimated at nearly $21,000 per citizen (United Nations 2002, 2004, 2021c).

Since Saudi Arabia is essentially a country located in a desert, water is a precious resource. More than 50% of fresh water for the country is from desalinization plants located near the Red Sea. The remainder originates from groundwater and some surface water located in the western mountains of the country (United Nations 2021c).

13.6.1 Environmental Regulatory Overview of Saudi Arabia

Environmental issues that Saudi Arabia faces are closely tied to its oil industry, which accounts for most of its economy. Environmental issues in Saudi Arabia are viewed towards the development of the county's oil and gas reserves. Saudi Arabia first passed the General Environmental Regulations and Rules for Implementation in 2001 and formally adopted environmental laws for ambient air, drinking water, wastewater discharge, solid and hazardous waste, emissions from mobile sources, noise, and pollution prevention and preparedness in 2012. The Presidency of Meteorology and Environment (PME) is responsible for implementing environmental regulations in Saudi Arabia (Chakibi 2013).

13.6.2 Solid and Hazardous Waste

Saudi Arabia defines wastes as substances which have been discarded or neglected and can no longer be put to beneficial use. Saudi Arabia defines hazardous waste as a type of waste that has characteristics that render it hazardous to human health and includes waste that are (Saudi Arabia Presidency of Meteorology and Environment 2012):

• Toxic	• Flammable	• Explosive	• Radioactive
• Corrosive	• Reactive	• Infectious	• Listed wastes

Land Pollution Regulations of Asia 265

Saudi Arabia also has a hazardous waste category of listed wastes that include any wastes generated by the following (Saudi Arabia Presidency of Meteorology and Environment 2012):

- Medical wastes
- Biocide or phyto-pharmaceutical processes
- Wood preservatives
- Waste mineral oil
- Wastes from heat treatment and steel tempering containing cyanides
- Waste oil
- PCBs
- Waste paints, inks dyes, pigments, lacquers, and varnish
- Waste resins, latex, plasticizers, glues, and adhesives
- Wastes generated from research and development and teaching arising from activities known or suspected to be harmful to humans
- Waste photographic chemicals
- Wastes from metal treatment processes

Saudi Arabia also considers wastes hazardous if they contain any of the following constituents (Saudi Arabia Presidency of Meteorology and Environment 2012):

• Metal carbonyls	• Beryllium	• Hexavalent chromium
• Copper	• Zinc	• Arsenic
• Selenium	• Cadmium	• Antimony
• Tellurium	• Mercury	• Thallium
• Lead	• Inorganic fluorine	• Inorganic cyanide
• Acidic solutions	• Basic solutions	• Organic phosphorus
• Organic cyanide	• Asbestos	• Phenols
• Ether compounds	• Halogenated organics	• Organic solvents
• Organic halogens	• Dibenzo-furans	• Dibenzo-p-dioxins

13.6.3 REMEDIATION STANDARDS

In Saudi Arabia, new developments or expansions of any facility require that the owner conduct a comprehensive environmental impact assessment. The process begins with notification of the Presidency of Meteorology and Environment and retaining a qualified consultant approved by the regulatory agency. Each project is graded as Category I, II, or III by the regulatory authority according to potential environmental impact, with Category I as the least potential impact and Category III as the most potential environmental impact (Saudi Arabia Presidency of Meteorology and Environment 2012). Each category must address the following (Saudi Arabia Presidency of Meteorology and Environment 2012):

- Air quality
- The marine and coastal environment

266   Environmental Compliance Handbook

- Surface and underground water
- Flora and fauna
- Land use
- Urban development
- General scenic view

After the EIA is conducted, a risk evaluation is conducted to evaluate whether there are any unreasonable risks posed by the presence of contamination discovered during the investigation process. Cleanup criterion are set by the regulatory authority based on the EIA and risk evaluation (Saudi Arabia Presidency of Meteorology and Environment 2012).

13.6.4 SUMMARY OF ENVIRONMENTAL REGULATIONS OF SAUDI ARABIA

Saudi Arabia's published environmental laws are comprehensive and in line with other developed countries. However, information on enforcement and contaminated site remediation is not available and is not readily shared with the outside world. Therefore, it's difficult to evaluate the current status of environmental compliance and sustainability. However, Saudi Arabia does face serious environmental issues related to air and especially water because Saudi Arabia is a desert country and water is scarce and can't afford to be polluted.

13.7 INDONESIA

Indonesia is the largest country in southeast Asia and is made up of 17,508 islands of which 6,000 are uninhabited. Its islands are spread between the Indian and Pacific Oceans that link the continent of Asia with Australia. Indonesia has five major islands, Sumatra, Java, Kalimantan, Sulawesi, and Irian Jaya (United Nations 2021d). The land area of Indonesia is characterized as tropical rainforest. Since most of the soil originates from volcanic activity, the soils in Indonesia and considered very fertile. The total land area of Indonesia is 1,904,569 square kilometers and ranks 14th largest in the world. Indonesia shares borders with Malaysia, Papua New Guinea, and East Timor (United Nations 2021d).

The population on Indonesia is estimated at 261 million residents, making it the fourth most populous nation in the world. The population density of Indonesia is 147 individuals per square kilometer, which is roughly five times more densely populated than the United States. Indonesia is home to over 500 languages and dialects. The capital of Indonesia is Jakarta, with an estimated population of 9.6 million residents. The gross domestic product of Indonesia is $687 billion, and the per capita standard of living is $3,000, which is approximately 20 times less than the United States. Trading partners with Indonesia include Singapore, Japan, and the United States (United Nations 2021d). The significant environmental issues on land in Indonesia include deforestation, air pollution, acid rain, and surface water pollution. Air pollution is especially significant in urban areas, especially Jakarta. The surrounding marine environment has been heavily impacted by poorly planned coastal developments,

Land Pollution Regulations of Asia

overfishing, and ocean acidification, which has placed 95% of Indonesia's reefs under serious threat (United Nations 2021d; Conservation International 2021).

13.7.1 ENVIRONMENTAL REGULATORY OVERVIEW OF INDONESIA

In Indonesia, the Ministry of Environment is responsible for administering and enforcing environmental laws and is also responsible for national environmental policy and planning (Indonesia Ministry of Environment 2021). The Indonesia Ministry of Environment is collaborating with USEPA to improve air quality, water quality, disposal of wastes, and cleanup sites of environmental contamination in Indonesia (USEPA 2021c).

13.7.2 SOLID AND HAZARDOUS WASTE

Management of waste in Indonesia is a mounting challenge because of a large population of over 250 million people, lack of infrastructure, and lack of financial resources. The Ministry of Environment published a new regulatory framework in 2014 in an attempt to streamline the regulations and make existing regulations stricter (Indonesia Ministry of Environment 2021).

The revised solid waste regulations also attempt to improve waste disposal methods, specifically to increase the amount of solid water disposed in licensed landfills instead of in streets, rivers and streams, and open burning of waste. Previous to the new solid waste regulations, solid waste was disposed of in the following manner as measured in 2008 (Lokahita 2017):

- 68% in landfills
- 16% small-scale incineration
- 7% composting
- 5% open burning
- 4% dumped in rivers and streams

Since the regulations were updated, Indonesia has made improvements, but these improvements in many circumstances have been met with a negative impact because of a rapid increase in population that overshadows many of the attempted improvements (Lokahita 2017).

13.7.3 REMEDIATION STANDARDS

Remediation standards for contaminated sites in Indonesia generally rely on the Dutch Intervention Values and USEPA. The USEPA is collaborating with the Ministry of Environment of Indonesia to assist and educate Indonesia so that at some point it will be able to build a framework of regulations to address legacy sites of environmental contamination. USEPA is currently collaborating with Indonesia on four sites and is working together with the Ministry of Environment to improve emergency response and cleanup procedures (USEPA 2021c).

13.7.4 SUMMARY OF ENVIRONMENTAL REGULATIONS IN INDONESIA

As stated by USEPA, Indonesia is a key factor in the global environmental arena, with significant importance in ecological resources but with significant challenges to protect and restore its environment (USEPA 2021c). Lack of infrastructure, large population, lack of technology, lack of financial resources, and lack of effective regulations and enforcement have all contributed to environmental deterioration in Indonesia that has led to:

- Poor water quality
- Poor air quality
- Poor waste disposal practices
- Deterioration of the marine environment and reef system
- Poor health
- Disease
- Extinction of species

13.8 MALAYSIA

Located near the equator, Malaysia is a country of two parts in southeastern Asia. The land area of Malaysia is 330,603 square kilometers, which ranks 66th in size. Malaysia is bordered by Thailand to the north and Indonesia to the south. Malaysia is a wet and tropical climate and averages approximately 3 meters of rain annually. The population of Malaysia slightly exceeds 31 million residents, and it has a GDP of 314.5 billion and a per capita income of nearly $10,000. The capital of Malaysia is Kuala Lumpur, with its population of nearly 2 million residents, located on the western peninsula region south of Thailand (United Nations 2021a).

13.8.1 ENVIRONMENTAL REGULATORY OVERVIEW OF MALAYSIA

Environmental issues in Malaysia include air pollution, water pollution from raw sewage from humans and livestock and industrial sources, deforestation, and loss of habitat. Industrial sources of water pollution originate from Malaysia's main industries, tin mining, natural rubber productions, and palm oil production (United Nations 2021e).

13.8.2 SOLID AND HAZARDOUS WASTE

Not unlike Indonesia, Malaysia struggles with appropriate disposal of solid and hazardous waste. Although the problem is not at the magnitude present in Indonesia because Malaysia's population is only 12% of Indonesia's, the trends and habits are similar (United Nations 2021e).

13.8.3 REMEDIATION STANDARDS

Malaysia has set priorities for improving and restoring its environment but achieving success will require changing cultural norms, time, and foreign investment.

Land Pollution Regulations of Asia

Currently, Malaysia is focusing on air and water pollution because those are the two environmental issues that are the highest priority for improving human health (Malaysia Department of Environment 2015; United Nations 2021e). Efforts to improve air quality include (Malaysia Department of Environment 2021a, 2021b):

- Upgrading transportation infrastructure improve to traffic flow and reduce vehicle emissions
- Restricting emissions from power-generating plants and industry through a comprehensive permitting process
- Eliminating the open burning of garbage and solid waste

Efforts to improve water quality include (Malaysia Department of Environment 2021c, 2021d):

- Improve zoning and land use planning
- Upgrade treatment plants
- Require industrial, agriculture, and livestock to pretreat water before discharge

13.8.4 SUMMARY OF ENVIRONMENTAL REGULATIONS OF MALAYSIA

Malaysia suffers from many of the same environmental issues as Indonesia. Those issues of particular concern include urban air pollution, rapid increases in population, providing clean water, sanitation, soil and groundwater contamination, deforestation, and erosion.

13.9 SUMMARY AND CONCLUSION

Although many countries of Asia have a robust set of environmental regulations, many of which are based on the USEPA model, Asia has perhaps the largest disparity in those countries that have effective environmental regulations and those that do not. Those that are effective include Japan and Korea. Those that are least effective are Indonesia, Malaysia, and India. China is somewhere toward the middle, but China is quickly improving and addressing air, water, and land environmental regulations.

13.10 REFERENCES

Asia Society. 2018. Korean History and Geography. https://asiasociety.org/education/korean-history-and-political-geography. (Accessed June 30, 2021).

Broughton, E. 2005. The Bhopal Disaster and Its Aftermath: A Review. *Journal of Environmental Health.* Vol. 4. No. 6. United States Institute of Health. Washington, DC. www.ncbi.nlm.nih.gov/pmc/articles/PMC1142333. (Accessed June 30, 2021).

Chakibi, S. 2013. Saudi Arabia Environmental Laws. *Journal of Environmental Health and Safety.* http://ehsjournal.org/sanaa-chibi/saudiarabialaws. (Accessed June 30, 2021).

China Ministry of Ecology and Environment (CMEE). 2021. *Law of the People's Republic of China on the Prevention and Control of Environmental Pollution by Solid Waste.* Beijing, China. www.china.org.cn/english/environment/34424.htm. (Accessed June 30, 2021).

China Ministry of Environment Protection (CMEP). 2014. *Five Standards of Environmental Protection*. MEP Notice No. 14 of 2014. CMEP. Beijing, China. 98p.

China Ministry of Environmental Protection (CMEP). 2021. New Law on Soil Pollution Prevention. www.china.org.cn/china/2018-09/01/content61417828. (Accessed June 30, 2021).

China Research Academy of Environmental Sciences. 2013. *Groundwater Challenges and Environmental Management in China*. Ministry of Environmental Protection. Beijing, China. 28p.

Conservation International. 2021. Indonesia: A Vast and Beautiful Country—At the Crossroads. https://conservation.org/where/pages/indonesia. (Accessed June 30, 2021).

India Ministry of Environment, Forestry, and Climate Change. 2021a. Solid and Hazardous Waste Regulations. www.envfor.nic.in/decision/solid-and-hazardous-waste. (Accessed June 30, 2021).

India Ministry of Environment, Forestry, and Climate Change. 2021b. Environmental Impact Assessment. www.envfor.nic.in/decision/environmental-impact-assessment. (Accessed June 30, 2021).

Indonesia Ministry of Environment. 2021. Solid Wastes. www.aecon.org/indonesia. (Accessed June 30, 2021).

Japan Ministry of the Environment. 2021a. The Basic Environmental Law and Basic Environmental Plan. www.env.go.jp/en/laws/policy/basic_lp.html. (Accessed June 30, 2021).

Japan Ministry of the Environment. 2021b. Soil Contamination Countermeasures Act of 2002. www.env.go.jp/en/soilact. (Accessed June 30, 2021).

Japan Ministry of the Environment. 2021c. *History and Current State of Waste Management in Japan*. Waste Management and Recycling Department. Tokyo. Japan. 31p. (Accessed June 30, 2021).

Japan Ministry of the Environment. 2021d. Waste and Recycling Standards for Verification. www.env.go.jp/en/recycle/manage/sy.html. (Accessed June 30, 2021).

Japan Ministry of the Environment. 2021e. Soil Pollution Control Values for Soil Leachate. www.env.go.jp/en/soil/leachate/html. (Accessed June 30, 2021).

Japan Ministry of the Environment. 2021f. Ambient Water Quality Standard for Groundwater and Criteria for Implementation of Soil Pollution Control Measures. Water Pollution Control Law. www.env.go.jp/en/spcl/html. (Accessed June 30, 2021).

Kichira, S. and Sugai, M. 1986. *The Origins of Environmental Destruction: The Ashio Copper Mine Pollution Case*. United Nations University. Shinyousha, Tokyo. 167 p. www.unu.edu.ashiocoppermine. (Accessed June 30, 2021).

Korea Ministry of the Environment. 2021a. Wastes Control Act of 2016: General Provisions. www.law.go.kr/DRF/law/wastecontrolact. (Accessed June 30, 2021).

Korea Ministry of the Environment. 2021b. Hazardous Waste Management Flowchart. www.eng.me.go.kr./eng/web/index.do?menuld=379. (Accessed June 30, 2021).

Korea Ministry of the Environment. 2021c. Korea Waste Prevention and Recycling Policy. www.eng.me.go.kr/eng/web/index.do?menuld=141. (Accessed June 30, 2021).

Korea Ministry of the Environment. 2021d. Liability, Compensation and Relief System for Damages from Environmental Pollution. www.eng/me/go.kr/eng/web/law/Actno.166922. (Accessed June 30, 2021).

Korea Ministry of the Environment. 2021e. Soil and Groundwater Contamination Prevention and Restoration. www.eng.me.go.kr/eng/web/index.do?menuld=311. (Accessed June 30, 2021).

Lokahita, B. 2017. *Indonesia Municipal Solid Waste: Regulations and Common Practice*. Tokyo Institute of Technology. www.academia.edu/23109258/indonesia/solidwasteregs. (Accessed June 30, 2021).

Malaysia Department of Environment. 2015. Environmental Issues and Challenges—Malaysian Scenario. www.doe.gov.my/environmentalissues. (Accessed June 30, 2021).

Land Pollution Regulations of Asia

Malaysia Department of Environment. 2021a. Malaysia Ambient Air Quality Standards. www. doe.gov.my/portelv1/info-umum/english-ambient-air-standards. (Accessed June 30, 2021).

Malaysia Department of Environment. 2021b. Malaysian Air Pollution Index. Official Portal of Department of State. www.doe.gov.my/portelv1/info-umum/englisk-api. (Accessed June 30, 2021).

Malaysia Department of Environment. 2021c. Water Pollution and Water Quality Standards. www.doe.gov.my/waterquality. (Accessed June 30, 2021).

Malaysia Department of Environment. 2021d. Malaysia Sewerage and Industrial Effluent Discharge Standards. www.doe.gov.my/discharge-standards/malaysia.htm. (Accessed June 30, 2021).

Marsh, W. B. and M. M. Kaufman. 2015. *Physical Geography*. University of Cambridge Press. Cambridge, UK. 647 pages.

Saudi Arabia Presidency of Meteorology and Environment. 2012. General Environmental Regulations. www.pme.gov.sa. (Accessed June 30, 2021).

United Nations. 2002. *Saudi Arabia Country Profile. Johannesburg Summit*. United Nations. New York, NY. 59p.

United Nations. 2004. *Kingdom of Saudi Arabia. Public Administration and Country Profile*. Department of Economic and Social Affairs (UNDESA). New York. NY. 16p.

United Nations. 2012. *Country Profile Series: Republic of Korea*. New York, NY. 92p. www. un.org.esa/agenda21/natlinfo/rekorea.pdf. (Accessed June 30, 2021).

United Nations. 2016. Human Development Report. United Nations Development Programme. www.hdr.undr.org/sites/2017. (Accessed June 30, 2021).

United Nations. 2021a. World Population Prospects. United Nations Department of Economic and Social Affairs. Population Division. https://esa.un.org/unpd/wpp/data. (Accessed July 1, 2021).

United Nations. 2021b. Country Profile for China. www.china.unfpa.org/em/united-nations-population-china-2. (Accessed June 30, 2021).

United Nations. 2021c. The Demographic Profile of Saudi Arabia with Population Trends. www.unescwa.org/sites/saudiarbiaprofile. (Accessed June 30, 2021).

United Nations. 2021d. Country Profile and Review of Indonesia. www.un.org/esa/earth summit/indon-cp.htm. (Accessed June 30, 2021).

United Nations. 2021e. Country Profile and Review of Malaysia. www.un.org/esa/earth summit/mal.htm. (Accessed June 30, 2021).

United States Census Bureau. 2021. United States Trade in Goods with China 2017. www. census.gov/foreign-trade/balance. (Accessed June 30, 2021).

United States Department of Commerce. 2021. Foreign Trade with China 2017. www. commerce.gov/tags/china. (Accessed June 30, 2021).

United States Department of State. 2021a. Office of the United States Trade Representative. (https://ustr.gov/nafta). (Accessed June 30, 2021).

United States Department of State. 2021b. United States Relations with Japan. www.state. gov/p/eap/ci/ja/. (Accessed June 30, 2021).

United States Department of State. 2021c. Profile and Statistics of Saudi Arabia. www.state. gov/saudiarabia. (Accessed June 30, 2021).

United States Environmental Protection Agency (USEPA). 2021a. EPA Collaboration with China. www.epa.gov.international-cooperation/epa-collaboration-china. (Accessed June 30, 2021).

United States Environmental Protection Agency (USEPA). 2021b. EPA Collaboration with the Republic of Korea. www.epa.gov.international-cooporation-korea. (Accessed June 30, 2021).

United States Environmental Protection Agency (USEPA). 2021c. EPA Collaboration with Indonesia. www.epa.gov.nternational-cooporation-Indonesia. (Accessed June 30, 2021).

Wencong, W. 2012. *China Unveils Plan to Control Hazardous Waste*. China Daily. Beijing, China. www.chinadaily.com.cn/china/2012-11/02content15867364.htm. (Accessed June 0, 2021).

World Bank. 2011. Environmental Assessment, Country Data: India. https://data.worldbank.org/countrydata.india.html. (Accessed June 30, 2021).

World Bank. 2016. India in Profile: Economy and Standard of Living. https://data.worldbank.org/countrydata.india.html. (Accessed June 30, 2021).

World Health Organization. 2021a. World Health Organization Profile of India. www.who.int/countries/ind/en. (Accessed October 1, 2018).

World Health Organization. 2021b. United Nations Global Analysis and Assessment of Sanitation and Drinking Water: Indonesia. www.who.org/drinkingwater/Indonesia. (Accessed June 30, 2021).

14 Land Pollution Regulations of Oceania

14.1 OCEANIA

Oceania is a region made up of thousands of islands in the Central and South Pacific Ocean. The two largest land masses that make up Oceania are Australia and the micro-continent of New Zealand. Oceania also includes three island regions: Melanesia, Micronesia, and Polynesia (National Geographic Society 2021a). From a geographic perspective, Oceania is divided into three region types (National Geographic Society 2021a):

- Continental Type—this type includes Australia and New Zealand
- High Islands—Also called volcanic islands, including Melanesia, Mount Yasur, and Vanuatu
- Low Islands—Also called coral islands, including Micronesia and Polynesia

We will evaluate the environmental regulations of Australia and New Zealand.

14.2 AUSTRALIA

Australia is the sixth-largest country of the world, with a land mass of 7,686,850 square kilometers. Australia is generally considered a desert. However, there is a small area located along its northeastern coast that is rainforest (United Nations 1997). Australia's population is estimated at 24.9 million, with most of the population living along the eastern coast and the western city of Perth (United Nations 2021a). Australia's largest city is Sydney, located in the southeastern portion of the country adjacent to the Pacific Ocean. Australia has no land borders with other countries. The Indian Ocean is located along its western shore, and the Pacific Ocean is located along its eastern shore (United Nations 2021a).

The average per capita income in Australia is estimated at $67,442, which ranks second-highest in the world, just behind Switzerland (World Bank 2021). The gross domestic product of Australia is estimated at $1.69 trillion and ranks 27th in the world. Economic drivers in Australia include mining and tourism, along with lesser amounts of farming and livestock. Mining in Australia consists of coal, natural gas, gold, aluminum, iron and other heavy metals, and diamonds and opals (World Bank 2021).

From an environmental perspective, Australia is perhaps best known for the Great Barrier Reef, which is composed of over 3,000 individual reef systems, coral cays, and hundreds of tropical islands along its northwestern coast (United Nations 2021b). The Great Barrier Reef contains the largest collection of coral reefs in the world,

DOI: 10.1201/9781003150107-14

with 400 different types of coral and 1,500 species of fish. The Great Barrier Reef was designated a United Nations World Heritage Site (UNESCO) in 1981 and covers an area of 348,000 square kilometers (United Nations 2021b).

14.2.1 Environmental Regulatory Overview of Australia

Environmental issues in Australia are wide ranging and can be summarized as the following (Australia Department of Environment and Energy 2021a; The National Geographic Society 2021a, 2021b):

- Invasive species. Specifically, the introduction of the rabbit.
- Endangered species, including the Tasmanian devil and southern cassowary.
- Extinction. More than 50 species of birds, mammals, and marsupials have gone extinct, including the Tasmanian wolf.
- Ocean dumping.
- Land degradation. Caused by land clearing, overgrazing, and forest clearing.
- Soil erosion from poor land use practices.
- Poor water quality from poor industry practices and land use planning.
- Salinization from poor water quality irrigation techniques.
- Urbanization.
- Pollution due to mining, including heavy metals, gems, and coal.
- Climate change. Reefs in the Great Barrier Reef System are dying off from warmer ocean waters as a result of climate change.

14.2.2 Solid and Hazardous Waste

In 2009, Australia set goals for reducing the amount of wastes generated nationwide. The goal was to create less waste and improve resources (Australia Department of Environment 2009). The need to revise its national policy and improve regulations was because Australia saw an increase in the amount of waste generated nationwide by 34.7% over a 4-year period from 2002 to 2006, and the amount of hazardous waste doubled during this same period of time (Australia Department of Environment 2009).

Since 2009, Australia is now recycling approximately 60% of its solid waste. The waste generated per person in Australia in 2006 was estimated at 2.783 metric tons and at 2.705 metric tons in 2015, which represents a slight decrease nationwide in the amount of solid waste produced (Australia Department of Environment 2021a).

In Australia, a solid waste is defined as material or products that are unwanted or have been discarded, rejected, or abandoned. Wastes typically arise from three dominant waste streams (Australia Department of Environment and Energy 2021b, 2021c):

- Domestic and municipal
- Industrial
- Construction and demolition

Determining whether a solid waste is hazardous in Australia is similar to in the United States and many other countries. In Australia, a solid waste is hazardous if it is a per-listed waste or has the following characteristics (Australia Department of Environment 2021b, 2021c):

- Explosive
- Ignitable
- Oxidizing agents
- Reactive
- Medical
- Radioactive
- Flammable
- Gas
- Corrosive
- Infectious
- Toxic
- Characteristic

A waste is a characteristic hazardous waste if a leachate test of the waste identifies any of the compounds listed at www.env.gov.au/protection/waste-class.pdf (Australia Department of Environment 2021c). Soil, sediment, and water remediation values for select compounds and pathways are listed at www.environment.gov.au/site-cleanup.pdf (Australia Department of Environment and Conservation 2010).

14.2.3 REMEDIATION STANDARDS

Cleanup or restoration of agricultural land in Australia is perhaps the most significant issue, not industry or urbanization. Approximately 50% of agricultural land in Australia is considered severely degraded because of rising salinity from irrigation, erosion, and vegetative loss. The result has been a reduced capacity for crops, which in turn then requires more land for agricultural and continued land degradation. This cycle of agriculture kills farmland, contaminates drinking water, and destroys natural ecosystems if it continues without drastic restoration efforts and modified farming techniques (Coolaustralia.org 2021).

14.2.4 SUMMARY OF ENVIRONMENTAL REGULATIONS OF AUSTRALIA

Australia has a comprehensive set of environmental regulations and an enforcement system and culture that are very engaged in protecting human health and the environment. Australia has some disadvantages when it comes to its dry climate, poor soil, and desert geography in that water is more a precious resource than other parts of the world since it is of limited quantity. This, as we discussed earlier in this section, is why Australia must protect its water from contamination. In addition, off Australia's northeast coast is the largest reef system in the world, the Great Barrier Reef. The Great Barrier Reef is experiencing an unprecedented die-off of coral due to warming of ocean water believed to be directly linked to climate change.

14.3 NEW ZEALAND

New Zealand is an island country 1,200 kilometers southeast of the coast of Australia in the South Pacific Ocean. New Zealand is composed of nearly 600 islands, two of

276 Environmental Compliance Handbook

which are large and make up most of its land mass: the North Island and the South Island.

New Zealand is composed of 267,710 square kilometers of land, making it the 75th-largest country in the world. The population of New Zealand is 4.8 million. New Zealand's largest city is Auckland, with an estimated population of 1.6 million (United Nations 2021c). The North Island of New Zealand is characterized as volcanic in nature, and the South Island is characterized as mountainous in nature. New Zealand's economy is the 53rd-largest in the world when measured by gross domestic product, which was $200.7 billion in 2016. The per capita income of New Zealand is $37,860 as measured in 2016. New Zealand's economy is considered one of the most globalized and depends greatly upon international trade. Major industries in New Zealand are aluminum mining, food processing, metal fabrication, and wood and paper products. Trading partners include Australia, United States, China, Japan, Germany, and South Korea (United Nations 2021c).

14.3.1 Environmental Regulatory Overview of New Zealand

Environmental issues facing New Zealand include deforestation, soil erosion, loss of habitat, and surface and groundwater quality degradation. Much of the development in New Zealand has been for farming, which has impacted the environment perhaps more than any other type of human activity, including industrial impacts. New Zealand is actively involved in sustainability issues and openly embraced efforts to deal with climate change beginning as early as 1991 (New Zealand Ministry for the Environment 2021a). In New Zealand, the Resources Management Act of 1991 is the main law that regulates discharges into the environment, including air and water. Much of the responsibility for managing discharges and permits has been handed down to district and regional councils. Therefore, just as in the United States, there are differences from district to district in how these issues are handled, but the basic goals, objectives, and specifics are the same (New Zealand Ministry for the Environment 2021a).

Agriculture is the largest land use in the lowlands of New Zealand, particularly dairy and livestock, which has been linked to increased concentrations of nitrogen in soil, surface water, and groundwater. Availability of water has also declined due to increased dairy farming, which uses more water in New Zealand than any other activity (New Zealand Ministry of the Environment 2021b). Wastewater discharges in New Zealand require what is termed a resource consent from the district regulatory authority. The resource consent is a discharge permit that makes the discharge subject to facility-specific discharge conditions that take into account the quality of the wastewater and the receiving body of water (New Zealand Ministry for the Environment 2021b).

14.3.2 Solid and Hazardous Waste

In New Zealand, a waste is considered hazardous if it presents some degree of physical, chemical, or biological hazard to people or the environment (New Zealand

Land Pollution Regulations of Oceania

Ministry for the Environment 2021c). Hazardous waste can be gases, liquids, or solids that are:

- Explosive
- Infectious
- Oxidizing agents
- Reactive
- Toxic
- Radioactive

- Flammable
- Ignitable
- Corrosive
- Medical
- Eco-toxic
- Characteristic

Similar to many other countries, hazardous wastes are tracked from the generator to the transporter to the final disposal site using a manifest system (New Zealand Ministry for the Environment 2021c).

New Zealand has developed a waste list or what is termed the "L Code," which is a method that reflects typical waste streams in New Zealand. The L-Code provides guidance on identifying wastes in a consistent manner and serves as the basis for record-keeping systems, especially for hazardous wastes. The L-Code is as follows (New Zealand Ministry for the Environment 2021d):

1. Wastes resulting from exploration, mining quarrying, and physical and chemical treatment of minerals.
2. Waste from agriculture, horticulture, aquaculture, forestry, hunting and fishing, and food preparation and processing.
3. Waste from wood processing and the production of panels and furniture, pulp, paper, and cardboard.
4. Wastes from leather, fur, and textile industries.
5. Wastes from petroleum refining, natural gas purification, and pyrolytic treatment of coal.
6. Wastes from inorganic chemical processes.
7. Wastes from organic chemical processes.
8. Wastes from the manufacture, formulation, supply and use of coating (paints, varnishes and vitreous enamels), adhesives, sealants, and printing inks.
9. Wastes from the photographic industry.
10. Wastes from thermal processes.
11. Wastes from chemical surface treatment and coating of metals and other material: non-ferrous hydro-metallurgy.
12. Wastes from shaping and physical and mechanical surface treatment of metals and plastics.
13. Oil wastes and wastes of liquid fuels.
14. Waste organic solvents, refrigerants, and propellants.
15. Waste packaging; adsorbents, wiping cloths, filter materials and protective clothing not otherwise specified.
16. Wastes not otherwise specified.
17. Construction and demolition wastes (including excavated soil from contaminated sites).

18. Wastes from human or animal health care and/or related research.
19. Wastes from waste management facilities, off-site waste water treatment plants, preparation of drinking water, and water use for industrial applications.
20. Municipal wastes (household waste and similar commercial, industrial, and institutional wastes), including separately collected fractions.

New Zealand has an extensive reduce, reuse, recycle program and also generates 80% of its electricity from renewable resources that include wind, hydro, and geothermal. The program for reducing waste follows a circular economy approach that is based on three principles (New Zealand Ministry for the Environment 2021e):

- Design out waste and pollution
- Keep products and material in use
- Regenerate natural systems

14.3.3 REMEDIATION STANDARDS

A Contaminated Sites Remediation Fund (CSRF) was established as part of the Resource Management Act of 1991 that supports regional councils to achieve goals for contaminated land management. Investigation and remediation are conducted using a prioritizing method to determine which sites are funded under the program. The New Zealand Ministry for the Environment assess each application and selects ten sites that are of the highest priority for funding. Once a site is remediated to an acceptable risk, it is removed from the list and replaced with a new site in need (New Zealand Ministry for the Environment 2021f).

New Zealand has developed detailed and comprehensive guidelines for conducting environmental investigations at sites of potential concern with the objective to achieve a nationally consistent approach to evaluate whether a risk to human health and environment exists (New Zealand Ministry for the Environment 2021g). New Zealand uses a risk screening system (RSS) that guides the investigation so that each site is investigated consistently and to appropriately rank a site by risks posed to protect human health and the environment. New Zealand utilizes the USEPA Integrated Risk Information System (IRIS) and the United States Agency for Toxic Substance and Disease Registry to develop minimum risk levels (New Zealand Ministry for the Environment 2021h).

14.3.4 SUMMARY OF ENVIRONMENTAL REGULATIONS OF NEW ZEALAND

The data certainly demonstrate that New Zealand has a comprehensive, effective, and forward-thinking environmental regulation program and a population that is engaged in protecting its air, water, and land. This is evident in New Zealand's efforts to reduce the amount of solid waste that is generated and that 80% of its electricity is generated from renewable resources. Challenges for New Zealand include striking a balance with its agricultural activities and environmental degradation of its surface and groundwater and continued population increase and urbanization.

Land Pollution Regulations of Oceania 279

14.4 SUMMARY AND CONCLUSION

Australia and New Zealand are the largest landmasses within Oceania. Geographically, they are very different. Australia is generally hot and dry, with the exception of the northeastern portion of the country, which is tropical in nature and has one of the oldest rainforests on Earth, called the Daintree (United Nations 1997). Soils are generally considered poor as well.

Australia has numerous mining operations for precious gems and natural resources, such as iron. The combination of its geography, climate, and mining activities, along with the fact that 90% of its population is located along its coastline, means contamination concentrates nearest the urban regions and near industrial and mining locations.

The geography of New Zealand is mountainous, and its climate is wetter and average temperatures cooler than Australia. These factors generally result in less severe environmental impacts. New Zealand's largest impacts are the result of agriculture, which has degraded its surface water and groundwater resources.

14.5 REFERENCES

Australia Department of Environment. 2009. *National Waste Policy*. Environmental Protection Heritage Council. Canberra, Australia. 22p. www.nepc.gov.au/node/849. (Accessed July 4, 2021).

Australia Department of Environment and Energy. 2021a. Publications and Resources. www.environment.gov.au/about-us/publications. (Accessed July 4, 2021).

Australia Department of Environment and Energy. 2021b. Solid Waste Classification System. www.env.gov.au/protection/waste-class.pdf. (Accessed July 4, 2021).

Australia Department of Environment and Energy. 2021c. The Waste and Recycling Industry in Australia. Report to Australian Government. Canberra, Australia. 152p.

Australia Department of Environmental and Conservation. 2010. *Assessment Levels for Soil, Sediment and Water*. Department of Environment and Conservation. Canberra. Australia. 56p. www.environment.gov.au/site-cleanup.pdf. (Accessed July 4, 2021).

Coolaustralia.org. 2021. Land Pollution. www.coolaustralia.org/land-pollution-primary. (Accessed July 4, 2021).

National Geographic Society. 2021a. Australia and Oceania: Physical Geography. www.nationalgeographic.org/encyclopedia/oceania-physical-geography. (Accessed July 4, 2021).

National Geographic Society. 2021b. Corals Are Dying on the Great Barrier Reef. https://news.nationalgeographic.com/2016/03/160321-coral-bleaching-great-barrier-reef.htm. (Accessed July 4, 2021).

New Zealand Ministry for the Environment. 2021a. Environmental Management in New Zealand. www.mfe.nz/publications/environmental-reporting. (Accessed July 4, 2021).

New Zealand Ministry for the Environment. 2021b. Wastewater Discharges. www.mfe.nz/publications/wastewater-discharges. (Accessed July 4, 2021).

New Zealand Ministry for the Environment. 2021c. Hazardous Waste Management in New Zealand. www.oag.govt.nz/2007/cg-2005-06/part11. (Accessed July 4, 2021).

New Zealand Ministry for the Environment. 2021d. Waste List. www.mfe.govt.nz/waste/guidance-and-resources. (Accessed July 4, 2021).

New Zealand Ministry for the Environment. 2021e. Waste and Resources. www.mfe.govt.nz/waste. (Accessed July 4, 2021).

New Zealand Ministry for the Environment. 2021f. Contaminated Sites Remediation Fund. www.mfe.govt.nz/more/funding/contaminated-stes. (Accessed July 4, 2021).

New Zealand Ministry for the Environment. 2021g. Contaminated Land Management Guideline. Revised 2016. www.mfe.govy.nz/sites/land. (Accessed July 4, 2021).

New Zealand Ministry for the Environment. 2021h. Contaminated Land Management Risk Screening System. www.mfe.govt.nz/sites/screening-system. (Accessed July 4, 2021).

United Nations. 1997. Australia Country Profile. Department of Policy Coordination. New York. http://www.un.org/esa/earthsummit/astra-cp.htm. (Accessed November 12, 2021).

United Nations. 2021a. World Population Prospects. United Nations Department of Economic and Social Affairs. Population Division. https://esa.un.org/unpd/wpp/data. (Accessed July 4, 2021).

United Nations. 2021b. Great Barrier Reef: UNESO World Heritage Site. www.whc.unesco,org/en/list/154. (Accessed July 4, 2021).

United Nations. 2021c. Country Profile of New Zealand. www.un.org/esa/earthsummit/NZ.htm. (Accessed July 4, 2021).

World Bank. 2021. Australia in Profile. Gross Domestic Product, Economy and Standard of Living. www.data.worldbank.org/countries/australia. (Accessed July 4, 2021).

15 Land Pollution Regulations of South America

15.1 INTRODUCTION

South America forms the southern landmass of the Americas, generally considered south of Panama. South America is composed of three basic types of terrain: the Andes Mountains, which span nearly the entire western boundary; the interior Amazon River Basin; and the Guiana Highlands in the east (United Nations 2021a). South America has a very diverse climate, with one of the wettest climates on Earth in Columbia to the driest desert in the world, the Atacama in Chile (United Nations 2021b, 2021c, 2021d).

South America is composed of 12 countries, has a total estimated population of 430 million, and represents only 5.6% of the total human population on Earth. The land area of South America is 17,840,000 square kilometers, making it the fourth-largest continent behind Asia, Africa, and North America. The population density of South America is 21.4 per square kilometer (United Nations 2021a). Most people live along the eastern and western coasts of the continent.

15.2 ARGENTINA

Argentina is bordered by the Atlantic Ocean to the east, Andes Mountains and Chile to the west, Bolivia and Paraguay to the north, and Brazil and Uruguay to the northeast. It is the eighth-largest country in the world, with 2,780,400 million square kilometers of land. Argentina has a population of 42.1 million making it the 32nd most populous country. The population density of Argentina is 16.2 people per square kilometer. The largest city in Argentina is Buenos Aires, which is also the capital of the country and has a population of over 15 million (United Nations 2021b). Argentina has a diverse economy that consists of manufacturing, agriculture, forestry, fishing, construction, transport, and utilities. The gross domestic product of Argentina was over $639 billion in 2017, which translates into a per capita income of $3,500 per individual, which is well below that of many other nations (World Bank 2021a).

15.2.1 ENVIRONMENTAL REGULATORY OVERVIEW OF ARGENTINA

Environmental issues facing Argentina include air pollution, water pollution, deforestation, salinity, and lack of environmental controls and enforcement (United Nations 2021b).

DOI: 10.1201/9781003150107-15

Air pollution in Argentina is most acute in its capital city of Buenos Aires. However, there are very limited data on current and historical air pollution levels in Buenos Aires and throughout Argentina. Argentina is struggling with cleanup of the 64-kilometer-long Matanzas-Riachuelo River, which flows through Argentina's most populous city, Buenos Aires, and is considered one of the most polluted rivers on Earth. The river has seen contamination for over 200 years from agriculture, livestock, tanneries, and industrial pollutants from over 15,000 industrial facilities that have discharged to the river and are located within the river drainage basin (World Bank 2013). Cleanup of the river was to begin in 2018 with assistance from the World Bank, Argentina Secretariat for the Environment and Sustainable Development, and several municipalities along and near the river at a cost of several billion US dollars (World Bank 2021b).

Cleanup for the Matanzas-Riachuelo River is expected to continue for decades before being complete. This situation along the Matanzas-Riachuelo River is but one example of the challenges that plague the development, implementation, and enforcement of water resources management in Argentina. Management and changing water management techniques in Argentina are further hindered because multiple institutions at the national, provincial, and river basin level have jurisdiction (World Bank 2021c).

Identified issues that must be addressed include (World Bank 2018b, 2021b).

- Incomplete or outdated legal and regulatory framework
- Limited capacity in water management at the central and provincial levels
- Outdated procedures for water resources planning
- Lack of integrated national water resources information system
- Deficient water resources monitoring system
- Serious water pollution problems
- High risk of flooding in urban and rural areas
- Lack of flood risk reduction strategies
- Lack of appropriate incentives for conservation and pollution reduction

15.2.2 Solid and Hazardous Waste

Solid and hazardous waste is regulated in Argentina by Law 24.051 that addresses the generation, transport, handling, treatment, and final disposal of hazardous waste. However, specific details on waste characterization, infrastructure, and management structure on a national level, such as the management of water resources, are lacking. In Argentina, a waste is hazardous if it can damage living beings or contaminate land, air, water, or the environment. Any facility that generates a hazardous waste must register with National Registry of Hazardous Waste Generators and Operators, which is required to operate any business that generates a hazardous waste (Argentina Secretariat for the Environment 2021). Just as in the United States and several countries of the world, the hazardous waste tracking system in Argentina is a "cradle to grave" system (Argentina Secretariat for the Environment 2021).

15.2.3 REMEDIATION STANDARDS

As with water and solid and hazardous waste, remediation standards on a national basis are lacking in Argentina and are plagued by several impediments that include overlapping jurisdictions, competing interests, lack of reliable cleanup standards, and lack of infrastructure (United Nations 2021b).

15.2.4 SUMMARY OF ENVIRONMENTAL REGULATIONS OF ARGENTINA

Argentina is a rather large country for its population. Argentina's significant environmental issues appear to be concentrated in and immediately around its largest city of Buenos Aires and involve air, water, and land pollution. Investigation and cleanup of one of the world's most polluted rivers will likely be a benchmark achievement or failure for Argentina's future in protecting human health and the environment. Cleanup of a 64-kilometer river in an urban environment in any country, including the United States, is a monumental task and is perhaps too much to ask for a country that is struggling with social and economic issues as well.

15.3 BRAZIL

Brazil occupies 48% of the continent of South America and has a total land area of 8,514,215 square kilometers. The only countries larger in land area are Russia, Canada, China, and the United States. Much of the climate of Brazil is tropical, with its southern portion being temperate. The largest river in Brazil, and the second longest in the world, is the Amazon River, with a length of 6,992 kilometers (United Nations 2021c).

Brazil has the eighth-largest economy in the world, with a gross domestic product of $2.138 trillion. Brazil has the second-largest economy in the Americas behind the United States. Significant natural resources include gold, uranium, iron, and timber. Main trading partners with Brazil include China, the United States, Argentina, and Japan. Brazil's main industries include textiles, chemicals, cement, lumber, iron ore, tin, steel, aircraft, motor vehicles, and other machinery equipment (United Nations 2021c).

Brazil undertook an ambitious program to reduce its dependence on imported oil, which accounted for 70% of its historical needs. By 2007, Brazil became self-sufficient in oil and is now one of the world's leading producers of hydroelectric power. In fact, hydroelectric power now generates 90% of Brazil's electricity needs. This was accomplished by the Itaipu Dam on the Parana River and the Tucurui Dam in Para in northern Brazil. Brazil is also expanding its use of nuclear power, with two nuclear power plants currently in operation and plans for 10 more (United Nations 2021c).

The population of Brazil is estimated at 207 million, making it the second-highest in population in the Americas only behind the United States. The two largest cites in Brazil are Sao Paulo, with an estimated population of 21.5 million, and Rio de Janeiro, with a population of 12.8 million. The per capita income of Brazil is approximately $15,500, or nearly one quarter of the per capita income of the United States and about the same as China (United Nations 2021c).

284 Environmental Compliance Handbook

15.3.1 Environmental Regulatory Overview of Brazil

The Amazon rainforest is the largest tropical rainforest in the world spanning 5.5 million square kilometers, 60% of which is located in Brazil. Rainforests worldwide once covered 15% of the land on Earth, but due to anthropogenic activities, only 6% are left, and they are decreasing each year. Current environmentally related threats to the Amazon rainforest include (National Geographic Society 2021):

- Deforestation. Deforestation in the Amazon rainforest varies between the nine countries that are within or border the rainforest. Types of deforestation within the Amazon basin include:

  - Clearing the forest by burning to make room for cattle pastures for large ranches and local farmers, who clear smaller, but perhaps more numerous, patches for agriculture
  - Mining, especially for gold
  - Urbanization from expending cities on the fringe

- Loss of biodiversity. Species lose their habitat or can no longer survive in the small fragments of forest left behind after clearing, especially if the patches of forest are not interconnected.
- Habitat degradation. Fragmentation by road building and other anthropogenic activities leads to habitat degradation that places additional negative pressures on indigenous species.
- Modified global climate. Deforestation releases CO_2 into the atmosphere and also prevents CO_2 removal and the release of oxygen into the atmosphere by the trees that are deforested.
- Interruption or loss of the water cycle. Deforestation reduces the critical water cycling services that trees provide by increasing the amount of erosion and decreasing the amount of evapotranspiration which in turn disrupts and decreases subsequent rain events.
- Social effects. Deforestation ultimately leads to degradation of the ecosystem and contributes to overall decline of long-term economic balance.
- Logging. Logging is another example of deforestation that destroys habitat.
- Hydroelectric plants—damming rivers in the Amazon rainforest flood.
- Poaching. Poaching of mammals, birds, reptiles, and even insects lowers the natural population numbers in species and places additional stress on Amazon wildlife.
- Pollution. Anthropogenic activities introduce a plethora of pollutants into the rainforest ecosystem through the air, water, and land that include but are not limited to organic and inorganic compounds, heavy metals, particulate matter, suspended solids, erosion, noise, increased atmospheric carbon dioxide, decreased oxygen, and habitat loss.

USEPA has been working with Brazil on a number of pollution issues including air, water, land, and other (USEPA 2021a).

15.3.2 Solid and Hazardous Waste

Brazil amended its 1998 waste law in 2010 (Brazil Ministry of Environment 2021a) and brought together in one law a set of principles, goals, instruments, targets, and actions toward the integrated management of environmentally sound solid waste management. The Brazilian Waste Law outlines concepts and provisions similar to that of the United States and includes (Brazil Ministry of the Environment 2021a):

- Solid waste management
- Recycling
- Reuse
- Reduce
- Sustainability
- Pollution prevention
- Polluters pay principle
- Shared responsibility, or cradle to grave concept

Hazardous wastes in Brazil follow those of most other countries and include (Brazil Ministry of the Environment 2021a):

- Explosive
- Flammable
- Oxidizing
- Corrosive
- Reactive
- Ignitable
- Infectious
- Toxic
- Characteristic
- Medical

Radioactive waste is addressed separately (Brazil Ministry of the Environment 2021a). The volume of hazardous waste generation in Brazil from 2006 to 2015 increased by 30%. The significant challenges for Brazil recently are the management of medical waste and e-waste, both of which appear to need improvement (Macedo and Sant'ana 2015).

15.3.3 Remediation Standards

Much like the rest of the world, Brazil uses risk-based standards for remediation. Remediation thus far in Brazil has generally been for its larger sites that pose the highest risk and usually involve water and river systems (Brazil Ministry of Environment 2021b).

15.3.4 Summary of Environmental Regulations of Brazil

Brazil is the largest and most populated country in South America and the country that has 60% of the Amazon rainforest. Brazil's most significant environmental issues are:

- Sanitation, especially in the poor areas within its major cities
- Cleaning up river systems, which is related to sanitation

286 Environmental Compliance Handbook

- Air pollution in larger urban areas
- Protecting the rainforest

The sanitation issue is one that plagues many of countries in South America and Mexico, portions of Africa, and Asia. Unfortunately, the longer it takes to finally address the sanitation issue, the harder and more costly it becomes, and disease continues to spread (United Nations 2015)

Brazil's environmental regulations seem robust but are not providing the desired results because of several factors:

- Lack of financial resources
- Overlapping responsibilities between agencies
- Other social and economic priorities
- Lack of political will and corruption, especially with respect to addressing the larger more complex environmental issues such as protecting the rainforest and sanitation

15.4 CHILE

Chile is located along the extreme western boundary of South America. Chile shares land borders with Argentina, Peru, and Bolivia. Chile's western border is the Pacific Ocean.

From north to south, Chile extends 4,270 kilometers but averages only 170 kilometers east to west, making it a long and narrow nation. The land area of Chile is 756,096 square kilometers. The Andes Mountains run along the entire eastern border of Chile. Chile is prone to earthquakes. In fact, Chile has experienced 28 earthquakes in the 20th century of magnitude 6.9 on the Richter scale or greater (USGS 2018; United Nations 2021d). Because Chile is so long and narrow in the north-south direction and goes from sea level to nearly 7,000 meters in elevation in the Andes Mountains, the climatic regions vary significantly and often over very short distances.

The population of Chile is 18.4 million. Santiago is Chile's largest city and is the capital, with a population of 6.3 million, which represents 33% of Chile's entire population. The gross domestic product of Chile was $247.03 billion in 2016. Per capita income is estimated at $15,059. Chile is considered one of South America's leading economies (World Bank 2015).

15.4.1 Environmental Regulatory Overview of Chile

The main statutory environmental framework in Chile is contained in the Environmental Law (No. 19.300, 1994 and as amended in 2010), which introduced the environmental impact assessment system, which set out the process for obtaining an environmental license. The 1994 law is known as the General Environmental Framework Law, which established the National Environment Commission (CONAMA), reporting directly to the president's office through the Ministry of General Secretariat of the Presidency. CONAMA coordinates government environmental policies, prepares environmental regulations, and fosters integration of environmental concerns in other policy (United Nations 2015).

Land Pollution Regulations, South America 287

The amendments to the 1994 law enacted in 2010 created the Ministry for the Environment. The Ministry for the Environment, which elevated environmental issues to a cabinet level rank, is charged with assisting the president of the Republic of Chile in the design and implementation of policies, plans, and programs for the protection of the environment, including (United States Library of Congress 2010):

- Biological diversity
- Renewable natural resources
- Hydraulic resources
- Sustainable development
- Environmental policy
- Regulatory framework

The environmental impact assessment system requires all investment projects and productive activities to undergo a detailed process to determine the real effects of proposed operations would have on the environment (Chile Ministry of the Environment 2021a).

USEPA has been collaborating with Chile since 2004 on issues relating to (USEPA 2017; Chile Ministry of the Environment 2021a):

- Risk management related to waste and contaminated sites
- Environmental enforcement
- Environmental forensics
- Compliance with air pollution standards
- Training
- Technical consultation
- Public participation
- Environmental education

Chile's most significant environmental issues center around deforestation and the subsequent erosion that it causes and air pollution in its capital city of Santiago. The city of Santiago experiences increased air pollution due to automobile exhaust and wood burning that is difficult to abate because the city is surrounded by mountains (Chile Ministry of the Environment 2021a). Many households use wood for fuel because it's more economical than other forms of energy. Other environmental issues in Chile include degraded water quality, especially in and around Santiago and from mining operations. Unfortunately, most of Chile's mining is in the northern desert region of the country where any water quality issues are heightened because of the relative scarcity of water resources (Chile Ministry of the Environment 2021a).

15.4.2 Solid and Hazardous Waste

Environmental regulations in Chile on solid and hazardous waste are similar to those of the United States and cover the following (Chile Ministry of the Environment 2021b):

- Solid waste
- Hazardous waste

288 Environmental Compliance Handbook

- Storage and handling of hazardous substances
- Transporting solid and hazardous waste
- Disposal

Hazardous wastes in Chile follow those of the United States and include (Chile Ministry of the Environment 2021b):

- Explosive
- Infectious
- Corrosive
- Medical

- Ignitable
- Oxidizing
- Characteristic
- Certain residues

- Flammable
- Toxic
- Reactive

Certain identified processes that produces waste are also considered hazardous and are similar to listed wastes in the United States. Some of these include the following (Chile Ministry of the Environment 2021b):

- Textiles
- Used absorbents
- Pesticides

- Petroleum
- Refrigerants
- Herbicides

- Solvents
- Propellants
- Adhesives and glues

15.4.3 REMEDIATION STANDARDS

Remediation of sites of environmental contamination in Chile has focused on mining and abandoned mines. USEPA has been actively collaborating with the Chile Ministry of the Environment, focusing on (USEPA 2017):

- Management of environmental aspects of mining
- Abandoned mine risk evaluation
- Remediation of contaminated mining sites
- Site investigation and characterization
- Conducting risk assessments
- Mapping
- Selection of remedial options
- Enforcement measures at mining sites

Additional remedial focus has been on improving air quality, especially in the urban area of Santiago through a Chilean Ministry of Environment and USEPA partnership designed to improve air quality, protect the climate, and provide public health benefits (USEPA 2017).

15.4.4 SUMMARY OF ENVIRONMENTAL REGULATIONS OF CHILE

Chile is a small country with a relatively low population located in an extremely diverse climate and geographic region. This tends to complicate regulation of the environment because there are so many different, separate and distinct, and

Land Pollution Regulations, South America

sometimes unrelated geologic and hydrologic environments. Therefore, Chile's geology and geography heavily influence its efforts to protect and improve environmental regulations.

Air quality regulations appear to be the most advanced in Chile compared with water, solid and hazardous waste, and remediation regulations. However, availability of clean water will likely be more of a pressing issue in the future due to climate change.

Chile's environmental regulations first became serious in 1994 and then were revised significantly in 2010. Together with cooperation and collaboration with the USEPA, it does not appear that environmental matters will largely be ignored in the future. The expectation is that future environmental regulations in Chile will be very similar to those of the United States.

15.5 PERU

Peru is located in western South America. Peru's west coast extends along the Pacific Ocean for 2,414 kilometers. Columbia and Ecuador are located to the north, Brazil to the east, and Bolivia and Chile to the south. Peru covers a land area of 1,279,999 square kilometers, making it the 20th-largest country by area in the world. Peru has an estimated population of 31 million people. The gross domestic product of Peru was $470 billion in 2017, which translates into a per capita income of $13,735. The economy of Peru has significantly improved over the last 20 years. As evidence of strong economic growth, poverty in Peru has been reduced from nearly 60% in 2004 to just below 26% in 2012 (World Bank 2021d).

The Peruvian economy is 43% service (including tourism) and 47% industry and manufacturing. Major exports include metals (copper, zinc, gold, mercury), chemicals, pharmaceuticals, machinery, textiles, and fish meal. Peru ranks second in the world for copper production. Copper accounts for 60% of Peru's exports (World Bank 2021d). The capital of Peru is Lima. It is located along the western coast with the Pacific Ocean. The population of Lima and associated surrounding communities is estimated at over 12 million residents, which represents 40% of the entire population of Peru (World Bank 2021d).

The geography of Peru is unique in that the Andean mountain range, which runs north to south, divides the country from its coastal region to the west from the Amazon Basin to the east (United Nations 2005). The Andean mountain range acts as a divide in geographical and climatic terms.

15.5.1 Environmental Regulatory Overview of Peru

The principal environmental issues of Peru include water pollution, soil erosion, soil pollution, and deforestation. Air pollution is significant in Lima, where pollutant concentrations sometimes reach unhealthy levels (WHO 2021). Peru is susceptible to high levels of soil erosion as deforestation occurs, and this is often exacerbated, especially in mountainous regions with increased erosion. Water pollution and lack of sanitation are also issues, especially in rural areas, where an estimated 62% of the population has access to clean water and access to sewage is as low as 22% (United Nations 2005).

The USEPA has been collaborating with Peru on several environmental issues since 2009 that include air, water, waste, land degradation, remedial technologies and management, and others (USEPA 2021b).

The Peruvian Ministry of Environment is a ministry of the Cabinet of Peru created in 2008. Its function is to oversee the environmental sector with the authority to design, establish, and execute government policies concerning the environmental management and strategic development of natural resources (Peru Ministry of the Environment 2021). Peru, by means of Law 30327, established an integrated permitting process titled "Global Environmental Certification" (IntegrAmbiente) that integrates all necessary environmental permits into one permit, similar to the process in China (Peru Ministry of the Environment 2021).

15.5.2 SOLID AND HAZARDOUS WASTE

Solid waste management in Peru was essentially non-existent until the Ministry of the Environment was created in 2008. Environmental concerns in Peru are not considered a national priority, including solid waste management. However, there are only nine officially recognized solid waste landfills in Peru, and five are located in and around the capital city of Lima. This means that many parts of the country still rely on informal disposal of solid waste in dumps and other poorly managed and poorly constructed locations (Peru Ministry of the Environment 2021; Peace Corp 2016).

USEPA is currently collaborating with Peru on many environmental issues, including upgrading infrastructure for solid and hazardous waste management (USEPA 2021b).

15.5.3 REMEDIATION STANDARDS

Remediation of contaminated sites in Peru is improving but still has a long way to go to be considered effective. USEPA has been assisting Peru in establishing an effective program through the following activities (USEPA 2021b):

- Management of environmental aspects of mining
- Mine risk evaluation
- Remediation of contaminated mining sites
- Site investigation and characterization
- Conducting risk assessments
- Mapping
- Selection of remedial options
- Enforcement measures at mining sites
- Managing mercury

Mercury is utilized at many gold mines in Peru to assist in extracting gold from gold-bearing ore. Releases of mercury are common, along with other heavy metals, including iron, manganese, lead, copper, and zinc (USEPA 2021b). These heavy metals have not only impacted the soil but have significantly degraded surface water quality and are a significant source of air pollution, especially near major mines,

Land Pollution Regulations, South America 291

including the city of La Oroya, located in the Andes Mountains 176 kilometers northeast of Lima (USEPA 2021b).

15.5.4 SUMMARY OF ENVIRONMENTAL REGULATIONS OF PERU

Although Peru does not have a significant population, the environmental concerns due to mining, especially copper and gold, and lack of infrastructure are significant. An additional item of note is that the environment has only been a national priority since 2008 when the Ministry of the Environment was established. Peru is making progress, but the environmental damage in erosion, deforestation for farming and mining activities, air pollution, and water quality degradation and the huge future costs of environmental remediation of legacy sites will certainly present Peru with huge economic costs well into the future.

15.6 SUMMARY AND CONCLUSION

Environmental challenges faced by the countries of South America are not unlike those of other countries that we have evaluated on other continents. Some of the unique challenges include protecting the Amazon rainforest predominantly located in Brazil. In addition, the two largest cities in Brazil, Sao Paulo, with an estimated population of 21.5 million, and Rio de Janeiro, with a population of 12.8 million, face significant challenges in providing sanitation and safe drinking water to their citizens. In addition, air pollution is significant, as is land pollution, which contributes to air and water pollution and compounds the challenge.

Chile stands out as the one country we have evaluated in South America that has embraced solving its environmental challenges in a progressive and proactive approach. Although Chile may not have the financial resources that the United States has to solve environmental issues, it is still considered a model country for South America. This is especially significant when one considers that Chile does not have geography or an abundance of water resources on its side to easily address environmental issues.

In summary, the countries of South America face some of the same issues that are present in the United States and involve challenges related to poor water quality and sanitation and poor farming techniques that include deforestation, erosion, pesticide and herbicide application, and overuse of fertilizers, which significantly impact surface water and the oceans.

15.7 REFERENCES

Argentina Secretariat for the Environment. 2021. National Definition of Waste Under Law 24.051. www.mrecic.gov.ar/waste. (Accessed July 8, 2021).

Brazil Ministry of the Environment. 2021a. National Solid Waste Law of 2010. www.wiego. gov.br/wast-law.htm. (Accessed July 8, 2021).

Brazil Ministry of the Environment. 2021b. Remediation in Brazil. www.brazilgovnews.gov. br/ministry-of-the-environment.htm. (Accessed July 8, 2021).

Chile Ministry of the Environment. 2021a. Overview and USEPA Cooperation. http://protal. mma.gob.cl/. (Accessed July 8, 2021).

Chile Ministry of the Environment. 2021b. Chile Division of Solid and Hazardous Waste. http://portal.mma.gob.cl.waste/. (Accessed July 8, 2021).

Macedo, A. A. and Sant'ana, L. P. 2015. Managing Emerging Hazardous Wastes in Brazil. Unisanta Science and Technology. ISSN 2317–1316. P. 51–54.

National Geographic Society. 2021. Threats to the Amazon Rainforest. www.nationalgeographic.com/environment/habitats/rainforest-threats. (Accessed July 8, 2021).

Peace Corps Peru. 2016. Trash Talking: Solid Waste Management in Peru. https://psu2.wordpress.com/2016/11/03/trash-talking-Peru. (Accessed July 8, 2021).

Peru Ministry of the Environment. 2021. Solid Waste. www.gob.pe/ambiento. (Accessed July 8, 2021).

United Nations. 2005. *Republic of Peru: Country Profile*. Department of Economic and Social Affairs. New York, NY. 18p.

United Nations. 2015. *International Decade for Action "Water for Life"*. United Nations Department of Economic and Social Affairs (UNDESA). New York. NY. www.un.org/waterforlifedecade/quality.shtml. (Accessed July 8, 2021).

United Nations. 2021a. World Population Prospects. United Nations Department of Economic and Social Affairs. Population Division. https://esa.un.org/unpd/wpp/data. (Accessed July 8, 2021).

United Nations. 2021b. Country Profile for Argentina. www.un.org/esa/earthsummit/AR.htm. (Accessed July 8, 2021).

United Nations. 2021c. Country Profile for Brazil. www.un.org/esa/earthsummit/brzl-cp.htm. (Accessed July 8, 2021).

United Nations. 2021d. Country Profile for Chile. www.un.org/esa/earthsummit.ch-cp.htm. (Accessed July 8, 2021).

United States Environmental Protection Agency (USEPA). 2017. EPA Collaboration with Chile. www.epa.gov.international-cooperation/epa-collaboration-chile. (Accessed July 8, 2021).

United States Environmental Protection Agency (USEPA). 2021a. EPA Collaboration with Brazil. www.epa.gov.nternational-cooporation-Brazil. (Accessed July 8, 2021).

United States Environmental Protection Agency (USEPA). 2021b. EPA Collaboration with Peru. www.epa.gov.nternational-cooporation-Peru. (Accessed July 8, 2021).

United States Geological Survey (USGS). 2018. Earthquakes. https://earthquakes.usgs.gov/earthquakes/map. (Accessed July 8, 2021).

United States Library of Congress. 2010. Chile: New Law Creates Ministry for the Environment. www.loc.gov/law/foreign-news/article/chile-new-law. (Accessed July 8, 2021).

World Bank. 2013. Matanza-Riachuelo River Cleanup, Argentina. Blacksmith Institute. http://worldbank.org/projects/argentine/matanza-riachuela-river. (Accessed July 8, 2021).

World Bank. 2015. Economy Profile for Chile. https://data.worldbank/country/chile. (Accessed July 8, 2021).

World Bank. 2021a. The World Bank in Argentina. www.worldbank.org/en.country/argentina. (Accessed July 8, 2021).

World Bank. 2021b. *Significant Advances in the Recovery of the Matanza-Riachuelo River Basin*. World Bank. Argentina. 6p.

World Bank. 2021c. Argentina: Policy for Sustainable Development in the 21st Century. http://openknowledge.wolrdbank.org/handle/10986/14980. (Accessed July 8, 2021).

World Bank. 2021d. Overview of World Bank in Peru. www.worldbank.org/en/country/peru/overview. (Accessed July 8, 2021).

World Health Organization. 2021. Peru: Country Profile. http://who.int/gho/countries/per/country_profiles/per.htm. (Accessed July 8, 2021).

16 Current Status of Land Pollution and Regulations of the World

16.1 INTRODUCTION

In this chapter, we will evaluate the current status of land pollution and environmental regulations that focus on the land throughout the world. We will first examine Antarctica and then the world's oceans. Evaluating Antarctica and the world's oceans provides an important perspective on pollution throughout the world. We will then evaluate and rank the effectiveness of environmental regulations that deal with land for each country examined.

16.2 ANTARCTICA

When evaluating environmental regulations of the world, one must not exclude an entire continent from that discussion. Therefore, we shall and must discuss environmental regulations and pollution in Antarctica. Antarctica is a continent on the bottom of the world with a total land area of 14 million square kilometers, with 98% of its surface covered in ice. Antarctica shares no border with any other country (see Figure 16.1).

There are no permanent settlements in Antarctica. In 1959, Antarctica was established as an international area dedicated to science and research. Several nations, including Argentina, France, Japan, the United Kingdom, and the United States, signed the Antarctic Treaty in 1959. The treaty contains 14 articles that establish key elements related to the peaceful international coordination on the continent, some of which include (Antarctic and Southern Ocean Coalition 2021):

- Banning military intervention
- Peaceful scientific research
- International information exchange
- No territorial sovereignty

16.2.1 POLLUTION IN ANTARCTICA

Some pollution issues in Antarctica include (McConnell et al. 2014; NASA 2014):

- Ozone depletion
- Increased atmospheric CO_2 concentrations
- Solid and hazardous waste generated from research activities and tourism

DOI: 10.1201/9781003150107-16

293

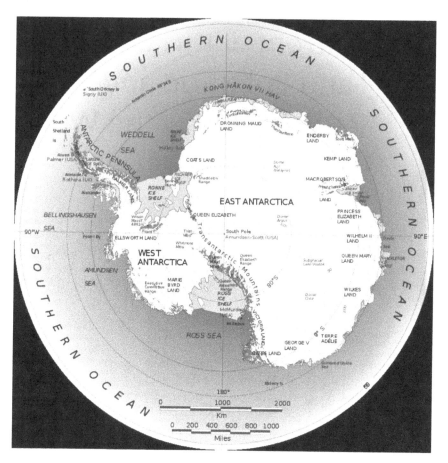

FIGURE 16.1 Map of Antarctica (from National Aeronautics and Space Administration 2021. Map of Antarctica. https://nasa.gov/images/contnet/600827main_24.jpg [accessed November 12, 2021], 2021).

- Marine pollution from anthropogenic sources such as spills from ships and cargo vessels
- Litter
- Lead
- Invasive species such as dandelions

16.2.2 Environmental Regulations of Antarctica

Environmental regulations in Antarctica do exist and are traced back to the Antarctic Treaty of 1959. All activities in Antarctica must be conducted in accordance with the regulations set forth by the Norwegian Polar Institute. For any activity, the following must be conducted (Norwegian Polar Institute 2021):

- Notification of planned activity
- Preparation of an environmental impact assessment

Current Status of Land Pollution 295

- Preparation of contingency plans
- Obtaining insurance
- List of experience and equipment
- Preparation of environmental emergency plan
- Waste handling and disposal plan
- Provisions for preventing introduction of alien species
- Preparation of a final report

16.2.3 Summary of Environmental Regulations in Antarctica

It should not come as a surprise after reading this chapter that there are environmental impacts even in Antarctica from anthropogenic sources. This may present an interesting and perhaps conflicting phenomenon to some of you because Antarctica is predominantly devoid of human settlement. It is essentially a wilderness of continental size. So how can Antarctica be polluted, and if Antarctica is polluted, what does that say about the rest of the Earth? Antarctica is perhaps the best example that pollution does not respect country borders. Evaluating Antarctica has placed a spotlight on the migration of pollution on Earth. One migration pathway that is common for pollution follows Fick's first law of molecular diffusion, which is on full display for us to observe and quantify in Antarctica.

16.3 OCEANS

Oceans cover 70% of the surface of the Earth, or 360 million square kilometers. In many respects, the oceans of the world are the last resting place for most anthropogenic pollution. As an example, an estimated 7 million metric tons of plastic is dumped into the world's oceans every year (see Figure 16.2) (NOAA 2021a). This is just a portion of the pollution that enters the oceans every year. NOAA (2021a) estimates that billions of pounds of trash and other pollutants enter the oceans each year.

FIGURE 16.2 Plastic and other garbage on a beach in Hawaii (from NOAA. *Ocean Pollution*. www.noaa.gov/resource-collections/ocean-pollution [accessed February 7, 2021], 2021a).

16.3.1 Ocean Pollution

The oceans have been the dumping ground for not just our garbage but for some of our most toxic wastes for hundreds of years. The oceans have been treated as a human privy and garbage disposal. Only in the past few decades has the dumping of wastes in the oceans come under some environmental regulations. However, polluting the oceans still continues to this day. The oceans receive pollution from direct discharge or dumping, from deposition from the atmosphere, from rivers and streams that discharge into the oceans, and directly from the land itself.

The oceans were once believed to be so vast and deep that the effects of pollution would never directly threaten them. This has been proven false. The rate of accumulation of pollution is proportional to human population and increasing energy requirements. The more people, the more pollution ends up in the oceans. The most common types of pollution in the oceans include plastic and other forms of garbage, heavy metals, organic pollutants such as PCBs and oil, radioactive substances and waste, biological pollutants, pesticides, herbicides, fertilizers, heavy metals, and untreated raw sewerage (NOAA 2021a).

Pollution in the oceans does not all originate from direct discharge but is also delivered through the atmosphere from increased concentrations of carbon dioxide in the atmosphere that is partially adsorbed by the oceans and causes acidification of the surface layers (NOAA 2021b). Absorption of carbon dioxide by the oceans has been estimated to be as high as 25%, which represents millions of tons of carbon dioxide absorbed by the oceans every year. The additional carbon dioxide absorbed by the oceans reacts with ocean water and forms an acid called carbonic acid that is lowering the pH of the oceans worldwide. In fact, the oceans are acidifying faster than they have in some 300 million years. The effect of the acidification has been linked to bleaching of coral reefs, including the Great Barrier Reef off the northeastern coast of Australia (Natural Resource Defense Council 2021).

An additional form of pollution in the ocean that is not commonly mentioned is noise. Sound waves travel father and faster in the oceans than they do in the atmosphere. Increased human-induced noise in the ocean is harming numerous marine species because they use sound to navigate, find food, and mate. These crucial activities for marine life survival are having widespread impacts (Natural Resource Defense Council 2021).

16.3.2 Environmental Regulations of the Oceans

The United States has enacted several laws over the last several decades aimed at protecting the Great Lakes and the oceans from pollution. As we learned in Chapter 3, the most significant federal laws addressing waterways were first enacted in the United States as far back as 1899 and include (USEPA 2021):

- Rivers and Harbors and Refuse Act of 1899
- Coastal Zone Protection Act of 1972
- Fisheries Conservation and Management Act of 1976
- Global Climate Protection Act of 1987
- Marine Protection Act of 1972

Current Status of Land Pollution

- Marine Mammal Protection Act of 2015
- Ocean Dumping Act of 1988
- Oil Spill Protection Act of 1990

16.3.3 SUMMARY OF ENVIRONMENTAL REGULATIONS OF THE OCEANS

For centuries, the oceans have been subject to the freedom of the seas doctrine, which is a principle recognized in the 17th century that limited national rights and jurisdiction over the oceans to a narrow belt of sea surrounding a nation's coastline, typically 19.3 kilometers (United Nations 2021). Since 1982, the United Nations has been actively involved in establishing protections for the oceans through the adoption of the Law of Sea Convention which (United Nations 2021):

- Established freedom of navigation rights
- Set territorial sea boundaries 19.3 kilometers offshore
- Set exclusive economic zones of up to 321 kilometers offshore
- Set rules for extending continental shelf rights up to 563 kilometers offshore
- Created the International Seabed Authority
- Created other conflict-resolution mechanisms

The United Nations is also involved in protection of the marine environment and biodiversity and reducing pollution from marine shipping, especially shipping in the polar regions with the adoption of the Polar Code in 2014. As of 2015, there were nearly 90,000 commercial shipping vessels that carry cargo that were registered worldwide that transported nearly 10 billion metric tons of cargo (United Nations 2021).

16.4 SUMMARY OF ENVIRONMENTAL REGULATIONS AND WATER POLLUTION OF THE WORLD

We have now completed a review of environmental regulations of more than 50 countries and every continent including a brief assessment of Antarctica and the oceans. With respect to pollution, we have learned that perhaps the most significant points are the following:

- No country, continent, or ocean is pollution free
- Pollution caused by humans is everywhere
- Significant environmental degradation caused by humans has touched every country
- Pollution does not respect country borders
- Pollution will continue to migrate in the air and water and on land
- The oceans have been treated as a human privy and garbage disposal of the world

With respect to environmental regulations, the most significant information we have learned includes the following:

- Each country has environmental regulations
- No two countries are exactly the same

- Each country has unique attributes, whether these are related to its climate, geography, geology, social and economic concerns, or politics, and these affect the ability of a country to protect human health and the environment and influence many aspects of environmental regulations
- The platform of environmental regulations of each country is similar to that of the United States or the European Union
- Although some countries are doing a better job than others, each country struggles with protecting human health and the environment for various reasons, two of which seem to be universal: a growing population and urban expansion
- Every country has been significantly affected by pollution of the air, water, and land

The largest polluter of air worldwide is motor vehicles, including automobiles, diesel trucks, trains, buses, marine vessels, and aircraft. The largest polluter of water worldwide is biological pollutants, largely from human waste and agricultural activities that include pesticides and herbicides, fertilizers, and erosion. The largest polluting activity of land worldwide is agriculture (WHO 2013, 2015, 2021; WHO and USEPA 2012).

The final resting place for much of the pollution of the world is the oceans.

16.4.1 Evaluating and Ranking Each Country's Regulatory Effectiveness for Land

To evaluate the effectiveness of each country's environmental regulations, we must examine several factors, assign integers that appropriately reflect where each country is in the process, and provide an enlightened point of view as to where each country can improve.

This evaluation involves assessing several attributes that we have discussed and highlighted throughout this chapter and includes whether each country:

1. Has the appropriate environmental regulations for air, water, and land
2. Has appropriately implemented and enforced environmental regulations
3. Has the appropriate infrastructure to support the implementation and enforcement of environmental regulations
4. Has the economic and financial resources to support continued pollution controls
5. Has other social issues that impede or override environmental concerns such as poverty, corruption, violence, unstable political issues, population, and urban constraints
6. Has sustainability initiatives, including pollution prevention, recycling, and waste reduction
7. Has geographical, geological, and climate factors and other natural conditions such as natural disasters that impede environmental performance

To make this evaluation, we must first consider how each country has chosen to protect the air, the water, and the land under the first criterion listed previously.

Current Status of Land Pollution

Therefore, the regulations that are most important are associated with protecting the air, water, and land and include:

- Air quality regulations from both stationary and mobile sources
- Water quality regulations for effluent discharges from industrial sources and public-owned treatment works and septic tank discharges
- Drinking water quality regulations
- Solid and hazardous waste regulations relating to characterization, generation, transport, and disposal
- Soil, water, and sediment cleanup regulations, including investigation and polluters pay principles
- Regulating different land-use designations such as residential, commercial, industrial, and agricultural

The remaining criteria (those listed as 2 through 7) are straightforward qualitative evaluations. Table 16.1 presents the assigned values for each criterion for each country ranked from most effective to least effective. Each criterion in Table 16.1 corresponds to the list of 1 through 7. Each criterion is assigned an integer of 0 through 3 that corresponds to the following qualitative assessment category:

- A value of zero (0) indicates that country lacks appropriate controls
- A value of one (1) indicates that the country has poor controls
- A value of two (2) indicates that the country has adequate or average controls
- A value of three (3) indicates that the country has good controls

For comparison purposes, the per capita gross domestic product is listed for each country (United States Central Intelligence Agency 2021), and a ratio of GDP to the total score of each country is presented in the last column.

There should be very few surprises when each country is ranked. The countries with the most room for improvement include Turkey, Egypt, Malaysia, Russia, Saudi Arabia, and India. These countries have a host of challenges that include lack of regulations or enforcement of existing regulations, lack of infrastructure, financial hardship, and population pressures and may have to deal with corruption issues. The countries in the middle that include Mexico, Indonesia, Brazil, Argentina, South Africa, Kenya, Tanzania, Peru, China, and Chile are developing nations that are struggling to learn and enforce their environmental regulations, have population pressures, lack infrastructure, and are struggling to deal with cleanup of existing sites of contamination. The countries that rank the highest include Japan, United States, Canada, Australia, New Zealand, the European Union, Switzerland, Norway, and South Korea, which are all developed nations with financial resources to address environmental issues but still struggle with issues such as population and urbanization and air and water quality but have all made significant progress compared to other nations.

Examining the final column in Table 16.1 provides perhaps the most telling information. The countries that rank the highest have a ratio of GDP to total score of less than 2.5. The countries that have the most room for improvement all have a ratio

TABLE 16.1
Country Comparison of Environmental Regulation Effectiveness

Country	Criteria							Total Score	GDP Rank*	GDP/ Score
	1	2	3	4	5	6	7			
Turkey	1	1	0	0	0	0	2	4	69	17.25
Egypt	1	1	1	0	0	1	1	5	105	21
Malaysia	1	1	0	1	1	0	1	5	64	16
Russia	1	1	1	1	1	0	1	6	65	10.8
Saudi Arabia	1	1	1	1	1	0	1	6	20	3.3
India	1	1	0	1	1	1	1	6	132	22
Mexico	1	1	2	1	1	1	1	8	82	10.25
Indonesia	2	1	1	2	1	1	1	9	108	12
Brazil	1	1	1	2	1	1	2	9	95	10.5
Argentina	1	1	1	2	1	1	2	9	95	10.5
South Africa	2	1	2	1	1	2	1	10	103	10.3
Kenya	2	2	1	1	1	1	2	10	154	15.4
Tanzania	2	2	1	1	1	1	2	10	156	15.6
Peru	1	1	1	2	2	1	2	10	104	10.4
China	2	1	2	3	2	1	2	13	94	7.2
Chile	2	2	2	2	3	1	1	13	74	5.7
Japan	3	3	3	3	2	2	1	17	38	2.2
United States	2	3	3	3	2	3	2	18	19	1.05
Canada	2	3	3	3	3	2	2	18	32	1.7
Australia	2	3	3	3	3	2	2	18	27	1.5
New Zealand	3	3	3	3	3	2	2	19	44	2.3
European Union	3	3	3	3	3	3	2	20	43	2.1
Switzerland	3	3	3	3	3	3	2	20	17	0.85
Norway	3	3	3	3	3	3	2	20	12	0.6
South Korea	3	3	3	3	2	3	3	20	42	2.1

* United States Central Intelligence Agency. 2021. *The World Fact Book.* www.cia.gov/library/publications/the-world-factbook.gos/NA.html (accessed February 7, 2021).

greater than 10, except for Saudi Arabia. Saudi Arabia ranks poorly for environmental controls but high for per capita GDP, which indicates that Saudi Arabia has financial resources to improve its environmental controls but chooses to focus its financial resources elsewhere. Egypt and India are the only countries evaluated that have a total score to GDP ratio greater than 20. China and Chile are unique when examining their ranking and numerical scores. China and Chile rank between low and high performance. Knowing that China has been improving its environmental controls

Current Status of Land Pollution

indicates that China is transitioning from low to high. Chile currently appears to be more stagnant.

16.4.2 Environmental Challenges of Each Country

Some of the challenges that each country faces that have significantly influenced its score and have been described in this chapter include:

- Turkey is currently facing financial and economic hardship, along with a lack of robust environmental standards and enforcement.
- Egypt is experiencing political unrest, a growing population and urbanization, lack of infrastructure, and water shortages.
- Malaysia lacks infrastructure, environmental enforcement, and financial resources and is seeing rapid population growth.
- Russia lacks infrastructure, environmental enforcement, financial resources, and political framework and struggles with corruption.
- Saudi Arabia lacks robust regulations and enforcement, and it has some lack of infrastructure but does have financial resources.
- India has high population pressure and lacks infrastructure and enforcement of regulations.
- Mexico lacks infrastructure, especially with water treatment, and struggles with corruption, lack of complete regulations (i.e., stormwater), and enforcement.
- Indonesia has pressure from a large population, newer regulations, deforestation, and lack of infrastructure.
- Brazil has population pressure, especially in urban areas; lack of infrastructure; lack of progress in cleaning up contaminated sites; water treatment problems; deforestation; and corruption.
- Argentina lacks infrastructure, which places pressure on enforcement, financial resources, water treatment, and cleanup of contaminated sites.
- South Africa has a degree of political unrest and financial pressures, climate change, and water issues.
- Kenya has financial pressures and lack of infrastructure.
- Tanzania has lack of financial and infrastructure resources.
- Peru has a lack of infrastructure to deal with water contamination issues, especially from mining, and lacks adequate financial resources.
- China lacks enforcement measures (but improvements are in the making) and has population challenges and infrastructure to deal with pollution control measures, especially in the more undeveloped western regions.
- Chile has urban population challenges and water issues in the desert regions of the northern portion of the country.
- Japan has geographical and geologic challenges (i.e., it is an island nation prone to earthquakes), population pressure, and lack of landfill space.
- The United States faces lack of regulation of agricultural areas and has significant issues related to chemical use of pesticides and herbicides, erosion, deforestation, pollution from septic systems, and invasive species.

302 Environmental Compliance Handbook

- Canada lacks some stormwater controls and faces habitat destruction from mining, deforestation, and petroleum development.
- Australia faces water issues, population pressure, and habitat destruction, especially the Great Barrier Reef, which it can do little to correct.
- New Zealand faces water quality challenges, especially from agriculture and livestock sources.
- The European Union faces continued challenges from cleaning up historical sites of contamination and population pressure but has done well with regulations and sustainability measures.
- Switzerland has erosion challenges and pollution from agriculture and livestock.
- Norway faces climate change challenges and pollution from agriculture and livestock.
- South Korea faces population pressures and air quality challenges, especially in its capital city of Seoul.

A final point to consider is the objectivity of the evaluation conducted through the selection of countries included in this chapter. As stated at the beginning of this chapter, the countries selected include 70% of the human population and 75% of the land area. The countries not selected include countries such as Somalia, Algeria, Libya, Sudan, Democratic Republic of Congo, Angola, Zimbabwe, Botswana, Mozambique, Congo, Syria, Iran, Pakistan, Afghanistan, Venezuela, and many former Soviet-bloc countries. These countries were not evaluated because of the more severe social and economic issues, poverty, sickness, and violence they are experiencing, which have degraded the natural environment and directly threaten human health. Therefore, when considering environmental degradation and threats to human health and the environment from pollution, the countries examined in this chapter are biased toward countries with better environmental performance.

16.5 SUMMARY AND CONCLUSION

Antarctica and the oceans have played and will play a significant role in measuring future sustainability efforts and the survival of our species and many other species on Earth. We have learned thus far that Antarctica and the oceans have become polluted from direct discharge and from migration of contaminates through the air, water, and land. We have also learned that pollution migrates following fundamental laws of physics and that fact is evident by examining the impacts observed in Antarctica and the oceans. We have also learned that pollution does not respect boundaries, whether they be political or media specific.

The oceans were the origin of life on Earth and remain the prime source of nutrients and food that sustain life on our planet but have become polluted by our own actions. If not for any other reason than our own survival, we must be diligent in assessing the risks posed by pollutants, enact appropriate protections, and act on those protections to prevent further degradation on a global scale.

To summarize this chapter is to state that we must now accept the reality that humans have adversely impacted the entire Earth and that our efforts to improve our environment since the enactment of environmental regulations in the United States and worldwide have simply not been enough. Earth scientists are now convinced that we have moved the needle of geologic time into a new period called the Anthropocene (Smithsonian 2018; Zalasiewicz et al. 2018; International Union of Geological Sciences 2016). This is significant because the definition of a geologic age is a time period that affects the entire Earth, will be recorded in Earth history, and cannot be reversed.

Now that we have described pollution and its behavior and examined how numerous countries have enacted environmental regulations and struggled with their own pollution issues, we will turn our attention to how to maintain compliance with environmental regulations. The next chapter outlines an approach to achieve and maintain environmental compliance anywhere in the world by using a rather simple approach that can be summed up in just a few words: "no matter where you are, it's all the same."

16.6 REFERENCES

Antarctic and Southern Ocean Coalition. 2021. Antarctic Environmental Protection. www. asoc.org/advocacy/antarctic-environmental-protection. (Accessed February 7, 2021).

International Union of Geological Sciences (IUGS). 2016. *The Anthropocene Epoch: Adding Humans to the Chart of Geologic Time*. International Geological Congress. Cape Town. South Africa. www.35igc.org/verso/5/scientific-programme. (Accessed February 7, 2021).

McConnell, J. R., Maselli, O. J., Sigl, M., Vallelongo, P., Neumann, T. Anschutz, H., Bales, R. C., Curran, M. A., Edwards, R., Das, S. B., Kipfstuhl, S., Layman, L. and Thomas, E. R. 2014. Antarctic-wide Array of High-Resolution Ice Core Records Reveals Pervasive Lead Pollution Began in 1889 and Persist Today. *Scientific Reports*. Vol. 4. No. 5848. 5p.

National Aeronautics and Space Administration (NASA). 2014. Lead Pollution Beats Explorers to the South Pole. www.nasa.gov/content/goddard/lead-pollution-in-antarctica. (Accessed February 7, 2021).

National Aeronautics and Space Administration (NASA). 2021. Map of Antarctica. https:// nasa.gov/images/contnet/600827main_24.jpg. (Accessed February 7, 2021).

National Oceanic and Atmospheric Administration (NOAA). 2021a. Ocean Pollution. www. noaa.gov/resource-collections/ocean-pollution. (Accessed February 7, 2021).

National Oceanic and Atmospheric Administration (NOAA). 2021b. DSCOVR: Deep Space Climate Observatory. www.nesdic.naoo.gov/drcovr-deep-space-climate-observatory. (Accessed February 7, 2021).

Natural Resources Defense Council (NRDC). 2021. Ocean Pollution: The Dirty Facts. https:// nrdc.org/stories/ocean-pollution-dirty-facts. (Accessed February 7, 2021).

Norwegian Polar Institute. 2021. Regulations for Activities in Antarctica. www.nplar.no/en/ regulations/the-antarctic. (Accessed February 7, 2021).

Smithsonian. 2018. The Age of Humans: Living in the Anthropocene. Smithsonian Statement on Climate Change. www.si.edu/newsdesk/releases/smithsonian-statement-climate-change. (Accessed February 7, 2021).

United Nations. 2021. Oceans and the Law of the Sea. www.un.org/en/sections/issues-depth/ oceans-and-law-sea. (Accessed February 7, 2021).

United States Central Intelligence Agency. 2021. The World Fact Book. www.cia.gov/library/ publications/the-world-factbook.gos/NA.html. (Accessed February 7, 2021).

United States Environmental Protection Agency (USEPA). 2021. Laws that Protect Our Oceans. www.epa.gov/beach-tech/laws-protect-our-oceans. (Accessed February 7, 2021).

World Health Organization and United States Environmental Protection Agency. 2012. *Animal Waste, Water Quality and Human Health*. IWA Publishing. London, UK. 489p.

World Health Organization (WHO). 2013. Water Quality and Health Strategy 2013–2020. www.who.int/water_sanitation_health/dwq/en/. (Accessed February 7, 2021).

World Health Organization (WHO). 2015. *Progress on Drinking Water and Sanitation*. World Health Organization. New York, NY. 90p.

World Health Organization (WHO). 2021. World Health Organization Global Urban Ambient Air Pollution Database. www.who.int/phe/health_topics/outdoorair/database.htm. (Accessed February 7, 2021).

Zalasiewicz, J. Waters, C., Summerhayes, C. and Williams, M. 2018. The Anthropocene. *Geology Today*. Vol. 34. No. 5. pp. 182–187.

17 Achieving and Maintaining Compliance with Land Regulations

17.1 INTRODUCTION

Conducting an environmental audit is critical for maintaining compliance with applicable laws and regulations that relate to the environment. Environmental compliance is sometimes perceived as just another set of rules and is simple: just follow the rules. However, the Earth is dynamic. There is change, ever more complex, as more information and science are discovered, learned, and shared. The weather changes constantly and is perhaps the best example of how to imagine environmental management and risk. It also changes, as does everything. Everything obeys the second law of thermodynamics, namely entropy, in that nature tends to become more complex and disordered as time marches on.

Understanding the natural setting of our urban and industrial areas and knowing the chemicals that are used and how they might cause harm to the environment and humans if they are released, what alternative chemicals are available, how to reduce chemical and energy use, preventing chemical releases to the environment, quantifying the costs involved with cleaning up the environment once a release occurs, and developing environmental stewardship are all subjects that should be taken into account when evaluating the environmental health of a facility, and it all begins with first conducting an environmental audit.

The significant lessons or principles of environmental compliance in this chapter are:

1. It all begins with conducting an environmental audit.
2. Being in compliance with environmental regulations does have sustainability implications.
3. Work yourself out of environmental regulations.
4. There is little we can control.
5. Limit chemical use when possible.
6. When possible, eliminate chemicals that are very toxic or have a high chemical risk factor.
7. Fully understand environmental permits.
8. Never accept a permit term that the facility can't achieve.
9. Always be accurate and truthful.
10. When in doubt, it's usually always better to report a spill or other incident where reporting may be required.

DOI: 10.1201/9781003150107-17

306   Environmental Compliance Handbook

11. The importance of waste characterization and points of generation.
12. Any operational changes may require notice and permit changes or a new permit.
13. Give yourself enough time for collecting additional compliance samples in case a data quality issue arises.
14. Keep well organized and communicate with management and employees regularly.
15. Housekeeping.
16. Signage.
17. Conduct spill drills.
18. Prepare a list of onsite chemicals with corresponding reportable quantities and recommended cleanup procedures.
19. Conduct regular inspections.
20. Proper preparation prevents poor performance.
21. In many instances, proving a negative is required when evaluating a release.
22. When a question arises, consult management, in-house environmental counsel, outside environmental counsel, or the regulatory authority, as appropriate.
23. Last, remember that "no matter where you are, it's all the same."

The central focus of this chapter will be conducting compliance audits. Conducting a compliance audit also provides an opportunity to evaluate a facility's environmental health and to identify sustainability opportunities.

17.2   USEPA AUDIT POLICY

In 1986, the USEPA published the Environmental Auditing Policy Statement (USEPA 1986). USEPA published the policy statement to encourage the use of conducting "self-assessments" by the regulated community to help achieve and maintain compliance with environmental laws and regulations and to identify and correct unregulated hazards. In addition, USEPA defined environmental audits as (USEPA 1986):

> a systematic, documented, periodic, and objective review of facility operations and practices related to meeting environmental requirements.

The policy also identified several objectives for environmental audits (USEPA 1986):

1. Verifying compliance with environmental requirements
2. Evaluating the effectiveness of in-place environmental management systems
3. Assessing risks from regulated and unregulated materials and practices

USEPA has published guidelines or protocols for conducting environmental audits in subject areas that include (USEPA 2021a):

- Comprehensive Environmental Response, Compensation and Liability Act
- Clean Water Act

Compliance with Land Regulations

- Stormwater under the CWA
- Wastewater under the CWA
- Federal Insecticide, Fungicide, and Rodenticide Act
- Emergency Planning and Community Right-to-Know
- Resource Conservation and Recovery Act
- Safe Drinking Water Act
- Toxic Substance and Control Act

The audit protocols listed previously are a checklist for conducting environmental audits on each subject listed and are 1,387 pages in length. Audit protocols are intended to assist the regulated community in developing programs at individual facilities to evaluate their compliance with environmental requirements under federal law. USEPA states that the protocols are intended solely as guidance in this effort. In addition, USEPA states that the regulated community's legal obligations are determined by the terms of applicable environmental facility-specific permits, as well as underlying statutes and applicable federal, state, and local law.

Therefore, when examining the audit protocols, the facility must also take into account state, county, municipal, or local regulations that apply in order to fully understand requirements that the facility must comply with before determining whether the facility is in full compliance. This again may require the services of in-house counsel, a qualified and experienced environmental attorney, an experienced and qualified environmental consultant, and/or all of these. Additionally, contacting the regulatory agency may be justified to clarify certain issues that may arise.

In 1995, USEPA published "Incentives for Self-Policing: Discovery, Disclosure, Correction, and Prevention of Violations," which both reaffirmed and expanded the 1986 policy (USEPA 2021b). USEPA again revised the policy in 2000 and in 2005 (USEPA 2021c). Under USEPA's environmental audit policy, gravity-based penalties for violations of USEPA-administered statutes are reduced or completely eliminated if the violations are voluntarily discovered, promptly disclosed to USEPA, and meet a number of other specified conditions. Those other specified conditions include the following (USEPA 2021c):

1. The violation must be systematically discovered, either through (a) an environmental audit or (b) a compliance management system reflecting the company's actions and methods in preventing, detecting, and correcting violations.
2. The violation must be discovered voluntarily and not by legally mandated sampling or monitoring required by statute, regulation, permit, judicial or administrative order, or consent agreement.
3. The company must fully disclose the specific violation in writing to the appropriate regulatory authority or USEPA within 21 days or within a shorter period of time if necessary (i.e., such as a spill exceeding the corresponding reportable quality in which case the reporting requirement is within 24 hours) after the company discovered that the violation has or may have occurred.

4. The company must discover and disclose the violation to USEPA prior to (a) the commencement of a federal, state, or local agency inspection or investigation or the issuance by the agency of an information request to the company; (b) notice of a citizen suit; (c) the filing of a compliant by a third party; (d) the reporting of the violation to USEPA (or other governmental agency) by a "whistleblower" employee; or (e) imminent discovery of the violation by a regulatory agency.
5. The company must correct the violation within 60 calendar days from the date of discovery, certify this in writing, and take appropriate action or measures as determined by USEPA to remedy any environmental or human harm due to the violation.
6. The company must agree in writing to take steps to prevent a recurrence of the violation.
7. The specific violation (or closely related violation) cannot have occurred previously within the past 3 years at the same facility, or within the past 5 years as part of a pattern of multiple facilities owned or operated by the same entity.
8. The violation cannot be one that (a) resulted in serious harm or may have presented an imminent and substantial endangerment to human health or the environment or (b) violates the specific terms of any judicial or administrative order or consent agreement.
9. The company must cooperate with USEPA and provide such information as is necessary and requested by USEPA to determine applicability of the policy.

USEPA has broad statutory authority to request relevant information on the environmental compliance status of regulated entities. However, USEPA believes routine requests for audit reports by the agency could inhibit auditing in the long run, decreasing both the quantity and quality of audits conducted. Therefore, as a matter of policy, USEPA will not routinely request environmental audit reports (USEPA 1986).

USEPA's authority to request an audit report, or relevant portions thereof, will be exercised on a case-by-case basis where the agency determines it is needed to accomplish a statutory mission or where the government deems it material to a criminal investigation. USEPA expects such requests to be limited, most likely focused on particular information needs rather than the entire report, and usually made where the information needed cannot be obtained from monitoring, reporting, or other data otherwise available to the agency. Examples would likely include situations where a company has placed its management practices at issue by raising them as a defense or state of mind or intent are a relevant element of inquiry, such as during a criminal investigation. This list is illustrative rather than exhaustive, since there doubtless will be other situations, not subject to prediction, in which audit reports rather than information may be required (USEPA 1986).

USEPA acknowledges regulated entities' need to self-evaluate environmental performance with some measure of privacy and encourages such activity. However, audit reports may not shield monitoring, compliance, or other information that would otherwise be reportable and/or accessible to USEPA, even if there is no explicit

Compliance with Land Regulations 309

"requirement" to generate the data. Thus this policy does not alter regulated entities' existing or future obligations to monitor, record, or report information required under environmental statutes, regulations, or permits or to allow USEPA access to that information. Nor does this policy alter USEPA's authority to request and receive any relevant information, including that contained in audit reports, under various environmental statutes, such as Clean Water Act section 308 and Clean Air Act sections 114 and 208 or in other administrative or judicial proceedings (USEPA 1986).

Regulated entities also should be aware that certain audit findings may by law have to be reported to governmental agencies. However, in addition to any such requirements, USEPA encourages regulated entities to notify appropriate state or federal officials of findings that suggest significant environmental or public health risks, even when not specifically required to do so (USEPA 1986).

Questions and reporting requirements related to this policy and conducting environmental audits should be directed to in-house counsel or to a qualified and experienced environmental attorney.

17.3 STARTING THE PROCESS

The first step in achieving environmental compliance is to conduct a compliance audit. Conducting a compliance audit will answer some very important questions concerning any operation. There are two basic questions that must be answered right at the beginning to determine if a compliance audit is advisable:

- Is the facility subject to any environmental regulations?
- Does the facility require an environmentally related permit for any of its operations?

Today, it is rare that a facility doesn't know the answer to those two basic questions, but for the sake of argument, let's assume that they do not know and want to do the right thing and evaluate whether they are subject to environmental regulations. Under this scenario, the following questions apply:

- Does the facility store or use any hazardous substances as defined by USEPA in Chapter 2? If the answer is no, then does the facility store of use any hazardous substances as defined by state or local regulations?
- Does the facility operate any equipment that generates any exhaust or has any air emissions?
- Does the facility use or discharge any water other than sanitary?
- Does the facility generate any solid waste?

If the answer is yes to any of these questions, the facility is likely subject to environmental regulations and may require one or more environmental permits. If questions persist, involvement of company management and in-house counsel, a qualified and experienced environmental attorney, an experienced and qualified environmental consultant, and/or all of these may be necessary. Additionally, contacting the regulatory agency may be justified to clarify specific questions.

310   Environmental Compliance Handbook

17.4 CONDUCTING AN ENVIRONMENTAL AUDIT

Conducting an environmental audit is the first step on a long and winding road toward environmental compliance and ultimately attaining sustainability and environmental stewardship.

Most large companies with several manufacturing locations conduct environmental audits on a regular basis. Some choose to conduct environmental audits more often than others, but most seem to choose conducting an environmental audit every year or every 2 years. Some companies conduct environmental audits with in-house staff to retain institutional knowledge. Some choose to retain an environmental consulting firm to conduct environmental audits because it's conducted by an independent third party. Some environmental audits are conducted without any prior notice, and some are conducted with prior notice. Whichever path is chosen, environmental audits are usually always conducted at the request of counsel.

17.4.1 CATEGORIES OF FINDINGS

When conducting an environmental audit, findings are typically weighted by priority, severity, or category and typically include:

1. Non-compliance
2. Best management practice (BMP)
3. Observation

An issue of non-compliance is the highest priority of any finding and would be considered a regulatory violation. USEPA (2021c) defines an issue of **environmental non-compliance** as not conforming with environmental law, regulations, standards, or other requirements such as an environmental permit. An example of a non-compliance issue would be an improper or unlabeled drum of hazardous waste.

A **best management practice** was first defined by USEPA in 1974 (USEPA 2021d) in the Clean Water Act and has since been adopted by the regulated community. Best management practices are defined for environmental purposes as specific practices that are capable of improving, preventing, and/or minimizing the potential of an occurrence or situation that can lead to environmental non-compliance and/or a release of a hazardous substance to the environment. A best management practice issue, when identified, does not mean that a non-compliant situation exists, but if left unattended, it could result in a non-compliance issue in the future. Examples of environmental best management practices include (Connecticut Department of Environmental Protection 2021):

* Storing products and wastes indoors
* Cleaning catch basins on a regular basis
* Ensuring that lids on dumpsters remain closed

An **observation** is an issue that could be improved but does not reach the threshold of a best management practice. For instance, a better method for keeping track of monitoring results or improving housekeeping are examples of observations.

Compliance with Land Regulations 311

17.4.2 Document List

If the environmental audit is announced prior to the audit date, a request for applicable documents to be available to be reviewed in advance audit will save time. As applicable to the facility being reviewed, the following list of documents should be available for review during the site visit.

1. Current site map
2. Current organization chart
3. Current process flow diagrams depicting: inputs, process units, waste streams, and other outputs
4. List of processes that have been shut down since the last audit
5. List of processes that are expected to start up with the next 12 to 24 months
6. SPCC plan
7. Spill reports for the last 3 years
8. All RCRA manifests since the last audit
9. All special waste notifications and shipping documents since the last audit
10. All analytical tests results and determination for special and hazardous waste
11. All analytical tests for each non-hazardous waste stream
12. Contingency plan
13. Personnel training records since the last audit
14. List of arrangements with local agencies
15. NPDES permit and most recent application
16. Discharge monitoring reports and any associated letters of explanation of permit limit exceedances since the last audit
17. Stormwater plan, intent to comply documents, and all monitoring reports and sample data
18. All notices of violation, noncompliance letters, or consent orders, including facility's noncompliance explanations and, if in significant noncompliance, the facility's plan to return to compliance
19. Water flow diagrams and water balances
20. Sketches of all wastewater treatment systems within the facility
21. Names of wastewater analytical laboratory, a list of all analytical methods used for wastewater analysis, and the quality assurance/quality control procedures used for wastewater sample collection and analysis
22. Air permit application modifications
23. All construction and operating permits for air sources
24. Copies of all visible emission observations taken by facility personnel or contractors for the past 3 years
25. Environmental management plan
26. Any other environmental permit to operate (associated with air, land, or water)
27. Waste shipment summary for the past 2 years
28. Toxic Release Inventory Reports (Form R) for the past 2 years

312 Environmental Compliance Handbook

For facilities not in the United States, and as applicable to the specific country, province or state, additional documents to be made available to the reviewers include:

1. Noise studies and noise compliance (e.g., Mexico, China, etc.)
2. Environmental impact assessment
3. Operating permits
4. European Union Registration, Evaluation, Authorization, and Restriction of Chemicals (REACH) certification status for those locations within the European Union

17.4.3 OPENING MEETING

The following sections describe the environmental review process in greater detail to ensure that a more complete review of the facility compliance status is conducted. In addition, this section also provides guidance for identifying and establishing pollution prevention opportunities to further minimize potential environmental impact in the future.

To improve the efficiency of conducting the environmental review, prior notice to the facility should be given so that all necessary information can be made readily available and appropriate plant personnel are in attendance for the review. Some companies may choose to conduct unannounced audits to emulate an agency inspection, which often are unannounced. Whichever method is used, it's recommended that the facility be reminded that an agency inspection can occur any time on any day and that proper preparation prevents poor performance. Therefore, organization is key and backup systems should be in place in case the facility person in charge of compliance is not present.

A list of documents is provided in Section 17.4.2 that should be available for review, as applicable to the facility.

An opening meeting should be scheduled in advance of the environmental audit. The purpose of the opening meeting is to:

1. Inform appropriate facility management of the purpose and objectives of the audit.
2. Coordinate logistics and review procedures.
3. Remind facility management with respect to:

   - Internal policy pertaining to environmental audits
   - Report privilege and distribution
   - Procedures to follow during agency inspections and emergency and spill events
   - Periodically review environmental permits
   - Pollution prevention initiatives
   - Preparation for environmental compliance requires an awareness and daily attention to such matters

4. Ensure that priority items are presented to facility management quickly so that appropriate corrective actions can be initiated promptly

Compliance with Land Regulations 313

5. Remind management and all personnel that compliance with environmental regulations is not optional
6. Remind management that the process for identified non-compliance items is that they must be addressed immediately and with the utmost priority
7. Remind management that controls must be evaluated and updated as necessary if non-compliance issues are identified to minimize the possibility of a reoccurrence

17.4.4 ENVIRONMENTAL AUDIT CHECKLIST

It is always very helpful to review and become familiarized with the overall facility layout and manufacturing process of the facility before the actual audit begins. This is usually accomplished by reviewing a diagram or figure of the general facility layout and manufacturing process flow and then reviewing a diagram or figure of the facility that identifies (1) locations of all air emission sources; (2) solid and hazardous waste stream sources and points of generation; (3) water and wastewater flow through the facility; and (4) areas where hazardous substances enter the facility, where they are stored and used within the facility, and where they are stored as wastes before offsite transportation for disposal.

The items listed in the checklist are intended to be used as a guide in conducting the environmental audit for environmental regulations that are focused on land. Some of the sections or listed items included in the checklist may not be applicable to a particular facility. Therefore, the checklist should be facility specific and should be used as applicable.

17.4.4.1 Hazardous Waste

Solid and hazardous waste regulations are generally considered the most complex and also vary significantly from facility to facility. To assist in conducting the solid and hazardous waste portion of the environmental review process, this section is divided into several subsections: (1) hazardous waste, (2) solid waste, (3) universal waste, (4) used oil, (5) company owned or operated landfills, (6) recycling, and (7) pollution prevention plans.

As an additional item to be noted, each facility's solid and hazardous waste requirements and permits may not be standardized under similar regulations. Therefore, federal, state, and local guidelines that affect each facility should be reviewed independently.

17.4.4.1.1 General Background Information

The following items should be considered first before conducting the detailed review of solid and hazardous waste to become familiarized with the current status of the facility with respect to status and compliance:

- Evaluate the types of solid waste that the facility generates
- Evaluate whether the facility generates hazardous waste
- Evaluate and record the facility EPA ID number
- Evaluate the current and previous facility generator status (e.g., permit, LQG, SQG, etc.)

314   Environmental Compliance Handbook

- Evaluate whether the facility has a permit to store waste onsite
- Review requirements (40 CFR, 264 and 265 [interim requirements]) and any state and local requirement if facility is permitted to store wastes onsite
- Evaluate whether the Boiler & Furnace Rule, 40 CFR 266 Subpart H, applies
- Verify that the correct EPA ID number is recorded on all documentation (i.e., manifests)
- Evaluate whether any processes have changed since the last audit that could influence the chemical or physical characteristics of any waste stream. If so, has the associated generated waste been recharacterized. If not, why?
- Review any correspondence from or to any regulatory agency
- Review whether any solid waste inspections have been conducted since the last environmental audit
- Review and evaluate whether the facility has received any solid waste NOVs since the last environmental audit
- Review and evaluate any corrective actions that have been undertaken to address any NOVs

17.4.4.1.2   Point of Generation Analysis

- Evaluate and verify that each hazardous waste stream is hazardous by either (1) listing in regulations, (2) laboratory analysis, or (3) knowledge of materials and processes used
- Evaluate whether waste needs to be reclassified or characterized

  1. Evaluate whether any process or manufacturing changes occurred that may impact the classification or chemical characterization of the waste (i.e., installation of new machinery).
  2. Evaluate whether any other changes have occurred that may impact the characterization of the waste (i.e., differences in parent material or re-formulation of bulk product inputs).
  3. Check to ensure that each hazardous waste stream is re-characterized at least every 3 years or sooner if there is any process changes that could influence the chemistry of the waste material.

- Review waste stream source diagram

17.4.4.1.3   Inventory of Hazardous Waste

- Ensure inventory of hazardous waste is correct
- Evaluate the appropriateness of generator status
- Check to make sure volumes and weights of each hazardous waste stream are correct and are verified

17.4.4.1.4   90 Day Storage

- Review inspection documentation
- Verify that no container has accumulated waste for more that 90 days, unless a 30-day extension was granted
- Verify each container and tank is labeled the words HAZARDOUS WASTE, the start accumulation date, and applicable waste codes

Compliance with Land Regulations

17.4.4.1.5 Treatment and Disposal
- Verify wastes are hauled by transporters with valid EPA identification numbers
- Verify wastes are taken to disposal facilities with valid hazardous waste permits
- Review annual waste shipment summaries
- Check to make sure the facility has a tracking procedure to ensure that appropriate actions are undertaken in case a waste manifest is not returned to the facility within the required timeframe

17.4.4.1.6 Biennial Reports
- Verify that the biennial report was complete and submitted by March 1
- Verify that copies of the report are kept for 3 years

17.4.4.1.7 Manifests
- Inspect for documentation accuracy
- Verify that manifests are used when shipping waste off-site
- Verify that exception reports are filed when a copy of the manifest is not received within the required timeframe of the waste being accepted by the initial transporter (45 days for SQG and 30 days for LQG)
- Check to make sure the facility has a tracking procedure to ensure that appropriate actions are undertaken if a waste manifest is not returned to the facility within the required timeframe
- Verify that manifests are kept for 3 years
- Verify that manifests are signed and dated correctly and clearly

17.4.4.1.8 Storage Areas
- Ensure internal communications system available
- Ensure telephone or two-way radio to summon emergency assistance
- Check for portable fire extinguishers
- Inspect spill control equipment
- Inspect decontamination equipment
- Check for fire hydrants
- Some of these may not be necessary depending on the waste
- Determine if equipment is tested and maintained
- Verify sufficient aisle space is maintained
- Verify arrangements with local fire, police, and emergency response teams if necessary

17.4.4.1.9 Personnel Training
- Verify that personnel complete classroom instruction
- Training must include contingency plan implementation, if a LQG
- Training must be completed within 6 months of employment/assignment
- Verify that annual review training is provided
- Verify that employees do not work unsupervised until training is completed
- Verify specifically that waste storage area managers and hazardous waste handlers have been trained

316 Environmental Compliance Handbook

17.4.4.1.10 Training Records

- Include job title and description for each employee by name
- Written description of how much training each position will receive
- Documentation of training received by name
- Determine if records on former employees are retained for 3 years
- Determine if records on current employees are maintained

17.4.4.1.11 Contingency Plans

- Verify that the contingency plan is designed to minimize hazards from fires, explosions, or releases, if a LQG
- Must describe actions to be taken in an emergency
- Must describe arrangements made local agencies as appropriate
- Must include the name, address, and phone number of the emergency coordinator and any alternates
- Must include a list of any emergency equipment, its location, and description
- Must include an evacuation plan if required
- Ensure that communications systems are satisfactory and are defined in case of an emergency
- Verify that revisions are maintained and submitted to local emergency services where required
- Verify that adequate isle space is available for unobstructed movement of emergency equipment in case of an emergency
- Verify that the plan is routinely reviewed and updated when regulations change, the plan fails, the facility changes, the emergency coordinators change or the emergency equipment changes
- Review personnel training records to ensure that facility personnel are able to respond effectively during an emergency. This should include but is not limited to (1) inspecting and maintaining equipment, (2) shutoff systems, (3) communication and alarms, (4) fire and explosions, (5) contamination, and (6) shutdown procedures

17.4.4.1.12 Emergency Coordinators

- Verify that at all times there is at least one employee at the facility or on call with responsibility for coordinating emergency response measures
- Verify that he/she is familiar with the facility and the contingency plan
- Verify that he/she has the authority to commit resources to carry out the contingency plan

17.4.4.1.13 Containers

- Verify that containers are not leaking, bulging, rusting, damaged, or dented
- Verify that the storage area is inspected at least weekly
- Verify that containers are empty
- Verify that containers are compatible with the waste
- Verify that containers are kept closed except when adding or removing waste
- Verify that containers are properly labeled and legible

Compliance with Land Regulations 317

17.4.4.1.14 Labels
- Verify that every waste container is labeled
- Verify that every waste container is labeled with appropriate accumulation dates, waste identification, and other required information
- Verify whether other labeling requirements apply

17.4.4.1.15 Satellite Accumulation
- Verify that the satellite accumulation point is at or near the point of generation
- Verify that the containers are in good condition, compatible with the waste, and kept closed except when adding or removing waste
- Verify that the containers are marked HAZARDOUS WASTE or other identified marking, as required
- Verify that when waste is accumulated in excess of quantity limitations that the date is marked on the container and within 3 days the container is moved to the 90-day storage area
- Verify proper containment is provided

17.4.4.1.16 Restricted Wastes
- Determine by analysis or process knowledge if wastes are restricted from land disposal
- Verify that all notifications and certifications has been made
- Verify that records are kept for 3 years

17.4.4.1.17 Record Keeping
- Ensure that records are organized and complete
- Evaluate backup record keeping systems
- Ensure that at least 3 years of written records are available for inspection

17.4.4.1.18 Agency Inspections
- When was last agency hazardous waste inspection?
- What areas of regulations were the focus?
- Were any violations discovered?
- Was the violation addressed after the inspection ended?
- Did the agency take any photographs?
- What triggered the inspection?
- What agency conducted the inspection?
- Did the agency issue any Notice of Violation?
- How were violations addressed?
- Were there any fines or other penalties?

17.4.4.2 Solid Waste

17.4.4.2.1 Non-Hazardous Waste Determination
- Verify that proper waste determinations are performed (at a minimum every 3 years or more every time the process changes that could result in input changes of waste chemical or physical characteristics)

318 Environmental Compliance Handbook

- Verify that non-hazardous wastes are actually non-hazardous. Ask yourself the question: How do I know its non-hazardous? If you don't know, test it
- Review solid waste flow diagram
- Review list of solid waste streams

17.4.4.2.2 Point of Generation Analysis

- Evaluate and verify that each waste stream is not hazardous by either (1) listing in regulations, (2) laboratory analysis, or (3) knowledge of materials and processes used
- Evaluate whether waste needs to be reclassified or characterized

  1. Evaluate whether any process or manufacturing changes occurred that may impact the classification or chemical characterization of the waste (i.e., installation of new machinery, new vendor, etc.).
  2. Evaluate whether any other changes have occurred that may impact the characterization of the waste (i.e., differences in parent material or re-formulation of bulk product inputs).
  3. Check to ensure that each hazardous waste stream is re-characterized at least every 3 years or sooner if there are any process changes that could influence the chemistry of the waste material.

- Review waste stream source diagram

17.4.4.2.3 Inventory of Solid Waste

- Ensure inventory of hazardous waste is correct
- Evaluate the appropriateness of generator status
- Check to make sure volumes and weights of each hazardous waste stream are correct and are verified

17.4.4.2.4 Storage

- Verify that materials are properly stored and labeled
- Verify that no applicable storage time frames are exceeded
- Ensure that storage containers are in good condition
- Ensure that wastes are not mixed together

17.4.4.2.5 Labels

- Verify that every waste container is labeled
- Verify that every waste container is labeled with appropriate accumulation dates, waste identification, and other required information
- Verify whether other labeling requirements apply

17.4.4.2.6 Transportation

- Verify that materials are hauled according to any applicable state and local requirements

17.4.4.2.7 Disposal

- Verify that the materials are disposed at an approved site
- Review special waste permits or waste disposal authorizations if applicable
- Review current waste profiles

Compliance with Land Regulations

17.4.4.2.8 *Record Keeping*
- Ensure that records are organized and complete
- Evaluate backup record-keeping systems
- Ensure that at least 3 years of written records are available for inspection

17.4.4.2.9 *Agency Inspections*
- When was last agency waste inspection?
- What areas of regulations were the focus?
- Were any violations discovered?
- Was the violation addressed after the inspection ended?
- Did the agency take any photographs?
- What triggered the inspection?
- What agency conducted the inspection?
- Did the agency issue any Notice of Violation?
- How were the violations addressed?
- Were there any fines or other penalties?

17.4.4.3 Used Oil
17.4.4.3.1 *Storage*
- Verify that the tanks, containers, and fill pipes are labeled USED OIL
- Verify that the tanks or containers are in satisfactory condition
- Ensure that all used oil is placed in appropriate containers (i.e., drip pans beneath machines are not an appropriate storage container)

17.4.4.3.2 *Labels*
- Verify that every waste container is labeled
- Verify that every waste container is labeled with appropriate accumulation dates, waste identification, and other required information
- Verify whether other labeling requirements apply

17.4.4.3.3 *Transportation*
- Verify that the used oil is transported only by transporters with a valid USEPA identification number

17.4.4.3.4 *Recycle*
- Verify the final destination of the used oil
- Inspect certificate of recycling

17.4.4.4 Universal Waste
17.4.4.4.1 *Identification*
- Verify batteries, pesticides, mercury-containing equipment, and lamps are being segregated and properly disposed at a licensed facility

17.4.4.4.2 *Storage*
- Verify that waste is not accumulated for more than the allowed time period
- Ensure that the storage container are appropriate and that the lids are closed

320 — Environmental Compliance Handbook

17.4.4.4.3 Labels
- Verify that all containers are properly marked or labeled

17.4.4.4.4 Training
- Verify that necessary employees receive required training

17.4.4.4.5 Transportation
- Verify that the materials are shipped properly and arrive at the final destination

17.4.4.5 Company Owned or Operated Landfills

17.4.4.5.1 Landfill Permit
- Review operational plan
- Review closure/post-closure plan
- Review water quality monitoring plan
- Discuss most recent permit renewal application
- Review list of acceptable/current waste streams

17.4.4.5.2 Compliance Documentation
- Inspect Landfill Operator Certificates, if required
- Review latest engineering reports
- Review water monitoring reports
- Review closure/post-closure financial assurance
- Obtain copy of any NOV since last review
- Review any agency inspection reports
- Discuss any other significant agency correspondence

17.4.4.5.3 Other Landfill-Related Documents
- Discuss any other significant agency correspondence
- Review closure/post-closure cost estimates
- Review remaining volume/life estimates
- Review recent topographic surveys

17.4.4.5.4 Beneficial Reuse Efforts
- Discuss facility's efforts towards beneficial reuse of landfill materials
- Discuss any state regulations, policies, or guidance regarding foundry waste streams
- Discuss any state regulations, policies, or guidance regarding beneficial reuse

17.4.4.5.5 Recycling and Pollution Prevention
- Evaluate whether the facility has a recycling plan or initiative
- Evaluate what items, products, and materials are or can be recycled
- Evaluate whether items, products, and materials are recycled internally or externally
- Evaluate whether the facility could benefit from a regulatory-sponsored recycling or pollution prevention program (i.e., USEPA's Climate Leaders Program)

Compliance with Land Regulations

- Evaluate the status of the facility's pollution prevention plan and awareness
- Evaluate whether the facility uses chlorinated solvents and enacts measures to eliminate the continued use if chlorinated solvents are used or stored at the facility
- Evaluate whether the facility uses other targeted compounds such as cadmium, mercury, chromium, and lead and evaluate measures to eliminate or reduce use of these compounds.
- Evaluate whether the facility uses large quantities of VOCs other than chlorinated solvents such as xylenes, toluene, MEK, and so on and evaluate initiating a pollution prevention strategy
- Evaluate the effectiveness of the facility's chemical ordering procedures as a method of pollution prevention and awareness

17.4.4.6 Spills

17.4.4.6.1 Liquid Inventory
- Evaluate volume of oil storage, including transformers
- Does volume exceed 1,320 gallons?

17.4.4.6.2 Spill History
- Review any spills since last review
- Review incident reports
- Review NRC or state spill reports
- Inspect spill areas
- Review corrective action
- Discuss historical spills

17.4.4.6.3 SPCC Plan
- Review SPCC applicability and current total quantity of regulated liquids stored onsite
- Review SPCC plan
- Review list of potential spill materials
- Review waste and material storage areas with diagram
- Review contingency plans
- Review training records
- Ensure that SPCC plan is signed by a PE
- Ensure that a Certification of Substantial Harm Determination is completed

17.4.4.6.4 Evaluation of Storage location
- Ensure storage locations are covered
- Are storage locations within the reach of a floor drain?
- Evaluate the integrity of the floor
- Evaluate whether the storage area needs additional engineering controls
- Is storage area restricted and fenced?
- Does storage area maintain appropriate signs and labels?

Environmental Compliance Handbook

17.4.4.6.5 Pre-Calculation of Reportable Quantities

- Ensure that the facility has pre-calculated reportable quantities for each liquid
- Double-check reportable quantities

17.4.4.7 PCBs

17.4.4.7.1 PCB Review and Record Keeping

- Evaluation and inventory of equipment potentially containing PCBs
- Record the type (liquid, dry, pole-mounted, pad, PCB, non-PCB), contents, capacity, age, and location of each transformer at the facility
- Review list of PCB equipment, including:

  Transformers (>50 ppm, >500 ppm, made before July 1, 1979)
  Evaluate ownership records of all transformers on facility property
  Capacitors (large, high voltages, made before July 1, 1979)
  Hydraulic presses (made before July 1, 1979)
  Oil-filled electrical switches (made 1935–1979)
  Oil-filled electrical motors (made before July 1, 1979)
  Oils, paints, inks, sealants (made 1935–1971)

17.4.4.7.2 Compliance Documentation

- Review analysis of oils for previous units
- Inspect PCB and non-PCB labeling
- Documents regarding storage and disposal of PCB equipment
- Review inspection reports of PCB equipment in use
- Review registration of PCB transformers

17.4.4.7.3 PCB Spill Review

- Evaluate whether any recent or historical PCB spills have occurred at the facility
- Obtain copy of any PCB or non-PCB spill reports

17.4.4.8 Storage Tanks

17.4.4.8.1 Review SPCC Plan

- Review SPCC and/or contingency plans for tank identification and documentation

  Review spill/release prevention and response procedures
  Review reporting procedures

17.4.4.8.2 Types and Location of Tanks

- Type and location of Tanks

  Review recordkeeping on the design, construction, installation, alterations, repairs
  Record material of construction, capacity, contents, and location of each AST and UST at the facility

Compliance with Land Regulations

Review contents and identification (labeling)
Review documented unloading procedures and training records
Procedures—are employees present during the unloading operations?

17.4.4.8.3 Containment

- Containment areas

Review containment for integrity, size, impermeability, and so on

17.4.4.8.4 Tank History

- Spill/leak history

Documentation, open issues

17.4.4.8.5 Spill Prevention

- Tank fail-safe measures

Review calibration and testing of alarms and gauges
Documented procedures and preventative maintenance
Recordkeeping

17.4.4.8.6 Record Keeping

- Inspections—PM and recordkeeping

Routine (weekly) housekeeping, levels, leaks, and so on
Formal (quarterly) same as previous, to include labeling, spill kits and PPE
adequacy and location, piping, tanks, containment, repairs
Integrity (internal and external) and leak testing (not to exceed every 5 years
and only applies to USTs)

17.4.4.8.7 Training

- Training records

Spill team, employees responsible for handling, loading, unloading, inspections

17.4.4.9 Hazardous Materials

- Review DOT requirements for hazardous material shipments

Registration requirements
Hazardous waste handler training documentation (every 3 years)
Management of hazardous materials for shipment
Record keeping
DOT security plan

17.4.4.10 Toxic Substance Control Act

17.4.4.10.1 Chemical Ordering Procedure

- Review written procedure
- Discuss environmental personnel involvement
- Discuss customer requirements on chemical use limitations

324 Environmental Compliance Handbook

- Ensure permitting issues and applicability are addressed
- Identify efforts to elimination of chlorinated solvents

17.4.4.10.2 New Chemicals Produced or Imported into the United States

- Review any new chemicals which are produced or imported into the United States since last review

17.4.4.10.3 Chemical Ordering Procedure Considerations

- Review any written procedures
- Ensure that the facility follows any written procedures
- Does company nurse or doctor sign off on any new chemical?
- Does environmental department sign off on any new chemical?
- Evaluate who can order new chemicals
- Review any new chemicals that are produced or imported into the United States
- Review training records for shipping and receiving personnel
- Evaluate whether a security plan for the facility has been developed, if required

17.4.4.11 Emergency Planning and Community Right-To-Know-Act

17.4.4.11.1 Program Review

- Hazard communication program

  Employee training on workplace chemicals, procedures, and recordkeeping
  Review MSDS management, updating, chemical listing, tracking, employee access/availability procedures

17.4.4.11.2 Notification

- Emergency planning notification

  Extremely hazardous substances threshold planning quantities
  Submission of emergency response and/or contingency plans to LEPC (verify if plans are up to date and current)

17.4.4.11.3 Reporting

- Tier I, II submission of reports

  Review other state requirements for supplemental reporting
  Hazardous chemical listing submissions (local fire department or fire marshal)

17.4.4.11.4 Form R Specific

- TRI form R

  Review applicability determination (used, manufactured, or processed) and submission of reports

17.4.4.12 Site Inspection

Following the review of written materials, a site inspection of the entire facility should always be conducted. It is recommended that the site inspection be structured in such

Compliance with Land Regulations

a way that it follows the manufacturing process as much as possible. Photographs are recommended to document the current environmental condition of the facility, especially those items that may require a corrective action. It is also recommended that a photograph be taken after any corrective action to document task completion and to demonstrate compliance.

It will be helpful to have a map of the facility while conducting the site inspection that shows the entire property, with buildings and operations clearly marked. In addition, it will be helpful to have waste storage areas, air emission points, points of generation of solid and hazardous waste, points of generation of waste water and waste water discharge points, and stormwater outfalls labeled on the map.

During the site inspection, the following items should be observed, evaluated, and noted:

- If there is an item observed during the site inspection or at any time that may present an immediate threat or imminent danger to human health or the environment, follow the organization's emergency notification procedures. Examples may include but would not be limited to the following:
  - Evidence of existing release of hazardous substances or petroleum products
  - Evidence of material threat of a release of hazardous substance or petroleum products

- Other areas to inspect, review, and/or observe:
  - General housekeeping
  - Any new equipment?
  - Any new processes?
  - Any new waste streams?
  - Review history of neighborhood complaints (documented)
  - Vehicle maintenance areas
  - Spill kits location, contents listing and adequacy
  - Emergency response phone list—postings, updated, controlled document, and so on
  - Chemical and waste storage areas
  - Labeling, storage, and housekeeping
  - Waste storage areas, piles, drop boxes, covered, leaks, proper contents
  - New product storage areas (oil and chemicals)
  - Universal waste storage practices and labeling
  - Drum management
  - Drum labeling, management, residue issues, secure when not in immediate use, empty drum storage
  - Secondary containment areas
  - Plant and property security—alarm system, guard service, fencing, gates, lighting, access control, and so on
  - Roof inspection
  - Drainage

326 Environmental Compliance Handbook

- Emission deposits
- Exhaust stack(s) deposits
- Roof vent(s) deposits
- Floor drains identified on plant drawings, purpose, discharge, potential discharge, drain blocker in the event of a spill
- Stormwater conveyance system preventative maintenance
- Erosion issues
- Catch basins
- Detention ponds
- Outfalls
- Outside processes, if any
- Equipment and product storage potential for contamination
- Utility supply and shutoff valves, identification, security, management, emergency procedures, and training
- Compressor discharge management
- Exhaust fan and vent deposits to exterior of the building
- Baghouse inspection, record keeping, and preventative maintenance
- Review facility's control of chemicals for chlorinated solvent content
- Secondary container labeling
- Neighboring properties identified and note (1) the potential for hazardous substance use or storage such as ASTs and drum storage areas, (2) stained soil, (3) topography, (4) drainage patterns, and (5) types of operations
- Property boundaries inspected, non-facility contributions, observations/ activity, any issues

17.4.4.13 Closing Meeting

A closing meeting should be conducted at the end of the environmental audit. The purpose of the closing meeting is to:

1. Inform facility management of the results of the environmental audit
2. Set a timeline for when the draft report will be completed
3. Establish a timeline and assign appropriate personnel for implementing corrective actions as a result of the findings of the environmental audit, if any
4. Remind facility management with respect to:

- Internal company policy pertaining to environmental audits
- Report privilege
- Report distribution
- Procedures to follow during agency inspections
- Procedures to follow during an emergency or spill event
- Periodically review environmental permits
- Review company policy for non-routine correspondence with a regulatory agency
- Review company policy for permit applications
- Reminding all employees that environmental compliance requires an awareness and daily attention

Compliance with Land Regulations

17.4.4.14 Report Preparation

A draft report should be prepared soon after the audit has been conducted, certainly no more than a month following the audit. The report should be a draft and marked as such when submitted for review. For items that require a corrective action that are discovered during the environmental review, it is recommended that they be separated into either best management practice–related issues, compliance-related issues, or observations at the end of each section of the report. An example environmental audit report outline is presented in the following:

- Fundamentals
  - Air
  - Water
  - Land
  - Living organisms
- Conducting the audit
  - The opening meeting
    - Air
    - Water
    - Hazardous waste
    - Universal waste
    - Non-hazardous waste
    - Spills
    - PCBs
    - Hazardous materials
    - Toxic Substance and Control Act
    - Emergency Planning and Community Right-To-Know-Act
    - Site inspection
  - The closing meeting
  - Summary and conclusions
  - Photographs

After the report has been reviewed and finalized, a system should be put in place to ensure that all the recommended items are properly addressed and correct, especially any identified compliance items. As a reminder, any identified compliance items are of the highest priority and must be addressed immediately.

Typically a spreadsheet is prepared for each item identified in the report as a compliance, BMP, or observation and typically includes the following:

- Name of person responsible for supervising or conducting corrective action
- Time period allowed to complete corrective action
- Actions to be taken to complete corrective action
- Completion sign off
- Completion date
- Management sign off

17.5 AGENCY INSPECTIONS

We first discussed the importance of environmental enforcement in Chapter 4. We learned that the United States has an established and effective enforcement policy and program. The United States has also consulted with many developing nations in providing assistance in establishing environmental regulations and enforcement programs. Much of the success that the United States has achieved has been, in part, through its environmental enforcement program.

The USEPA has also helped itself by promoting "self audits." Self audits have benefited the regulated community in many ways, including education and conducting corrective actions when issues are discovered. Self audits have benefited USEPA as well by shifting much of the responsibility to the regulated community. However, this is predicated on conducting inspections to ensure compliance.

Inspections by regulatory authorities vary from location to location and state to state. Some inspections may target one particular media such as air, water, or solid and hazardous waste, while others may cover several all at once, but that is rare. Usually, agency inspections will focus on one particular set of regulations.

Typically, an agency inspection will consist of the following:

1. Opening meeting. This is to discuss the reason behind the inspection (e.g., routine inspection or complaint). To discuss the facility layout, production processes, plant history, wastes generated and storage areas, air emission points and stacks, air emission controls, water distribution and water use, water treatment and discharge, and other environmental aspects.
2. Document review. The document review usually follows the opening meeting. Document review usually focuses on demonstrating compliance with applicable environmental permits such as air and water discharge. Document reviews may also include inspection of manifests and waste characterization. Waste characterization may also focus on non-hazardous waste streams to ensure that a non-hazardous waste has been properly characterized. Training records, logs, plans, reports, forms, and correspondence records are also routinely inspected.
3. Site inspection. From the information obtained from the document review, the agency inspector will have a much more educated perspective on operations and what regulations apply to the facility. The inspector will usually request that a site inspection be conducted to visually inspect the area or areas that the inspection has focused (air, water, solid and hazardous waste, etc.). The regulatory inspector may also ask if photographs can be taken during the site inspection. The regulatory inspector is typically looking to evaluate the consistency between the document review with site observations.
4. Closing meeting. The closing meeting will usually involve whether any violations have been identified and may also involve additional requests for information and documentation that was not readily available at the time of the inspection.

Compliance with Land Regulations 329

17.6 ENVIRONMENTAL AUDITS AND SUSTAINABILITY

Environmental audits should not be confused with sustainability. An environmental audit is a measure of compliance with environmental laws, regulations, standards, and other requirements. An environmental audit is not a measure of sustainability. However, an environmental audit is a good resource to begin evaluating sustainability. Sustainability is based on a simple principle that states that (USEPA 2021e):

> Everything we need for our survival and well-being depends, either directly or indirectly, on our natural environment.

USEPA defines **sustainability** as creating and maintaining conditions under which humans can exist in productive harmony to support present and future generations (USEPA 2021e). Note that the definition states "productive harmony." One may ask, in productive harmony with what? The answer is with the environment. Therefore, in order to achieve some level of sustainability, we must understand the environment and also understand the aspects of how facility operations impact the environment. Sustainability is then the outcome of analyzing the aspects of facility operations with the natural environment and developing and engineering methods to either minimize or eliminate harmful potential harmful impacts.

A good start for understanding facility operations that may negatively impact the environment is with a comprehensive environmental audit. The other key element in the definition of sustainability is understanding the natural environment.

17.7 SUMMARY OF FUNDAMENTAL CONCEPTS OF ENVIRONMENTAL COMPLIANCE

The primary purpose of environmental regulations is to protect human health and the environment. We should also have an appreciation of the sheer immensity and oftentimes overwhelming complexity of environmental regulations.

An important aspect of environmental regulations and conducting environmental audits is that environmental regulations focus on what is called "end of the pipe" or wastes that are generated. For instance, looking back over this chapter on how to conduct an environmental audit, the regulations focused on the type and amount of air emissions, water discharges, and solid and hazardous wastes that were generated.

Each facility should also have a keen awareness that any process or manufacturing changes or installation of new equipment may have a profound effect on the types, amounts, and chemical and physical characteristics of emissions, discharges, and wastes that are generated that may require a permit or permit modification and may render existing waste profiles obsolete and require associated wastes to be recharacterized.

In this chapter, we have boiled things down to a few points that will provide assistance and focus our efforts to guide us in achieving compliance with environmental regulations and beyond.

The significant lessons or principles of environmental compliance were:

1. **It all begins with conducting an environmental audit.**
   The first step to compliance is to conduct an environmental audit. An environmental audit is an assessment of whether a facility or organization is observing practices to minimize harm to the environment. USEPA defines an **environmental audit** as a systematic evaluation to determine the conformance to quantitative specifications in environmental laws, regulations, standards, permits, or other legally required documents (USEPA 2021a).
2. **Being in compliance with environmental regulations does have sustainability implications.**
   In this chapter, we explored what it means to look toward environmental stewardship and sustainability as part of environmental compliance with air regulations. Environmental stewardship and sustainability require going beyond environmental compliance and addressing deeper questions about how we may impact the environment. Being in compliance means conducting activities that are consistent with environmental regulations. It does not ask the question of whether you should be doing something else that will enhance your organizational goals. In other words, you have to tell yourself, "Just because you can doesn't always mean you should."
3. **Work yourself out of environmental regulations.**
   This is often difficult and challenging but is well worth the trouble and helps pave the way to environmental stewardship and sustainability.
4. **There is little we can control.**
   Environmental professionals for the most part do not control the location where manufacturing takes place or the type of manufacturing or products that are made. However, environmental professionals typically have input in the chemicals that are used and what management actions and engineering controls to put in place to prevent mismanagement and potentially causing harm to the environment if a release occurs.
5. **Limit chemical use when possible.**
   Remember, if there we no hazardous substances, there would be little need for environmental regulations or environmental professionals.
6. **When possible, eliminate chemicals that are very toxic or have a high chemical risk factor.**
   Elimination of chemicals provides the highest level of confidence in protecting human health and the environment. Focus should be placed on those chemicals that are highly toxic or have a high chemical risk factor. Some of these include the following:

   - Arsenic
   - Lead
   - Mercury
   - Cadmium
   - Volatile organic compounds, especially:

     - Benzene
     - Chlorinated or halogenated VOCs

Compliance with Land Regulations 331

- PCBs
- 1,4-dioxane
- Dioxin
- Cyanide
- PFAS compounds
- Chlordane
- DDT

7. **Fully understand environmental permits.**
   This sounds easy, but it is not, and many mistakes occur due to numerous reasons. Some include lack of understanding of a permit because it's too long, too complex, or too technical.

8. **Never accept a permit term that the facility can't achieve.**
   All too often, a permit term may seem easy to comply with, but under actual operating conditions, the facility can't achieve compliance. Therefore, always collect data to evaluate whether the permit term can be achieved before accepting the permit.

9. **Always be accurate and truthful.**
   There are instances when communicating with a regulatory agency in writing or even verbally where data can be misinterpreted or a mistake is made, such as a typo. In any instance, it's usually always better to be as accurate as possible and truthful. If a circumstance arises where interpretation of data can be questioned or any other circumstance where data can be challenged, get advice from a knowledgeable, qualified, and reliable source such as an environmental consulting firm, the local environmental authority, or counsel.

10. **When in doubt, it's usually always better to report a spill or other incident where reporting may be required.**
    If there is a spill or other incident that requires reporting to the environmental agency within a prescribed time after the incident, it is usually always beneficial to report the incident. The only time when reporting is not advised is when there is absolute certainty that the incident does not require reporting. Under these circumstances, detailed documentation should be kept to ensure that properly procedures were followed. In addition, know in advance the reporting procedures and company policies on reporting, and seek training so that when an incident does occur, everyone will be prepared.

11. **The importance of waste characterization and points of generation.**
    In general, a facility should not rely on generator knowledge to either characterize or partially characterize a waste stream. Reliance should be on laboratory analysis of a representative sample of the waste. In addition, the point of generation of a waste is equally important to ensure that the waste is properly characterized and representative.

12. **Any operational changes may require notice and permit changes or a new permit.**
    Each facility should also have a keen awareness that any process or manufacturing changes or installation of new equipment may have a profound

332 Environmental Compliance Handbook

effect on the types, amounts, and chemical and physical characteristics of emissions, discharges, and wastes that are generated that may require a permit or permit modification and may render existing waste profiles obsolete and require associated wastes to be recharacterized.

13. **Give yourself enough time for collecting additional compliance samples in case a data quality issue arises.**
Compliance sampling and monitoring should be conducted well in advance of reporting results to the regulatory authority in case a data quality issue is identified during any step of the process.

14. **Keep well organized and communicate with management and employees regularly.**
Maintaining well-organized files for compliance will ensure that there is less confusion and a shorter compliance audit or regulatory inspection. In addition, establishing a yearly compliance schedule by month and regular communication with management will help to ensure that ongoing compliance is achieved. Regular communication with employees can lead to identifying areas for improvement and can generate a better environmentally oriented culture.

15. **Housekeeping.**
Maintaining a clean and organized facility will assist in maintaining compliance.

16. **Signage.**
Proper labels are part of compliance requirements. However, expanding signage in key areas and in waste storage areas and compliance monitoring equipment will increase compliance and will educate employees on maintaining compliance.

17. **Conduct spill drills.**
Conducting spill drills will ensure proper response and will provide educational opportunities for everyone involved.

18. **Prepare a list of onsite chemicals with corresponding reportable quantities and recommended cleanup procedures.**
A spill incident is stressful, so calculate RQs in advance. This will greatly increase a proper response and reporting if required.

19. **Conduct regular inspections.**
It is recommended to conduct regular inspections beyond those that are required under environmental permits. Conducting regular inspections generally has many benefits that include better housekeeping and identifying potential compliance issues.

20. **Proper preparation prevents poor performance.**
Don't wait for a regulatory inspection to improve performance. A Notice of Violation is not something desirable.

21. **In many instances, proving a negative is required when evaluating a release.**
Evaluating the risk posed when a release of hazardous substances has occurred focuses on ensuring that the release will not adversely impact human health or the environment. This involves a higher level of inquiry because the perspective is to collect a robust body of scientific data that

Compliance with Land Regulations 333

demonstrate that the release will not result in harming humans or the environment with the highest degree of confidence.

22. **When a question arises, consult management, in-house environmental counsel, outside environmental counsel, an environmental consultant, or the regulatory authority, as appropriate.**
Environmental regulations are complex, dynamic, and sometimes difficult to interpret. Therefore, engaging others is often the best approach.

23. **Last, remember that "no matter where you are, it's all the same."**
We are learning that we live on one planet with no environmental boundaries and that contamination does not respect borders. Another point is that environmental regulations are built on the USEPA or EU platforms, and those platforms rely on basic principles that include:

1. Protect the air
2. Protect the water
3. Protect the land
4. Protect living organisms
5. Protect our cultural and historic places

17.8 REFERENCES

Connecticut Department of Environmental Protection. 2021. *Environmental Best Management Practices Guide.* Hartford, CT. 4p. www.ct.gov/dep/compliance-assistance. (Accessed October 2, 2021).

United States Environmental Protections Agency (USEPA). 1986. Environmental Auditing Policy Statement. *Federal Register.* Vol. 51. No. 131. Washington, DC. Wednesday, July 9, 1986. pp. 25004–25010.

United States Environmental Protection Agency (USEPA). 2021a. United States Environmental Protection Agency QA Glossary. www.epa.gov/emap/archive-emap/web/html. (Accessed October 2, 2021).

United States Environmental Protection Agency (USEPA). 2021b. Audit Protocols. www.epa. gov/compliance/audit-protocols. (Accessed October 2, 2021).

United States Environmental Protection Agency (USEPA). 2021c. Compliance. www.epa.gov/ compliance. (Accessed October 2, 2021).

United States Environmental Protection Agency (USEPA). 2021d. National Menu of Best Management Practices (BMPs) for Stormwater. www.epa.gov/npdes/national-menu-best-management-practices-bmps-stormwater. (Accessed October 2, 2021).

United States Environmental Protection Agency (USEPA). 2021e. What Is Sustainability? www.epa.gov/sustainability. (Accessed October 2, 2021).

Index

Numbers in **bold** indicate a table. Numbers in *italics* indicate a figure.

1-Bromopropane, 155
1,-Dichloroethylene
  maximum concentration for toxicity
    characteristic, **120**
1,1-Dichloroethane
  generic residential cleanup levels, **142**
  soil cleanup levels for New Jersey, Illinois, New
    York, California, and USEPA, **143**
1,1-Dichloroethene
  generic residential cleanup levels, **142**
  soil cleanup levels for New Jersey, Illinois, New
    York, California, and USEPA, **143**
1,1,1-Trichloroethane (TCA), 38
  generic residential cleanup levels, **142**
  high persistence of, 98
  Reich Farm Superfund Site, 138
  as VOC, 97
  Woburn, Massachusetts, 137
1,1,1,2-Tetrachloroethane
  generic residential cleanup levels, **142**
1,1,2-Trichloroethane
  generic residential cleanup levels, **142**
1,1,2,2-Tetrachloroethane
  generic residential cleanup levels, **142**
1,2-Dibromo-3-chloropropane
  generic residential cleanup levels, **142**
  soil cleanup levels for New Jersey, Illinois, New
    York, California, and USEPA, **143**
1,2-Dibromoethane
  generic residential cleanup levels, **142**
  soil cleanup levels for New Jersey, Illinois, New
    York, California, and USEPA, **143**
1,2-Dichlorobenzene
  generic residential cleanup levels, **142**
  soil cleanup levels for New Jersey, Illinois, New
    York, California, and USEPA, **143**
1,2-Dichloroethene (DCE), 38, 138
  high persistence in the environment, 98
  soil cleanup levels for New Jersey, Illinois, New
    York, California, and USEPA, **143**
  trans-1,2-Dichloroethene (trans-DCE) 38, 137
1,2-Dichloropropane
  generic residential cleanup levels, **142**
  soil cleanup levels for New Jersey, Illinois, New
    York, California, and USEPA, **143**
1,2,3,-Trichlorobenzene
  generic residential cleanup levels, **142**
1,2,3,-Trichloropropane (TCP), 63

1,2,4-Trichlorobenzene
  generic residential cleanup levels, **142**
1,3-Dichlorobenzene
  generic residential cleanup levels, **142**
1,3-Dichloropropane
  soil cleanup levels for New Jersey, Illinois, New
    York, California, and USEPA, **143**
1,4-Dichlorobenzene
  generic residential cleanup levels, **142**
  maximum concentration for toxicity
    characteristic, **120**
1,4-Dioxane ($C_4H_8O_2$), 64, 108
  importance of eliminating, 331
  generic residential cleanup levels, **142**
  soil cleanup levels for New Jersey, Illinois, New
    York, California, and USEPA, **143**
  USEPA's identification as high priority, 155
2-Chlorotoluene
  soil cleanup levels for New Jersey, Illinois, New
    York, California, and USEPA, **143**
2,2-Dichloropropane
  soil cleanup levels for New Jersey, Illinois, New
    York, California, and USEPA, **143**
2,3,7,8-Tetra-chlorodibenzo-p-dioxin (TCDD), 49
2,4D, 74
  as carcinogen, 48
2,4,5-TP (Silvex)
  maximum concentration for toxicity
    characteristic, **120**
2,4,5-Trichlorophenol
  maximum concentration for toxicity
    characteristic, **120**
4-Chlorotoluene
  soil cleanup levels for New Jersey, Illinois, New
    York, California, and USEPA, **143**

A

acenaphthene
  as common PAH compound, **41**
acenaphthylene
  as common PAH compound, **41**
acetone
  acetone cyanohydrin, **162**
  common uses of VOCs, **35**
  soil cleanup levels for New Jersey, Illinois, New
    York, California, and USEPA, **143**
acid derivatives, 45

336 Index

acidic solutions
  Saudi Arabia's regulation of, 265
acidification of ocean, 55–56, 267
acid rain, xviii, 57, 106
  EU legislation addressing, 209
  fog and, 80
  Indonesia, 266
  leaching from, 102
  sulfuric acid as, 107
  wet deposition, 95
acids and bases, 52–53, 106
  carbonic acid, 106, 296
  carboxylic acid group (-COOH), 83
  common, **52**
  corrosivity, 119
  organic, 82, 154
  in soil, 106
  sulfuric acid, 107
acrolein
  as extremely hazardous substance, **162**
acrylamide
  as extremely hazardous substance, **162**
acrylonitrile
  as extremely hazardous substance, **162**
actinolite (fibrous), 51
  See also, asbestos
adiponitrile
  as extremely hazardous substance, **162**
adverse health effect
  bacteria (*E. Coli*), 59
  definition of, 33
  particulate matter, 57–58
  perchlorates, 63
  radon, 53
  THMs, 39
Africa
  countries in which environmental regulations
    are examined, **190**
  DDT used in, 174
  European Commission and, 212
  land pollution regulations of, 229–243
  overview of, 229–231
  See also, Egypt; Kenya; South Africa; Tanzania
African Americans
  environmental justice and, 24
African elephant, 180
African giraffe, 180
agency inspections (compliance), 328
agriculture
  Argentina, 281, 282
  Australia, 275
  CERLA and, 139
  DDT used in, 48
  as most polluting activity on Earth, 5, 10, 27,
    62, 298
  Kenya, 237

New Zealand, 276, 277, 279, 302
Norway, 221, 302
percentage of land use, 1, *2*, 6
pesticides used in, 173, 186
South Korea, *260*
Switzerland, 302
Tanzania, 240
USDA and, 50, 172
aldicarb
  as extremely hazardous substance, **162**
aldrin
  Dutch intervention values for, **193**
  as extremely hazardous substance, **162**
algae
  fertilizers/nitrates and, 49–50, 105
  origin of oxygen with, 93
algaecide, 44
allyl alcohol
  as extremely hazardous substance, **162**
allylamine
  as extremely hazardous substance, **162**
alpha-naphthlthiourea (ANTU)
  as extremely hazardous substance, **162**
alpha radiation, 53
aluminum
  aluminum phosphide, **162**
  mining in Australia of, 273
  mining in New Zealand of, 276
  silicoaluminate catalyst, 45
amines, 44–45
  capacity to degrade in sunlight, 101
  capacity to dissolve in water, 100
aminopterin
  as extremely hazardous substance, **162**
ammonia
  amines compounds and, 45
  as common base, **52**, 53
  as extremely hazardous substance,
    162
  as hazardous waste, 118
  health effects of, 53
  as household cleaning agent, 53
ammonium perchlorate (NH_4ClO_4), 63
Antarctica, 293–295, 297
  countries in which environmental regulations
    are examined, **190**
  Fick's First Law of molecular diffusion
    observable in, 295, 302
  Norwegian Polar Institute protocols for, 294
Antarctic Environmental Protocol, 249
Antarctic Treaty of 1959, 294
anthracene
  benzo(a)anthracene, **41**
  composition of, 40
  Dutch intervention values for, **193**
  as common PAH compound, **41**

Index

337

antimony
  Kenya regulation of, 238
  Saudi Arabia regulation of, 265
  Tanzania regulation of, 241
anthrophyllite (fibrous), 51
  See also, asbestos
Anthropocene, xviii, 303
anthropogenic factors
  in contaminant degradation, 81
anthropogenic sources of pollution xviii, 1, 5
  Amazon rainforest destruction due to, 284
  in Antarctica, 295
  atmospheric contaminants, 92
  automobile exhaust, 57
  carbon dioxide, 55, 106
  carbon monoxide, 56, 107
  causes of, 15
  cultural eutrophication, 105
  heavy metals, 101, 102
  land pollution in relationship to, 192
  ships and cargo vessels as source of, 294
  sulfur dioxide, 57
Antiquities Act of 1906, 9, 22, 171, 176–177,
    178, 186
Appalachian Mountain area, 53
aquifers, 4, 37, 71, 84, *91*, *92*, 96, 97, 99
Argentina, 281–283
  Antarctic Treaty signatory, 293
  Buenos Aires, **7**, 282
  Chile sharing land borders in, 286
  countries in which environmental regulations
    are examined, **190**
  country comparison of environmental regulation
    effectiveness, 299, **300**
  environmental challenges faced by, 301
  Matanzas-Riachuelo River cleanup, 282
aroclors, 42
arsenate
  lead arsenate, 172
arsenic
  as carcinogen, 46
  Dutch intervention values for, **193**
  as hazardous waste, 122
  as heavy metal, 45, **46**
  as highly toxic chemical, 330
  Kenya regulation of, 238
  maximum concentration of contaminants for,
    120
  in pesticides/herbicides, 47
  Saudi Arabia regulation of, 265
  solubility in water, 101
  Tanzania regulation of, 241
asbestos, 50–52, 105
  actinolite, 50
  air-borne, 105
  amosite, 50

anthrophyllite, 51
  as carcinogen, 34, 52
  chrysotile, 50
  crocidolite, 51
  Dutch intervention values for, **194**
  Kenya regulation of, 238
  OSHA health hazard, 175
  Phase I ESA evaluation of, 165, 166
  Saudi Arabia regulation of, 265
  South Africa regulation of, 234
  South Korea regulation of mines, 263
  Tanzania regulation of, 241
  Toxic Substance Control Act/USEPA and,
    154
  USEPA's identification as high priority,
    155
  tremolite, 51
  "white" 51
asbestosis, 52
Asia
  environmental regulation based on USEPA
    model, 269
  environmental regulations examined, **190**
  DDT used in, 174
  land pollution regulations of, 247–269
  Russia as part of, 214
  sanitation in, 286
  Turkey as part of, 224
  See also, China; Japan; India; Indonesia;
    Malaysia; Saudi Arabia; South Korea
Asian citrus psyllid, 61
Asian Elephant, 180
Atomic Energy Act, 178
atrazine
  Dutch intervention values for, **194**
Australia, 273–275
  country comparison of environmental regulation
    effectiveness, **300**
  environmental challenges faced by, 299
  environmental regulation based on US, **17**
  environmental regulations examined, list of
    countries including **190**
  Great Barrier Reef, 56, 107, 296
  Indonesian geographic position in relationship
    to, 266
  New Zealand trade with, 276
  in Oceania, 273, 279
  South African regulation compared to, 234,
    236
  Sydney, **7**, 237
Austria
  environmental regulation based on US, **17**
  environmental regulations examined, list of
    countries including, **190**
  as part of European Union, 209
Azerbaijan, 224

338   Index

B

bacteria, 59–60, 108
  antibacterials as emerging contaminants, 62
  antibiotics to inhibit growth of, 63
  Asian citrus psyllid as carrier of, 61
  biotic degradation involving, 75
  cyanide produced by, 50*E. coli*, 32
  as environmental pollutant, 32
  land pollution and, 15
  as pest, 47
  in soil, 84
barium
  Dutch intervention values for, **193**
  heavy metal, 45, **46**, 101
  maximum concentration for toxicity
    characteristic, **120**
basic (chemical)
  ammonia as, 53
  pH greater than 7, 52
  solutions, Saudi Arabia regulation of,
    265
Basic Act for Environmental Pollution Control of
  1967 (Japan), 265
Belgium
  environmental regulation based on US, **17**
  environmental regulations examined, list of
    countries including, **190**
  as part of European Union, 209
benzenamine, 3-(trifluoromethyl)
  as extremely hazardous substance, **162**
benzene
  benzene ring, 36, *37*, 40, 43, 82
  biotic degradation signaled by, 82
  BTEX compounds, 36, 99
  chlorobenzene, **35**, 38, 138, **141**, **143**
  common uses of VOCs, **35**
  dioxin and, 49
  Dutch intervention values for, **193**
  ethyl benzene, 99, **193**
  generic residential cleanup levels, **141**
  high chemical risk factor, 330
  as light nonaqueous phase liquid (LNAPL)
    compound, 36
  maximum concentration for toxicity
    characteristic, **120**
  nitrobenzene, **120**
  soil cleanup levels for New Jersey, Illinois,
    New York, California, and USEPA, **143**
  trichlorobenzene, **142**, **144**
  as VOC compound, 99
benzenearsonic acid
  as extremely hazardous substance, **162**
benzene, toluene, ethyl benzene, and xylenes
  (BTEX), 36, 99
benzo[a]anthracene
  as PAH compound, **41**

benzo[a]pyrene
  as PAH compound, 40, **41**
benzo[b]fluoranthene
  as A1 carcinogen, 40
  as PAH compound, **41**
benzo[ghi]perylene
  as PAH compound, **41**
benzoic acid, **52**
benzo[k]fluoranthene
  as PAH compound, **41**
benzothrochloride
  as extremely hazardous substance, **162**
benzoylmethylecgonine (cocaine), 62, 63
beryllium
  Kenya regulation of, 238
  radioactivity of, 53
  Saudi Arabia regulation of, 265
  Tanzania regulation of, 241
Bhopal, India gas tragedy, 8, 156, 257, 259
bi or bis(2-ehtylhexyl) phthalate (DEHP), 43
bis(chloromethyl) ketone
  as extremely hazardous substance, **162**
bisphenol A (BPA), 44
Bolivia
  Argentina bordered by, 281
  Chile bordered by, 286
  environmental regulation compared to the
    US, **17**
  Peru bordered by, 289
boric acid, **52**
boron
  as extremely hazardous substance, **162**
Brazil, 283–286
  Amazon rainforest, 284
  Amazon River, 283
  Argentina border, 281
  country comparison of environmental regulation
    effectiveness, **300**
  environmental challenges faced by,
    299
  environmental regulation compared to the
    US, **17**
  environmental regulations examined, list of
    countries including, **190**
  Parana River, 283
  Rio de Janeiro, **7**, 291
  Sao Paulo, **7**, 291
  USEPA and, 284
  Waste Law, 285
bromine
  as halogen, 38, 82, 109
  extremely hazardous substance, **160**
  THMs as byproduct of chlorine and, 39
bromobenzene
  generic residential cleanup levels, **141**
  soil cleanup levels for New Jersey, Illinois,
    New York, California, and USEPA, **143**

Index 339

bromochloromethane
    generic residential cleanup levels, **141**
    soil cleanup levels for New Jersey, Illinois, New York, California, and USEPA, **143**
bromodichloromethane
    common uses of VOCs, **35**
    generic residential cleanup levels, **141**
    health impacts of, 39
    soil cleanup levels for New Jersey, Illinois, New York, California, and USEPA, **143**
    as trihalomethane VOC, 39, 96, 99
bromoform
    common uses of VOCs, **35**
    generic residential cleanup levels, **141**
    health impacts of, 39
    soil cleanup levels for New Jersey, Illinois, New York, California, and USEPA, **143**
    as trihalomethane VOC, 39, 96, 99
bromomethane
    common uses of VOCs, **35**
    generic residential cleanup levels, **141**
brownfield
    definition of, 163
    Brownfield Revitalization Act (BRA), 147, 163–167, 168
    Mexico's Waste Law mirroring, 203
Brown tree snake, 61
Bulgaria
    as EEA member, 212
    as EU country, 209
    environmental regulation based on US, **17**
    Turkey bordered by, 224
Bullard, Robert, 24
Bureau of Land Management (BLM) (Department of Interior, US) 167
    Mineral Leasing Act administered by, 178

C

cadmium
    cadmium stearate, **162**
    as carcinogen, 46
    as common urban heavy metal, 45, 46, 101
    Dutch intervention values for, **193**
    as hazardous waste, 121, 122, 238
    as highly toxic chemical, 330
    Kenya regulation of, 238
    maximum concentration for toxicity characteristic, **120**
    radioactivity of, 53
    River Rouge watershed, 46
    Saudi Arabia regulation of, 265
    in soil, 101
    Tanzania regulation of, 241
calcium arsenate
    as extremely hazardous substance, **162**

calcium hydroxide
    as common acid or base, **52**
    radioactivity of, 53
Canada, 197–201
    climate change impacts in, 55
    country comparison of environmental regulation effectiveness, 299, **300**
    environmental challenges faced by, 302
    environmental laws, list of, 199
    Environmental Protection Act (1999) 200
    environmental regulation based on US, **17**, 206
    environmental regulations examined, list of countries including, **190**
    Lake Ontario **7**
    Migratory Bird Treaty Act, 177
    PBA banned by, 44
    right to self-govern, 198
    Toronto, **7**
cancer
    asbestos as cause of, 51
    definition of, 34
    fertilizers not linked to, 50
    MTBE as potential cause of, 37
    PCBs linked to, 42
    PFOAs as cause of, 62
    pollutants and, 63, 95
    radon as cause of, 53
    radon as part of cancer therapy, 106
carbofuran
    Dutch intervention values for, **194**
    as extremely hazardous substance, **160**
carbon
    carbon disulfide, **35**, **162**
    carbon monoxide, 32, 56, 57, 107, 175
    carbon tetrachloride, **35**, 38, 57, **120**, **141**, **143**, 155
    China's pledge to reduce, 249
    hydrocarbons, *see* polynuclear aromatic hydrocarbons (PAHs)
    Kyoto Protocol, 249
    See also, carbon dioxide
carbon dioxide
    Brazil's attempt to reduce, 284
    carbonic acid and, 56, 106, 206
    composition of the atmosphere, 94, **95**
    emissions inside domestic residence, 107
    as greenhouse gas, 54–56, 106–107
    increase in atmosphere since 2021, *54*, 54–56
    ocean absorption of, 296
    Switzerland's attempt to reduce, 222
    as VOC, 31
carbonic acid, **52**
    impact on oceans of 56, 106, 296
carbon tetrachloride 155
    common uses of VOCs, **35**, 38
    generic residential cleanup levels, **141**

Index

maximum concentration for toxicity characteristic, **120**
as ozone-depleting substance, 57
soil cleanup levels for New Jersey, Illinois, New York, California, and USEPA, **143**
USEPA regulation of, 155
carcinogenicity, 46, 63
carcinogenic risk, **143–144**, 144
carcinogens (known, suspected, and considered)
1,2,3,-trichloropropane (TCP), 63
1,4 dioxane, 63
2,4D not classified as, 48
A1, 36, 39, 40, 51–53
A2, 39, 42, 43, 48
acids and bases, 53
asbestos, 51–52
benzene, 36
benzo(a)pyrene, 40
chromium VI, 47
cyanide not classified as, 50
DDT, 48
DEHP (suspected), 43
dioxins, 49
ethyl benzene, 36
five general categories of, 34
glyphosate (Roundup; suspected), 48
heavy metals, 46
lead, 47
PAHs, 40
PCBs, 42
perchlorates, 63
phthalates, 43
radon, 41
TCDD, 49
trihalomethanes, 39
vinyl chloride, 36
Carson, Rachel
Silent Spring, 8, 19, 23, 47–48, 174
cement making, 55
Brazil, 283
cesium
radioactivity of, 53
chemical
acute response to exposure to, 34
chronic response to exposure, 34
oxidation, 83
reduction, 83
routes of exposure, 33
Chemical Abstracts Service (CAS), 36
Chernobyl, Ukraine
nuclear accident, 215
Chile, 286–289
Atacama desert, 281
country comparison of environmental regulation effectiveness, 299, **300**
environmental challenges faced by, 300–301
environmental regulation based on US, **17**

environmental regulations examined, list of countries including, **190**
National Environment Commission (CONAMA), 286
Santiago, **7**, 286
USEPA collaboration with, 287
China, 247–253
Beijing, **7**, 247
Brown Marmorated Stink Bug, 60
country comparison of environmental regulation effectiveness, 299, **300**
Emerald Ash Borer, 60
environmental challenges faced by, 300–301
environmental regulation based on US, **17**
environmental regulations examined, list of countries including, **190**
Ministry of Ecology and Environment (CMEE), 249
Ministry of Environmental Protection (MEP), 251
Northern Snakehead, 60
Shanghai, **7**, 247, *248*
South China Tiger, 179
USEPA collaboration with, 249
Yongding River, **7**
China Ministry of Ecology and Environment (CMEE), 249
China Ministry of Environmental Protection (MEP), 251
chloracetic acid
as extremely hazardous substance, **162**
chlordane
as bioaccumulative contaminant, 87
Dutch intervention values for, **193**
as extremely hazardous substance, **162**
as highly toxic chemical, 331
maximum concentration for toxicity characteristic, **120**
as pesticide, 48
USEPA banning of, 48
chloride
acryloyl chloride, **162**
benzal chloride, **162**
benzyl chloride, **162**
carbachol chloride, **162**
chlormequat chloride, **162**
chromin chloride, **162**
Dutch intervention values for, **193**
floroacetyl chloride, **163**
hydrogen chloride, 53
methylene chloride, 38, 96, **142**, **144**, 155
polyvinyl chloride (PVC), 43
potassium chloride, 50
vinyl chloride, **35**, 38, 39, 98, 99, **120**, **193**, 214
chlorine
alkyl halide hydrolysis when replaced with, 83
chlorinated VOCs and, 38

Index

chlorine gas, 42
chlorine trifluoride, **162**
dioxins and, 49, 105
as extremely hazardous substance, **162**
halogen atoms and, 82
halogenated, 109
HFCs and, 57
hydrolysis and, 83
organochlorines, 49
ozone and, 557
PCBs and, 42, 100, 156
PCP and, 44
THMs formed by bromine and, 38, 39
toxicity of, 49, 109
trihalomethanes and, 39
chlorobenzene
common uses of VOCs, **35**
generic residential cleanup levels, **141**
as groundwater contaminant, 138
as halogenated VOC, 38
soil cleanup levels for New Jersey, Illinois, New York, California, and USEPA, **143**
chloroethane
generic residential cleanup levels, **142**
soil cleanup levels for New Jersey, Illinois, New York, California, and USEPA, **143**
chloroethyl chloroformate
as extremely hazardous substance, **162**
chloroform
as anesthetic and disinfectant, 39
common halogenated VOC, 38
common uses of VOCs, **35**
as extremely hazardous substance, **162**
generic residential cleanup levels, **142**
maximum concentration for toxicity characteristics, **120**
methyl, 57
soil cleanup levels for New Jersey, Illinois, New York, California, and USEPA, **143**
as trihalomethane VOC, 98, 99
chlorofluorocarbons (CFCs), 31, 56, 154
chloromethane
common detection of, 96
common uses of VOCs, **35**
generic residential cleanup levels, **142**
soil cleanup levels for New Jersey, Illinois, New York, California, and USEPA, **143**
chloromethyl methyl ether
as extremely hazardous substance, **162**
chlorophacinone
as extremely hazardous substance, **162**
chloroxuron
as extremely hazardous substance, **162**
chromium
Africa regulation of, 241
chromium III, 45
chromium VI, 45, 47

Dutch intervention values for, **193**
as hazardous waste, 121, 122
hexavalent chromium, 104, 138, 139, 154, 265
at industrial sites, 104
maximum concentration for toxicity characteristic, **120**
PG&E Hinkley site, 138–139
River Rouge watershed 46
in soil, 101
chrysene
as PAH compound, **41**
chrysotile, 50, 51
See also, asbestos
Clean Air Act of 1970, 22
Clean Air Act of 1990, 107
Clean Water Act (CWA) of 1972, 9, 22, 107
stormwater, 307
wastewater, 307
Coastal Zone Management Act of 1972, 9, 22, 296
cocaine, 62, 63
congeners, 42, 238, 241
contaminants, contamination
abiotic degradation of, 82–83
advection of, 76, 84, 85, 87, 88, 89, 91, 92
basic degradation concepts, 80–81
basic transport concepts, 76–81
biotic degradation, 82
chemical, 15, 30
chemical oxidation of, 83
chemical reduction of, 83
convection of, 76, 92
emerging, 63, 108
fate and transport of, 96–108
Fick's first law of diffusion, 76, 85, 295
flux density in reference to, 76, 85, 89
global assessments and standards for, 189–193
half reaction of, 83
household chemical, 30, *31*
hydrolysis, 83
land, 29–64, 189–193
need for regulation of, 29–32
oxidant, 83
reductant, 83
sites of, 19, 27
toxicity of, 32–34
transport and fate, principles of 72–76
transport concepts, basic, 76–81
transport and fate in atmosphere, 91–96
transport and fate by contaminant group, 96–109
transport and fate in groundwater, 87–91
transport and fate in soil of, 83–85
transport and fate in surface water, 85–87
copper
Ashio Copper Mine disaster, Japan, 254
Chile mining of, 289
Dutch intervention values for, **193**

as essential trace metal, 47
health impacts of high levels, 47
as heavy metal, 45, 47, 290
Kenya regulation of, 238
in plants, 101
Peru mining and export of, 289, 289, 291
Rouge River watershed, 46
Saudi Arabia regulation of, 265
South Korea regulation of, 228
Tanzania regulation of, 241
corrosivity
definition of, 119
cresol
Dutch intervention values for, **193**
as extremely hazardous substance,
162
m-Cresol, **120**
maximum concentration for toxicity
characteristic, **120**
o-Cresol, 120
p-Cresol, **120**
Croatia
as EEA member, 212
as EU country, 209
environmental regulations examined, list of
countries including, **190**
cryptosporidium, 59
cultural eutrophication, 105
cyanide, 50, 105
as common pollutant, 32
Dutch intervention values for, **193**
as extremely hazardous substance, **162**
as highly toxic chemical, 331
hydrogen cyanide, 105
inorganic, 241, 265
organic, 265
cyanide (complex)
Dutch intervention values for, **193**
cyanobacteria, 239
cyanogen chloride
WHO drinking water assessment, 175
cyanogen bromide
as extremely hazardous substance, **162**
cyano group, 50
acetone cyanohydrin, **162**
cyanophos
as extremely hazardous substance, **162**
Cyprus
as EU country, 209
environmental regulation based on US, **17**
environmental regulations examined, list of
countries including, **190**
Czech Republic
as EU country, 209
environmental regulation based on US, **17**
environmental regulations examined, list of
countries including, **190**

D

DDD
Dutch intervention values for, **193**
DDE
Dutch intervention values for, **193**
DDT, *see* dichlorodiphenyltrichloroethane
Deepwater Horizon oil spill, 8
Denmark
as EU country, 209
environmental regulation based on US, **17**
environmental regulations examined, list of
countries including, **190**
dibenzo-p-dioxin, 265
dibromochloromethane, 97
dichloroethane (DCE)
as commonly detected VOC, 96
common uses of VOCs, **35**
generic residential cleanup levels, **142**
maximum concentration for toxicity
characteristic, **120**
soil cleanup levels for New Jersey, Illinois, New
York, California, and USEPA, **143**
dichloroethyl ether
as extremely hazardous substance, **162**
dichloromethane
bromodichloromethane, 39, 96, **141**, **143**
common uses of VOCs, **35**
Dutch intervention values for, **193**
dichlorodiphenyltrichloroethane (DDT), 47–48
as bioaccumulative contaminant, 87
Carson's book on, 8, 19, 23, 47–48, 174
Dutch intervention values for, **193**
high chemical risk factor of, 331
impact on birds, 19, 23
toxicity, 119
dieldrin
Dutch intervention values for, **193**
diethyl chlorophosphate
as extremely hazardous substance, **162**
diethyl phthalate (DEP), 43
di-n-butyl phthalate (DBP), 43
dioxane, *see* 1,4-dioxane
dioxins, 49, 104–105
as bioaccumulative contaminant, 87, 105–106
dibenzo-p-dioxins, 265
Dutch intervention values for, **193**
as environmental pollutant, 32, 134
Form R required for, 161
high chemical risk factor of, 331
Japan's regulation of, 255
polychlorinated dibenzo-p-dioxin, congeners of
238, 241
USEPA restrictions on, 154
Times Beach, 134–135
Dutch Intervention Values (DIV), 192, **193–194**,
235, 252, 267

Index

343

E

eddies, 85
EEA, *see* European Environment Agency
Egypt, 241–243
  Cairo, **7**, 242
  country comparison of environmental regulation effectiveness, 299, **300**
  Egyptian Environmental Affairs Agency (EEAA), 242, 243
  environmental challenges faced by, 301
  Environmental Law, 242
  environmental regulation based on US, **17**
  environmental regulations examined, list of countries including, **190**
  Gulf of Aqaba, 264
  Nile Delta, 242
  Nile River, **7**, 242
  Suez Canal, 242
EHS, *see* extremely hazardous substances
EINETICS, *see* European Information and Observation Network
Emergency Planning and Community Right to Know Act (EPCRA), 147, 156, 159–161, 163
Endangered Species Act, 178
endrin
  as extremely hazardous substance, **162**
  maximum concentration for toxicity characteristic, **120**
Energy Policy Act, 184
entropy 9, 11, 18, 305
environmentalism
  1960s and 1970s in the US, 8, 19
  definition of, 19
  in Russia, 215
environmental audit, 2, 20, 305–333
  agency-led, 328
  checklist 313–326
  closing meeting 326–327
  compliance with land regulation and, 305–6
  conducting, 310–327
  document list, 311–312
  fundamental concepts of, 329–333
  opening meeting 312–313
  report preparation, 327
  Russia, 219
  sustainability and, 329
  USEPA audit policy 306–309
environmental enforcement, 25–26
environmental impact statements (EIS) 20
  definition of, 24
environmental justice, 24–25
environmental laws, United States 22–23
environmental non-compliance, 198, 310
Environmental Protection Act (EPA), 23–24

environmental regulations
  Antarctica, 295
  Argentina, 283
  Australia, 275
  Brazil, 285
  Chile, 288
  complexity and effectiveness of, 8–10
  Egypt, 243
  European Union, 214
  history of, 7–8
  Indonesia, 268
  Kenya, 237, 239
  list of US laws regarding, 8–9
  maintaining compliance with 11–12, 305–333
  Malaysia, 269
  New Zealand, 278
  Oceans, 297
  Peru, 291
  Russia, 219
  Saudi Arabia, 266
  South Africa, 236
  South Korea, 263
  Switzerland, 224
  Turkey, 225
  United States, 15–27, 186
  world, 10–11
EPCRA, *see* Emergency Planning and Community Right to Know Act
Escherichia coli (*E. coli*), 32, 59
esters, 43, 45, 100, 101
  compounds, 83
Estonia
  as EU country, 209
  environmental regulation based on US, **17**
  environmental regulations examined, list of countries including, **190**
ethylbenzene
  generic residential cleanup levels, **142**
  Soil Cleanup Levels for New Jersey, Illinois, New York, California, and USEPA, **144**
Europe
  environmental regulation based on US, **17**
  environmental regulations examined, list of countries including, **190**
  land pollution regulations of, 209–226
  See also, European Union; Norway; Switzerland; Russia; Turkey
European Commission, 210–211
European Environment Agency (EEA), 212–213
European Information and Observation Network (EINETICS), 212
European starling, 60
European Union, 209–214
  country comparison of environmental regulation effectiveness, 299, **300**
  Court of Auditors, 210
  Court of Justice, 210, 211

344 Index

economy, 247
environmental challenges faced by,302
European Environment Agency (EEA), 212–213
Institute for European Environmental Policy, 209
Registration, Evaluation, Authorization, and Restriction of Chemicals (REACH) certification, 312
regulatory overview of 209–214
eutrophication, 105
cultural, 105
extinction or endangerment of wildlife
of Australian animals, 274
Endangered Species Act, 178–180
of Indonesian animals, 268
list of endangered animals, 179–180
list of recently extinct animals, 180
sixth earth extinction, xviii
Exxon Valdez oil spill, 8, 164
extremely hazardous substances (EHS), 160
See also, hazardous waste

F

Federal Insecticide, Fungicide, and Rodenticide Act (FIFRA) of 1947, 9, 22, 171, 172–174, 186
Federal Power Act of 1935, 178
fertilizer, 49–50, 105
FIFRA, *see* Federal Insecticide, Fungicide, and Rodenticide Act (FIFRA) of 1947
Finland
as EU country, 209
environmental regulation based on US, **17**
environmental regulations examined, list of countries including **190**
Norway bordered by, 219
Fisheries Conservation and Management Act, 9, 22
F-Listed waste, 122
Florida
Everglades, 60
invasive species in, 60–61
Florida Grasshopper Sparrow, 179
Florida Key Deer, 179
fluoranthene
as PAH compound, **41**
fluorene
as PAH compound, **41**
fluoride
cyanuric fluoride, **162**
sulfur hexafluoride (SF6), 55
fluorine
as halogen, 38, 82, 83, 109
as hazardous waste, 238
as extremely hazardous substance, **163**

inorganic, 265
Saudi Arabia regulation of, 265
foam, 42
foaming agents, 57
fonofos
as extremely hazardous substance, **163**
Food Quality Protection Act of 1996, 184
formaldehyde
as extremely hazardous substance, **163**
phenolic resin formed by phenol combined with, 44
France
Antarctic Treaty signatory, 293
as EU country, 209
environmental regulation based on US, **17**
environmental regulations examined, list of countries including, **190**
impacted soil from WWII, *213*
Paris, **7**
Paris Summit of 1972, 209
Seine River **7**
fungi, 47
biotic degradation involving, 75
cyanide produced by, 50
fungicide
Federal Insecticide, Fungicide, and Rodenticide Act (FIFRA) of 1947, 9, 22, 171, 172–174, 186
PCP as, 44

G

Germany
Berlin, **7**
as EU country, 209
environmental regulation based on US, **17**
environmental regulations examined, list of countries including, **190**
New Zealand as trading partner with, 276
giardia, 59
Gorbachev, Mikhail, 215, 219
Great Lakes, 8, 19, 87
invasive species in, 60
protection of, 296
Greece
environmental regulation based on US, **17**
environmental regulations examined, list of countries including, **190**
greenhouse gases, 54–55, 106
agriculture as source of, 27
HFCs, 57
RCRA and, 115
sustainability efforts targeting, 221
groundwater/groundwater contamination
1,4-dioxane in, 63
advection in, 87
Canada, 199

Index

345

capture zone, 91
CERLA and, 139–141
China, 249, 250, 251–252
chromium in, 104
dissolved. metals in, 103
Dutch Intervention Values (DIV), 192, **193–194**
European Union, 214
fertilizer in, 105
Japan, 255–256
large scale contamination incidents, 127
LDRs and, 126
Love Canal, 135
Mexico, 204
MTBE found in, 37, 99
Netherlands, 214
PAHs in, 100
PFAHs in, 108
Phase I ESA, 165, 166
Reich Farm Superfund Site, 138
Saudi Arabia, 264
soil protective of, 83
South Africa, 235–236
South Korea, 262–263
toxic wastes in, 119
transport and fate of contaminants in 71, 87–91
VOCs found in, 96–97
Wells G and H, 137
Gulf of Mexico, **6**
Deepwater Horizon oil spill, 8
Mississippi River flooding and, 80

H

half-life
of cyanide, 105
definition of, 80
of PAHs, 100
of permethrin and toxaphene, 104
of radon, 106
half reaction, 83
hazardous waste
Argentina, 282
Australia, 274
Brazil, 285
Canada, 200
characteristic, 118–122
Chile, 287
China, 250
contained-in rule for, 125
definition of, 118, 119
Egypt, 242
environmental audit checklist, 313
European Union, 212
extremely hazardous substances (EHS), 160
flood events and, 80, 254
hazardous chemical, definition of, 160

Hazardous and Solid Waste Amendments (HSWA), 125
hazardous materials (HAZMAT), 116, 150
Hazardous Materials Transportation Act (HMTA), 9, 22, 147–153
hazardous household products, 30
hazardous ranking system (HRS), 134, 139
hazardous waste label, 123, *123*
India, 257
Indonesia, 267
Japan, 254
Kenya, 237
land disposal restrictions (LDR) and, 126
listed hazardous waste, 122
Malaysia, 268
medical waste, 129–130
Mexico, 202
New Zealand, 276
Norway, 220
OSHA regulation of, 160
Peru, 290
RCRA regulation of, 115–130
Russia, 217
Saudi Arabia, 264
South Korea, 261
Subtitle-C, hazardous waste, 117
Subtitle-D, non-hazardous waste, 117
Switzerland, 223
Tanzania, 240
toxic hazardous waste, criteria for, 119
Turkey, 225
Uniform Hazardous Waste Manifest, 127
universal, 124
used oil, 124
heavy metals, 45–47, 101–104
heptotoxins, 34
hexachlorobenzene
maximum concentration for toxicity characteristic, **120**
hexacloroethane
maximum concentration for toxicity characteristic, **120**
hexavalent chromium, *see* chromium
herbicides, 47–48, 104
Canada regulation of, 200
Chile regulation of, 288
China use of, 250
as common environmental pollutant, 31
definition of, 47
glysophate and 2,4D, 48
as household product, 30
increase in use of, 104
Kenya high use of, 237
Kenya regulation of, 238
land pollution from, 5, 10
ocean pollution from, 296, 298
organchlorines, 49

346 Index

PCP, 44
Roundup, 48
in soil, 108
South Africa regulation of, 235
South Korea regulation of, 262
Tanzania use of, 240
TSCA regulation of, 153
United States lack of regulation for, 301
herbivores, 43
Hungary
environmental regulations examined, list of
countries including, **190**
as EU country, 210

I

Iceland
as EEA member, 212
environmental regulation based on US, **17**
ignitability
hazardous waste criteria, 119, 250 257, 258,
275, 277, 285, 288
India, 256–259
Bhopal gas tragedy, 8, 156, 257
country comparison of environmental regulation
effectiveness, 299, **300**
Delhi, **7**
environmental challenges faced by, 301
environmental regulation based on US, **17**
environmental regulations examined, list of
countries including, **190**
Indian Elephant, 180
Indonesia, 266–68
country comparison of environmental regulation
effectiveness, 299, **300**
environmental challenges faced by, 301
environmental regulation based on US, **17**
environmental regulations examined, list of
countries including, **190**
Jakarta, **7**, 266
major islands of, 266
Ministry of the Environment, 267
insecticide
anthracene, **41**
arsenic, **46**
chlorobenzene, **35**
DDT, 174
Insecticide Act of 1910, 9, 22, 171, 172, 186
Insecticide, Fungicide, and Rodenticide Act
(FIFRA) of 1947, 173–174, 186, 307
PCP, 44
permethrin, 104
toxaphene, 104
insects, 47, 61, 166, 284
insect repellants, 48, 62
isomer, 40
invasive species, 60–61
Global Invasive Species Program (GISP), 60

iodine
as halogen 38, 82, 109
perchlorate's impact on uptake of, 63
radioactivity of, 53
Ireland
as EU country, 210
environmental regulation based on US, **17**
environmental regulations examined, list of
countries including, **190**
Italy
as EU country, 210
environmental regulation based on US, **17**
environmental regulations examined, list of
countries including, **190**

J

Japan, 253–224
Ashio Copper Mine disaster, 254
country comparison of environmental regulation
effectiveness, 299, **300**
Emerald Ash Borer, 60
environmental challenges faced by, 301
environmental regulation based on US, **17**
environmental regulations examined, list of
countries including, **190**
Fukushima nuclear disaster, 253
Sea of Japan, 216, 259
Tokyo, **7**, 253
Japan Contamination Countermeasures Act of 2002, 255

K

Keep America Beautiful campaign, 8, 19
Kenya, 236–239
country comparison of environmental regulation
effectiveness, 299, **300**
environmental challenges faced by, 301
Environmental Management and Co-Ordination
Water Quality Act, 237
environmental regulation based on US, **17**
environmental regulations examined, list of
countries including, **190**
Mount Kenya, 236
Nairobi, 236, 237
Nairobi River, 236
plastic waste, 239
Tanzania compared to, 241
K-listed waste, 122
Kyoto Protocol of 1997, 217, 249
Kyrgyzstan
environmental regulation based on US, **17**

L

Lacey Act of 1990, 9, 22, 171, 176, 186
lakes and water bodies
China, 251
Clean Water Act, *see* Clean Water Act

Convention on the Protection and Use
of Trans-boundary Watercourses and
International Lakes of 1992, 217
dioxin detected in, 104
eddying effects in, 85
Lake Erie, Michigan, 8, 19
Lake Victoria, Africa, 239
major cities of the world, **7**
mercury in, 101
nitrates in, 49
PCBs, 100, 156
pH of, 81
phosphorus in, 50
PFASs in, 108
Saudi Arabia, 264
South Korea, 227, 228
Switzerland, 222
toxaphene used to kill fish in, 48
See also, Great Lakes; rivers; oceans and seas
land pollution
definition of, 192
global assessments and environmental standards
for, 192
Latvia
as EU country, 210
environmental regulation based on US, **17**
environmental regulations examined, list of
countries including, **190**
lead
in Antarctica, 294
Dutch intervention values for, **193**
as hazardous waste, 121, 122
as heavy metal, 45, **46**, 101, 102
high chemical risk factor of, 330
Kenya regulation of, 238
lead arsenate, 172
lead paint, 154, 165
lead paint abatement, 230
major sources of, 101, 102
maximum concentration for toxicity
characteristic, **120**
OSHA ruling on, 175
in Peru, 290
Saudi Arabia regulation of, 265
soil value, Norway, 221
soil value, United States, 221
as suspected human carcinogen, 47
Tanzania regulation of, 241
toxicity of waste containing, 119
LEPC, *see* local emergency planning committee
lindane
maximum concentration for toxicity
characteristic, **120**
Lithuania
as EU country, 210
environmental regulation based on US, **17**
environmental regulations examined, list of
countries including, **190**

LOAEL, *see* lowest observed adverse effect level
local emergency planning committee (LEPC),
159–160, 324
Love Canal, 8, 19, 134, 135–136
lowest observed adverse effect level
(LOAEL), 33
Luxembourg
as EU country, 210
environmental regulation based on US, **17**
environmental regulations examined, list of
countries including, **190**

M

Macedonia, 212
Malayan Tiger, 179
Malaysia, 268–269
country comparison of environmental regulation
effectiveness, 299, **300**
environmental challenges faced by, 301
environmental regulations examined, list of
countries including, **190**
Kuala Lumpur, 268
Malta
as EU country, 210
environmental regulation based on US, **17**
environmental regulations examined, list of
countries including, **190**
manganese, 290
m-Cresol
maximum concentration for toxicity
characteristic, **120**
mercury
as bioaccumulative contaminant, 87
dimethyl mercury, **162**
as common heavy metal, 45, **46**
Dutch intervention values for, **193**
health impacts of, 47
high chemical risk factor of, 330
Kenya regulation of, 238
latency period of, 34
maximum concentration for toxicity
characteristic, **120**
methylation of, 101
methyl mercury, 101
M-Listed waste, 122
Peru's mining and export of, 289, 290
Saudi Arabia regulation of, 265
South Korean regulation of, 262
Tanzania regulation of, 241
toxicity of waste containing, 119
as universal hazardous waste, 124
Marine Mammal Protection Act of 2015, 9,
22, 297
Marine Protection Act of 1972, 9, 22
methoxychlor
maximum concentration for toxicity
characteristic, **120**

methylamine, 44
  as common base, **52**
methylation, 101
methyl bromine, 57
methyl chloroform, 57
methylene chloride, 38, 96
  generic residential cleanup levels, **142**
methyl ethyl ketone (MEK)
  common uses for VOCs, **35**
  maximum concentration for toxicity
    characteristic, **120**
  soil cleanup levels for New Jersey, Illinois, New
    York, California, and USEPA, **143**
methyl isocyanate (MIC), 156
  see also Bhopal, India gas tragedy
methyl parathion
  USEPA banning of, 186
methyl-tert-butyl ether (MTBE), 37
  generic residential cleanup levels, **142**
  soil cleanup levels for New Jersey, Illinois, New
    York, California, and USEPA, **144**
Mexico, 201–206
  country comparison of environmental regulation
    effectiveness, 299, **300**
  environmental challenges faced by, 301
  environmental regulation based on US, **17**
  environmental regulations examined, list of
    countries including, **190**
  Federal Attorney Generalship of Environmental
    Protection (PROFEPA), 202
  Mexico City, **7**, 202
  Secretariat of the Environment and Natural
    Resources (SEMARNAT), 202
Migratory Bird Treaty, 177
Mineral Leasing Act, 177
Mississippi River, **6**
Missouri
  Times Beach, 8, 19, 134
Missouri River, **6**
mobility (of contaminant), 4, 72, 104, 109, 139
Montenegro, 212
mutagens, 34, 49

N

naphthalene
  Dutch intervention values for, **193**
  as VOC, **35**
  as PAH compound, 40, **41**
  phthalates/phthalic acid derived from, 43
National Environmental Policy Act (NEPA),
  19–20, 23–24
National Forest Management Act, 180
National Historic Preservation Act, 178
National Oceanic and Atmospheric Administration
  (NOAA), 295
National Parks Act, 177

n-Butylbenzene
  generic residential cleanup levels, **141**
NEPA, see National Environmental Policy Act
Netherlands
  as EU country, 210
  environmental regulation based on US, **17**
  environmental regulations examined, list of
    countries including, 190
New Zealand, 275–278
  Auckland, 276
  country comparison of environmental regulation
    effectiveness, 299, **300**
  environmental challenges faced by,
    302
  environmental regulation based on
    US, **17**
  environmental regulations examined, list of
    countries including, **190**
nickel
  allergic reaction to, 47
  as heavy metal, 45, **46**
  Dutch intervention values for, **193**
nitrate
  as common fertilizer, 31, 105
  in drinking water, 49
nitrobenzene
  maximum concentration for toxicity
    characteristic, **120**
nitrogen
  in amines, 44
  as carcinogen, 50
  as common fertilizer, 49–50, 105
  composition of atmosphere, **95**
  eutrophication and, 105
  ground-level ozone and, 57, 107
  New Zealand agricultural sector and, 276
nonpoint source (NPS) water pollution, 223
North America
  environmental regulation based on US, **17**
  environmental regulations examined, list of
    countries including, **190**
  land pollution regulation of, 107–206
  See also Canada; Mexico; United States
Norway, 219–221
  country comparison of environmental regulation
    effectiveness, 299, **300**
  as EEA member, 212
  environmental challenges faced by, 302
  environmental regulation based on US, **17**
  environmental regulations examined, list of
    countries including, **190**
  Norwegian Climate and Pollution Agency, 220
  Norwegian Directorate for Nature
    Management, 220
  Norwegian Environment Agency (NEA), 220
  Norwegian Polar Institute, 294
  sustainability efforts, 221

Index

349

Norway rat, 61
Norwegian Climate and Pollution Agency, 220
Norwegian Directorate for Nature
    Management, 220
Norwegian Environment Agency (NEA), 220
Norwegian Polar Institute, 294
NPS, *see* nonpoint source (NPS) pollution
Nuclear Waste Policy Act, 183

O

Occupational Safety and Health Act (OSHA) of
    1970, 10
  asbestos, 51
  carbon monoxide, 56
  safety hazards in the workplace, 174–176
Oceania
  environmental regulation based on US, **17**
  environmental regulations examined, list of
    countries including, **190**
  land pollution regulations of, 273–279
  See also, Australia; New Zealand
oceans and seas
  acidification, 107
  Aegean Sea, 224
  Antarctica and, 302
  Aral Sea, 215
  Atlantic Ocean, **6, 7**, 61, 137, 219, 231, 281
  Black Sea, 217, 224
  carbon dioxide absorbed by, 55–56
  dumping, 274, 296
  East China Sea, 247, 259
  freedom of the seas doctrine, 297
  Indian Ocean, 231, 236, 239, 240, 256, 266
  Law of the Sea Convention, 217, 297
  major cities of the world associated with, **6, 7**
  Mediterranean Sea, 224, 241
  Ocean Dumping Act of 1988, 10, 297
  Pacific Ocean, **6, 7**, 181, 212, 253, 266, 273,
    225, 286, 289
  Polar Code, 297
  pollution of, 106, 295–297
  as pollution sinks, 4, 5, 10, 72, 189, 298
  Red Sea, 264
  Sea of Japan, 216, 259
  US cities associated with, **6**
  Yellow Sea, 247, 259
odor
  contaminant as, 199
  hydrogen cyanide, 50
  naphthalene, 40
  sodium cyanide, 50
ODSs, *see* ozone-depleting substances
Office of Environmental Justice, 24
Oil Spill Protection Act of 1990, 10, 22
OSHA, *see* Occupational Safety and Health Act
oxidant, 83

ozone, 56–57, 107
  CFCs impact on, 91
  China's policy on, 249
  composition of atmosphere, **95**
  depletion in Antarctica, 293
  EU environmental legislation addressing, 209
  ground level, 56, 57, 107
  Kyoto Protocol and, 249
  Montreal Protocol on Substances that Deplete
    the Ozone Layer of 1987, 217
  ozone-depleting substances (ODSs), 56
  ozone layer, thinning of, 209
  Vienna Convention for the Protection of the
    Ozone Layer of 1985, 217
ozone-depleting substances (ODSs), 56

P

PAHs, *see* Polynuclear or polycyclic aromatic
    hydrocarbons
Pakistan
  environmental regulation based on US, **17**
palladium
  radioactivity of, 53
parasites, 59, 108
  definition of, 59
  permethrin used to kill, 48
Paris, France, *see* France
Paris green, 172
  Paris Summit of 1972, 209
parks
PCBs, *see* polychlorinated biphenyls
particulate matter (PM), 57–59, 108
  dry deposition, 95
  PAHs attaching to, 100
  PCBs attaching to, 100
permethrin, 48, 104
persistence (of contaminant), 4, 72
  of chlorinated VOCs, 139
  USEPA's recommendations, 155
Peru, 289–291
  country comparison of environmental regulation
    effectiveness, 299, **300**
  environmental challenges faced by, 301
  environmental regulations examined, list of
    countries including, **190**
  La Oroya, 291
  Lima, 289
  Ministry of Environment, 290
PFOA or PFAS, *see* polyfluoroalkyl substances
PG&E Hinkley site, 138–139
pH, *see* acids and bases
phenanthrene
  as PAH compound, 40, **41**
phenols, 43–44
  Dutch intervention values for, **193**
  Kenya regulation of, 238

Index

solubility of, 100
Saudi Arabia regulation of, 265
South Africa regulation of, 235
Tanzania regulation of, 241
phthalates, 43
solubility of, 100
phyto-phamaceutical processes 265
phytoremediation 101
P-Listed waste, 122
plutonium
radioactivity of, 53
Poland
as EU country, 210
environmental regulation based on US, **17**
environmental regulations examined, list of
countries including, **190**
"polluters pay" principle 18, 27, 145, 212, 221,
223, 232, 255, 286
pollution prevention
Brazil, 285
Canada, 201
China, 250, 251
Convention for the Prevention of Pollution at
Sea, 217
Convention on the Prevention of Marine
Pollution by Dumping of Wastes and Other
Matter, 216
European Union, 212
Law of the People's Republic of China on the
Prevention and Control of Environmental
Pollution by Solid Waste, 250
Pollution Prevention Act (PPA) of 1990, 10, 22,
147, 161, 163
recycling and, 320–321
Russia, 218
Saudi Arabia, 264
TSCA, 155
polychlorinated biphenyls (PCBs), 40–42, 100
banning in the US, 40–41
basic physical properties of, 41–42
as bioaccumulative contaminant, 87, 100, 109
as carcinogen (suspected), 42
environmental audit, 322
evaluation and inventory of (hazardous waste
checklist), 322
high chemical risk factor of, 331
ocean polluted by, 296
regulation under TSCA, 157
in soil, 108
South Korean regulation of, 228
Times Beach, 135
toxicity of, 119
Valley of the Drums, 136
polychlorinated dibenzo-p-dioxin, congeners of
238, 241
polyfluoroalkyl substances (PFOA or
PFAS), 61

polynuclear or polycyclic aromatic hydrocarbons
(PAHs), 100
biotic degradation of, 82
common PAH compounds, **41**
sources of, 40
Portugal
as EU country, 210
environmental regulation based on US, **17**
environmental regulations examined, list of
countries including, **190**
potash, 50
potassium
as common environmental pollutant, 31
as common fertilizer, 49, 50, 105
Mineral Leasing Act and, 177
potassium hydroxide, **52**
potassium perchlorate, 63
radioactivity of, 53
Putin, Vladimir, 216, 219
pyrene
as PAH compound, **41**
pyridine
as common base, **52**
maximum concentration for toxicity
characteristic, **120**

R

radioactive
Canadian class of hazardous material, 200
as class of hazardous material, 150
compounds, 32, 53–54, 75
EU class of hazardous material, 213
low-level waste, 115, 130
Low-Level Radioactive Waste Policy Act of
1980, 9, 22, 130
Nuclear Waste Policy Act, 183
oceans polluted with, 296
radioactive decay, 53, 82
radon, 53, 106
Sea of Japan, waste in 216
radionuclides, 34
radium
radioactivity of, 53
radon, 154
radioactivity of, 53, 106
reactivity
contaminant, 92
definition of, 119
hazardous waste criteria, 118
potential, 152
reductant, 83
Reich Farm Superfund Site, 134, 137, 138
Resource, Conservation, and Recovery Act
(RCRA), 10, 115–129
contained-in rule or policy, 124–125
creation of, 27, 115

Index

definition of hazardous waste, 115, 118, 125
definition of solid waste, 115
definition of waste, 115
Hazardous and Solid Waste Amendments of
    1984, 127
hazardous waste manifests, 148
"listed wastes" under, 122
subtitle I, 127–129
underground storage tanks (UST), 127
used oil, 124
River Rouge watershed 46
rivers
    Amazon River Basin, 281
    Amazon River, Brazil, 283
    Amnok River, 259
    Cuyahoga River, Ohio, 8
    Delaware River, **6**
    Edo River, Japan 254
    Ganges River, India, 256–257
    Han River, **7**
    Hudson River, **6**
    Lerma River, **7**
    major rivers/cities in the United States, **6**
    major rivers/cities in the world, **7**
    Mississippi River, **6**
    Missouri River, **6**
    Moskove River, Russia, **7**
    Nairobi River, 236
    Nile River, **7**, 242
    River and Harbors Act of 1899, 9, 10, 19, 22,
        23, 296
    River Rouge, Michigan, 46
    Seine River **7**
    Simbacom River, **7**
    Spree River, **7**
    Thames River, **7**, 18
    Toms River, 137
    Tone River, Japan, 254
    Watarase River, Japan, 254
    Wild and Scenic Rivers Act of 1968, 10, 23
    Yamuna River, **7**
    Yongding River, **7**
    See also, watershed
Romania
    as EEA member, 212
    as EU country, 210
    environmental regulation based on US, **17**
    environmental regulations examined, list of
        countries including, **190**
Russia , 214–219
    Aral Sea, 215
    as Asia/Europe, 214, 247
    Chernobyl, 215
    country comparison of environmental regulation
        effectiveness, 299, **300**
    Emerald Ash Borer, 60
    environmental challenges faced by, 301

environmental regulation based on US, **17**
environmental regulations examined, list of
    countries including, **190**
glasnost, 215
Gorbachev, 215, 319
Ministry of Natural Resources (MNR), 216
Moscow, **7**, 214, 247
Moskove River, **7**
Putin, 216, 219
Russian State Committee on Environmental
    Protection, 215
Saint Petersburg, 214
Yeltsin, 216

S

Safe Water Drinking Act (SDWA) (US), 10,
    23, 307
    Canada (Ontario), 199
safe water drinking practices
    Africa, 230
    Brazil, 291
salts
    inorganic cyanides, 50
    perchlorates, 50
    road salts, 73
SARA, *see* Superfund Amendments and
    Reauthorization Act of 1986
Saudi Arabia, 263–266
    Arabian Desert, 264
    country comparison of environmental regulation
        effectiveness, 299, **300**
    environmental challenges faced by, 301
    environmental regulations examined, list of
        countries including, **190**
    Riyadh, **7**
sec-Butylbenzene
    generic residential cleanup levels, **141**
selenium
    Dutch intervention values for, **193**
    as hazardous waste, 121, 238, 241, 265
    health impacts of exposure to, 47
    as heavy metal, 45, **46**
    maximum concentration for toxicity
        characteristics, **120**
semi-volatile organic compounds (SVOCs),
    100–101
    common groups of organic compounds
        associated with, 43
    common uses of, 42, 47
    degradation pattern of, 82
    in soil, 108
Serbia, 212
silver
    arygia caused by exposure to, 47
    drinking water standards, **132**
    Dutch intervention values for, **193**

as hazardous waste, 121
as heavy metal, 45, **46**
maximum concentration for toxicity
characteristic, **120**
site inspection, 324–326
agency inspection, 328
items to document, 165
offsite inspection, 166
Slovakia
as EU country, 210
environmental regulation based on US, **17**
environmental regulations examined, list of
countries including, **190**
Slovenia
as EU country, 210
environmental regulation based on US, **17**
environmental regulations examined, list of
countries including, **190**
sodium
Mineral Leasing Act and, 177
sodium bicarbonate, **52**
sodium cyanide, 50
sodium hydroxide, **52**
sodium perchlorate, 63
smog
components of, 57, 107
urban, 56, 96
soil
acids and bases released into, 106, 107
ammonia naturally present in, 53
bacteria present in, 59
cleanup, 256
climatological factors affecting migration of
pollutants in, 79
contaminant release in, 75
contaminants in, 30, 72
cyanide compounds in, 105
diffusion of contaminants in, 84
dioxin in, 135
Dutch intervention values for, **193–194**
emerging contaminants in, 62
erosion, 4, 60, 72, 229, 249, 274, 276, 289
fumigant, **35**
heavy metals in, 101, 102, 103
as intermediate or final sink, 4, 71–72
mass flow of contaminants through, 84
migration of contaminants in, 78
PAHs in, 40, 100
particulate matter (PM) in, 108
PCBs in, 100, 108, 135
pesticides and herbicides in, 104
phosphorous occurring in, 49
recognized environmental condition (REC) in,
166, *166*
remediation, 213
specific climatological features affecting
migration of pollutants in, 79

specific physical factors affecting migration of
pollutants in, 78
specific vegetative factors that affect migration
of pollutants in, 80
SVOCs in, 101
toxaphene in, 104
transport and fate of contaminants in, 83–85
VOCs in, 96, 97
soil cleanup criteria, NJ, Illinois, NY, California,
USEPA 143, **144**
Soil Conservation Service, 167
soil intervention values, 214
Soil Screening Value, 235
South Africa, 231–236
country comparison of environmental regulation
effectiveness, 299, **300**
DDT used in, 174
environmental challenges faced by, 301
environmental regulation based on US, **17**
environmental regulations examined, list of
countries including 190
Johannesburg, **7**
National Environmental Management Act
(NEMA), 231–233
South America
DDT used in, 174
environmental regulation based on US, **17**
environmental regulations examined, list of
countries including 190
land pollution regulations of, 281–291
nutria, 60
See also, Argentina; Brazil; Chile; Peru
South Korea, 259–263
Amnok River, 259
country comparison of environmental regulation
effectiveness, 299, **300**
Han River, **7**
environmental challenges faced by, 302
environmental regulations examined, list of
countries including, **190**
Korean War, 259
Seoul, **7**, 259, 302
Spain
as EU country, 210
environmental regulation based on US, **17**
environmental regulations examined, list of
countries including, **190**
Spill Control and Countermeasures (SPCC) Plan,
311, 321, 322
state emergency response commission (SERC),
159–160
stormwater
Canada, 206, 302
Clean Water Act (CWA), 307
Mexico, 202
runoff, 86
United States, 202

Index

353

strontium
  radioactivity of, 53
sulfates
  as particulate matter, 57
sulfur dioxide, 57, 107
sulfur hexafluoride (SF6), 55
sulfuric acid, **52**, 57, 107
Superfund Amendments and Reauthorization Act
    (SARA) of 1986, 10, 23, 156–163
Superfund law, US 133, 232
  see also, Polluters pay
Superfund liability, 148
Superfund Sites, US, 19, 116, 133
  CERCLA and, *134*, 134–139
  PRPs and, 140
Surface Mining Control and Reclamation Act of
    1977, 10, 23
surface soil
  heavy metal concentrations in, 103, **104**
  pesticides in, 104
surface water, *2*, *86*
  as intermediate sink, 4, 71
  fertilizers in, 105
  heavy metals and, 101, 102, 103
  major urban areas of the US associated with, **6**
  PAHs and, 100
  pesticides in, 104
  PFAHs in, 108
  phosphorus and, 49
  photolysis in, 82
  pollutants and, 75
  transport and fate of contaminants in, 85–87
SVOCs, *see* semi-volatile organic compounds (SVOCs)
Sweden
  environmental regulation based on US, **17**
  environmental regulations examined, list of
    countries including, **190**
  as part of European Union, 210
  Stockholm, 48
  Stockholm Convention on Persistent Organic
    Pollutants, 217
  UN Environmental Program meeting of
    2001, 48
Switzerland, 221–224
  country comparison of environmental regulation
    effectiveness, 299, **300**
  environmental challenges faced by, 302
  environmental regulation based on US, **17**
  environmental regulations examined, list of
    countries including, **190**
  as member of EEA, 212

T

Tanzania, 239–241
  country comparison of environmental regulation
    effectiveness, 299, **300**

Dar es Salaam, 240
Dodoma, 240
  environmental challenges faced by, 301
  environmental regulations examined, list of
    countries including, **190**
TCDD (2,3,7,8-tetra- chlorodibenzo-p-dioxin) 49
teragens, 34
tert-Butylbenzene
  generic residential cleanup levels, **141**
  soil cleanup levels for New Jersey, Illinois, New
    York, California, and USEPA, **143**
tetrachloroethene
  carbon tetrachloroethene, **143**
  degradation of, *98*, 99
  generic residential cleanup levels, **142**
  as halogenated VOC, 38
  Reich Farm, 138
  Wells G and H Superfund site, 137
tetrachloroethylene
  maximum levels for toxicity characteristic, **120**
  USEPA's classification as high priority, 155
thermodynamics
  second law of, 11, 305
  See also, entropy
thiocyanate
  Dutch intervention values for, **193**
thorium
  radioactivity of, 53
Times Beach, Missouri, 8, 134–135
tin
  radioactivity of, 53
toluene
  biodegradability of, 99
  common uses for VOCs, **35**
  Dutch intervention values for, **193**
  generic residential cleanup levels for, **142**
  in groundwater, 97
  as LNAPL compound, 36
  soil cleanup levels for New Jersey, Illinois, New
    York, California, and USEPA, **143**
  See also, BTEX
toxaphene
  as insecticide, 48, 104
  maximum concentration for toxicity
    characteristic, **120**
toxicity
  characteristics of, 119–122
  of contamination, 32–34
  definition of, 32, 119
  FQPA requirements, 186
  heavy metals, 46
  high acute, 155
  high or higher, 4
  lowest observed adverse effect level
    (LOAEL), 33
  maximum concentration of contaminants for
    toxicity characteristic, **120**

persistent and mobile, 4, 72, 109
risk assessment for, 32
USEPA's role in evaluating, 172, 186
threshold effect value, 32
toxicity characteristic leaching procedure
(TCLP) test, 119
waste extract test (WET) and, 119
toxicity characteristic leaching procedure (TCLP)
test, 119
toxicity, mobility, persistence, 139
trichloroethylene (TCE)
high risk characteristics of, 155
maximum concentration of contaminants for
toxicity characteristic, **120**
trihalomethanes, 39, 99
Turkey, 224–225
Ankara, 203
country comparison of environmental regulation
effectiveness, 299, **300**
as EEA member, 212
environmental challenges faced by, 301
environmental regulation based on US, **17**
environmental regulations examined, list of
countries including, **190**
Istanbul, **7**, 224

U

United Kingdom (UK)
Antarctic Treaty signatory, 293
Brexit, 209
environmental regulation based on US, **17**
environmental regulations examined, list of
countries including, **190**
United States (US)
age of environmentalism in, 19
anti-pollution advertisement, *20*
Centers for Disease Control (CDC), 135
country comparison of environmental regulation
effectiveness, 299, **300**
environmental challenges faced by, 301
environmental laws, list of, 9–10, 22–23
environmental regulation, history of 7
forest land, *181*, *182*
history of land pollution and regulation in,
15–27
land-related environmental regulation in,
171–186
major urban areas and associated water
features, **6**
United States Code (USC), 21
See also, CAA; Carson; CERCLA; CWA;
NEPA; Superfund sites
United States Department of Agriculture
(USDA), 50
United States Environmental Protection Agency
(USEPA), 12

Africa and, 229
asbestos banned by, 51–52
audit policy, 306–309
audit process, 309
brownfield defined by, 163
chemicals restricted by, 154–155
Chile and, 287, 288
China and, 249
chlordane banned by, 48
CERCLA and, 27
Congress' authorization of, 17
creation of, 19, 20
dioxins listed by, 49
enforcement of environmental laws by, 25,
133
environmental regulations and, 12, **17**
fertilizers declared non-carcinogenic by, 50
hazardous waste defined by, 118
Indonesia and, 268
invasive species, Great Lakes, 60
NEPA and, 19, 20, 23, 24
Office of Environmental Justice, 25
ozone, 56
particulate matter, 58
perchlorates, 63
Peru and, 290
"polluters pay" principle and, 133
radon classed as carcinogen by, 53
regions, 21, *21*
as regulatory agency, 17
self-audits promoted by, 328
soil cleanup levels for New Jersey, Illinois, New
York, California, and, **144**
THMs regulated by, 39
uranium
radioactivity of, 53

V

Valley of the Drums, 136–137
volatile organic compounds (VOCs), 34–39,
96–100
chlorinated (CVOC), 37, 139, 155
common uses of 34, **35**
halogenated (HVOC), 37, 38, 155
in pesticides and herbicides, 47, 108
in smog, 57
trihalomethanes, 39
See also, semi-volatile organic compounds
(SVOCs)

W

Warren, North Carolina, 24
water
aquifers, 4, 37, 71, 84, *91*, *92*, 96, 97, 99
Clean Water Act, 9, 22, 72

Index

major cities in the US associated with water bodies, **6**
major cities in the world associated with water bodies, **7**
Roman Empire regulation of, 18
Safe Water Drinking Act (SDWA), 10, 23
See also, groundwater; surface water; stormwater
watersheds
Chesapeake, 60
industrial, 46
waterways, 19
Wells G and H, Woburn, 134, 137
wetland destruction, 3, 29
wetlands, 4, 71, 81, 166, 220, 237, 249, 257
Wild and Scenic Rivers Act of 1968, 10, 23
wildlife, *see* extinction or endangerment of wildlife
Woburn, Massachusetts 134, 137
World Wildlife Fund (WWF), 179–180

X

xylenes
common uses of VOCs, **35**
Dutch intervention values for, **193**
health impacts of exposure to, 37
as VOC compound, 99
See also, BTEX

Y

Yeltsin, Boris, 216
Yugoslavian Republic of Macedonia, 212

Z

Zambia, 239
zinc
corrosion control and, 133
drinking water standards, **132**
Dutch intervention values for, **193**
health impacts of, 47
as heavy metal, 45, **46**, 101
Kenya regulation of, 238
Peru mining and export of, 289, 290
Saudi Arabia regulation of, 265
Tanzania regulation of, 241

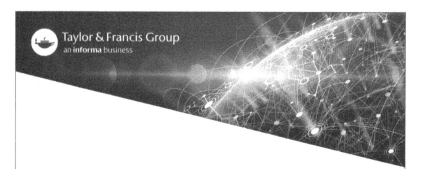